Silvio R.A. Salinas

Introduction to Statistical Physics

With 67 Illustrations

Springer

Silvio R.A. Salinas
Instituto de Fisica
Universidade de São Paolo
Caixa Postal 66318
05315-970 São Paolo
Brazil
ssalinas@if.usp.br

Series Editors

R. Stephen Berry
Department of Chemistry
University of Chicago
Chicago, IL 60637
USA

Joseph L. Birman
Department of Physics
City College of CUNY
New York, NY 10031
USA

Jeffery W. Lynn
Department of Physics
University of Maryland
College Park, MD 20742
USA

Mark P. Silverman
Department of Physics
Trinity College
Hartford, CT 06106
USA

H. Eugene Stanley
Center for Polymer Studies
Physics Department
Boston University
Boston, MA 02215
USA

Mikhail Voloshin
Theoretical Physics Institute
Tate Laboratory of Physics
The University of Minnesota
Minneapolis, MN 55455
USA

Library of Congress Cataloging-in-Publication Data
Salinas, Silvio R.A.
 Introduction to statistical physics / Silvio R.A. Salinas.
 p. cm. — (Graduate texts in contemporary physics)
 Includes bibliographical references and index.
 ISBN 978-1-4419-2884-9
 1. Statistical physics. I. Title. II. Series.
QC174.8 .S35 2000
530.13—dc21 00-059587

Printed on acid-free paper.

Translation from Introdução a Física Estatística (1997) published by Editora da Universidade de São Paolo, Brazil.

Production managed by Allan Abrams; manufacturing supervised by Jacqui Ashri.
Photocomposed copy prepared from the author's TeX files.

Printed in the United States of America.

9 8 7 6 5 4 3 2 1

Springer-Verlag New York Berlin Heidelberg
A member of BertelsmannSpringer Science+Business Media GmbH

Preface

This book is based on the class notes for an introductory course on statistical physics that I have taught a number of times to the senior undergraduate students in physics at the University of São Paulo. Most of the students were motivated and quite well prepared. Usually, they had already gone through the basic physics and calculus courses, and through some standard intermediate courses, including classical mechanics, electromagnetism and an introduction to quantum physics. The first 10 chapters of this book are a translation of the class notes that used to circulate among the students as Xerox copies (before the publication of the original version of the text in Portuguese by the University of São Paulo in 1997). I hope I have preserved the informal and introductory features of the class notes.

Owing to the increasing relevance of the concepts and techniques of statistical mechanics, the standard undergraduate programs in physics usually include at least one semester of an introductory course on the subject. However, there are no sharp distinctions between senior year undergraduates and beginning graduate students. The first 10 chapters of this book, with the exception of a few starred (more specialized) topics, but including some highlights of the final chapters, may very well be taught to senior undergraduates during just one semester. With less emphasis on the introductory points, and including most of the material from the final chapters, the book can as well be used for a one-semester graduate course. Besides the classical topics and examples in the area, which should be taught to all students, I have included a few examples from solid-state physics. Also, I have included some topics of more recent interest, such as an introduction to phase transitions and critical phenomena, and two chapters on kinetic

and stochastic methods dealing with non-equilibrium phenomena. The instructor will be able to select enough new material to stimulate the more advanced students.

Thermodynamics is a phenomenological branch of physics that organizes in a systematic way the empirical laws governing the thermal behavior of macroscopic bodies. One of the main goals of statistical physics consists of explaining the thermodynamic laws from considerations about the behavior of the enormous number of constituent particles of the macroscopic bodies. The microscopic world of particles in motion is governed by the laws of mechanics – to be more precise, by the laws of quantum mechanics, but to a good approximation, in many cases of interest, by classical mechanics itself (or by some simplified schemes, which give rise to very relevant statistical models). In principle, the behavior of the macroscopic bodies can be explained by a straightforward application of the laws of mechanics (either classical or quantum). However, as the number of particles is excessively large, of the order of the Avogadro number ($A \approx 6 \times 10^{23}$), this program is unfeasible, and maybe even useless. Indeed, the immense number of microscopic particles gives rise to new and distinct regularities, the statistical laws. Therefore, the basic ingredients of statistical physics are the laws of mechanics and the principles of the theory of probabilities.

With the exception of the last two chapters (15 and 16), this book is dedicated to the consideration of systems in thermodynamical equilibrium (characterized by macroscopic parameters that do not change with time). Statistical mechanics in equilibrium, as formulated in terms of Gibbs ensembles, is a well-established theory, with many important applications in condensed matter physics. The methods of equilibrium statistical mechanics, including the formulation and the treatment of microscopic models, have been used in many different contexts (from more abstract problems of theoretical physics, to more concrete situations of chemical or biological interest). Therefore, Gibbsian statistical mechanics should be a highly important element in the scientific education of all students in physics (and also in modern areas of chemistry, mathematics and biology).

Unfortunately, there is no satisfactory "Gibbsian description" of non-equilibrium phenomena. The historical investigations of Boltzmann on the origins of irreversible behavior (as we know, the laws of mechanics are time reversible!) are still very much up-to-date. In recent years, there has been quite a deal of interest in the applications of stochastic methods, including different types of numerical simulations, to analyze systems out of thermodynamical equilibrium. I then decided to write the final chapters, as an introduction to Boltzmann's kinetic method and to some stochastic techniques derived from the earlier treatments of Brownian motion.

The initial chapters of this book were written under the influence of an introductory text by F. Reif (see the bibliography) that was an important contribution to the innovation of the teaching of thermal physics. In Chap-

ter 1, the problem of the one-dimensional random walk is used to present the basic ideas and concepts of the theory of probabilities. Chapter 3, based on the classical text of H. B. Callen (see the bibliography) presents a summary of the Gibbsian formulation of classical thermodynamics. This is far from the more intuitive way of teaching thermodynamics, but it is certainly the most adequate way to recall the theory and make the necessary connections with statistical physics. Chapters 1 and 3 may be omitted by the students with a better background, with no difficulty for understanding the text. The formulation of equilibrium statistical mechanics is presented in Chapters 2, 4, 5, 6, and 7, which are the backbone of any introductory text. I have tried to use simple language, with many examples. The theory is formulated in the microcanonical ensemble, and some heuristic arguments are used to construct additional ensembles (which are in general much more useful). I make a point to introduce modern ideas (thermodynamic limit, equivalence between ensembles) through a sequence of simple models (gas of classical particles, Einstein solid, ideal paramagnet), keeping the mathematical manipulations at a relatively limited level. Some more specific sections (which may demand some extra mathematical effort) are marked by a star and may be deferred for a second reading. Some mathematical topics of frequent use are presented in the appendices. It should be emphasized that the collection of exercises at the end of each chapter is an essential part of an introductory course, and that some topics are clarified by these exercises (hints and answers to selected exercises may be found at www.fge.if.usp.br/~ssalinas).

Chapters 8, 9, and 10 deal with ideal quantum gases, a standard topic in all courses of statistical physics. The required level of quantum mechanics is elementary (for example, I avoid the language of second quantization). In Chapter 8, besides the introduction of the quantum statistics (of Bose–Einstein and Fermi–Dirac), I also discuss the Maxwell–Boltzmann statistics and the conditions for the validity of the classical limit. Chapter 9 refers to the ideal Fermi gas, with some examples from solid state physics (electronic specific heat, paramagnetism, and diamagnetism). The first part of Chapter 10 is dedicated to the ideal Bose gas, with emphasis on the phenomenon of Bose–Einstein condensation. In the second part of Chapter 10, I use the problem of black body radiation to illustrate the relationships among different branches of physics (from the spectral decomposition of the electromagnetic field, I obtain the classical Rayleigh–Jeans law; then, the quantization of the electromagnetic field is used to obtain Planck´s formula for the black body radiation).

In Chapter 11, I decided to include some additional examples from solid-state physics: phonons and magnons, with a discussion of spin waves and the Heisenberg model. All of these topics, which of course reflect my own preferences, are treated at a simple level, and many times I resort to the analysis of one-dimensional model systems. During a full semester of a regular undergraduate course in physics, the instructor should make an

effort to cover most of the topics up to the end of Chapter 10. Depending on the interests of students and instructor, some of the subsequent topics may be covered as special seminars.

In Chapters 12, 13, and 14, I offer an introduction to the modern theories of phase transitions and critical phenomena. In Chapter 12, I discuss some phenomenological aspects of critical phenomena (van der Waals and Curie-Weiss equations; order parameters; the Landau theory of second-order phase transitions). I also introduce the definition, and point out the universal character, of the main critical exponents. Chapter 13 is dedicated to the Ising model: exact solution in one dimension, mean-field approximations, and some comments on the Onsager solution of the Ising ferromagnet on the square lattice. The modern phenomenological scaling theories and the renormalization-group techniques, including a number of simple examples, are presented in Chapter 14. From the 1960s, the understanding of critical phenomena is one of the triumphs of equilibrium statistical physics, with repercussions in several areas of science.

In Chapters 15 and 16, I discuss some methods and techniques of non-equilibrium statistical physics. Chapter 15 is devoted to kinetic methods, including the deduction of Boltzmann's transport equation and the famous H-theorem. In order to give an illustration of the statistical character of Boltzmann's ideas, I discuss the well-known urn model proposed by Ehrenfest. In Chapter 16, I initially derive the Langevin and Fokker–Planck equations for Brownian motion, and then present a heuristic deduction of the master equation for stochastic Markovian processes. As an example, there is a discussion of Glauber's kinetic Ising model. I then use the master equation to justify the Monte Carlo simulations of statistical models (whose increasing relevance is directly related to the widespread availability of electronic computer facilities). Again, the choice of topics is a matter of personal taste and limitations: some important advances in non-equilibrium and non-linear problems have not been mentioned.

I thank many students and colleagues for pointing out obscure paragraphs (and mistakes) in the original version of this book. I have already mentioned the influence of earlier texts of statistical mechanics (Reif, Huang) and thermodynamics (the classical book by Callen). Also, I benefitted from very good teachers in São Paulo (Mario Schoenberg) and at Carnegie-Mellon University (Robert B. Griffiths). The publication of this English version of the text is entirely due to the efforts of Marcia Barbosa and H. Eugene Stanley, to whom I am mostly grateful.

São Paulo, Brazil, February 2000 *Silvio R.A. Salinas*

Contents

1

Introduction to Statistical Methods

We have chosen the problem of the random walk in one dimension to introduce some concepts and techniques of the theory of probabilities. We use this problem to analyze the main features of binomial and Gaussian distributions, to define mean (expected) values and standard deviations, and to investigate the role of large numbers. The simplest versions of the random walk problem are already related to models of physical interest, suggested by the diffusion of particles in a viscous medium.

We believe that there is no need to review the most elementary ideas of the theory of probabilities. It should be enough to be used to flip coins or to play dice or any game of cards. If we have a coin, the probability of coming up heads is 1/2 (there are two possibilities, and each one of them is "equally likely"). Given just one standard six-faced die, the probability of coming up face 3, for example, is 1/6. We know that there are six possibilities (events) that correspond to the full sampling space (to the set of "accessible microstates of the system," in the language of statistical physics), and that only one of these states is "face 3". We are assuming *a priori* that all accessible states are equally likely (and we simply suppose that we are able to recognize equally likely phenomena). As another example, suppose that we toss two distinct dice. What is the probability of coming up "face 3" and "face 5", for example? It easy to see that it is $2/36 = 1/12$, since there are 36 equally likely events, but only two of them correspond to "face 3" and "face 5". What is the probability of getting seven points if we add the dots on each face? Now, we see that there are six possibilities out of 36, which leads to a probability of 1/6. In general, if an experiment has N equally

FIGURE 1.1. One-dimensional random walk of steps of length l.

likely outcomes, and N_s of them lead to "success," we then say that the *a priori* probability of "success" is N_s/N.

1.1 The random walk in one dimension

Consider a particle in one dimension, performing successive displacements of the same length (l), with probability p of moving to the right, and probability $q = 1 - p$ of moving to the left. The problem consists in finding the probability $P_N(m)$ that the particle is in position $x = ml$ after performing N successive steps (m is an integer, such that $-N \leq m \leq N$), from the initial position $x = 0$. A three-dimensional version of this problem, with vector displacements, is related to the diffusion of a gas of particles under the action of intermolecular collisions.

The probability of a **given particular sequence** of N steps, with N_1 steps to the right, and $N_2 = N - N_1$ steps to the left, is given by the product

$$(p...p)\,(q...q) = p^{N_1} q^{N_2}.$$

Note that each step is an independent event, so the probability of a given sequence is just a simple product of individual probabilities. The **number of sequences** of this type (with N_1 steps to the right, and $N_2 = N - N_1$ steps to the left) is given by the combinatorial factor

$$\frac{N!}{N_1! N_2!}.$$

Therefore, the probability that the particle performs N_1 displacements to the right, out of a total of N displacements, is given by the well-known binomial distribution,

$$W_N(N_1) = \frac{N!}{N_1! N_2!} p^{N_1} q^{N_2}, \tag{1.1}$$

with $p + q = 1$ and $N_1 + N_2 = N$. Note that this probability is duly normalized. Indeed, we have

$$\sum_{N_1=0}^{N} W_N(N_1) = \sum_{N_1=0}^{N} \frac{N!}{N_1! N_2!} p^{N_1} q^{N_2} = (p+q)^N = 1.$$

Also, note that $0 \leq W_N(N_1) \leq 1$, for $0 \leq N_1 \leq N$, so that the probability is a positive integer, between 0 and 1. Since $m = N_1 - N_2$, the probability $P_N(m)$ is finally written as

$$P_N(m) = \frac{N!}{\left(\frac{N+m}{2}\right)! \left(\frac{N-m}{2}\right)!} p^{\frac{N+m}{2}} q^{\frac{N-m}{2}}, \tag{1.2}$$

with $p + q = 1$.

To make a connection with the phenomenon of diffusion, we now formulate the random walk problem in terms of a stochastic difference equation. Let us assume the same time interval τ between all successive steps (of the same length l), and interpret $P_N(m)$ as the probability of finding the particle at position $x = ml$, at time $t = N\tau$. Now, we can write an equation to calculate the probability $P_{N+1}(m)$. To reach the position $x = lm$, at the time $t = (N+1)\tau$, the particle has to be either at $x = (m+1)l$ or at $x = (m-1)l$, at the previous time $t = N\tau$. Thus, it is intuitive to write the recurrence relation

$$P_{N+1}(m) = pP_N(m-1) + qP_N(m+1). \tag{1.3}$$

The binomial distribution, given by equation (1.2), is a solution of this stochastic difference equation. Sequences of this kind, where the probability at a certain time depends only on the probabilities at the immediately previous time, are known as "Markov chains" and play an important role in many problems of physical interest. In the stochastic equation (1.3), the detailed dynamics of the physical problem has been replaced by some probabilistic assertions. Stochastic formulations of this nature play an increasingly important role to describe the behavior of physical systems out of equilibrium (see Chapter 16). Consider the particular case of equal probabilities for steps to the right and to the left, $p = q = 1/2$, and take the limit where τ and l are very small. We then write the continuous representation,

$$\frac{P_{N+1} - P_N}{\tau} \sim \frac{\partial P}{\partial t},$$

and

$$\frac{P(ml - l) + P(ml + l) - 2P(ml)}{l^2} \sim \frac{\partial^2 P}{\partial x^2}.$$

Therefore, in this limit we obtain the well-known diffusion differential equation,

$$\frac{\partial P}{\partial t} = D \frac{\partial^2 P}{\partial x^2}, \tag{1.4}$$

with the diffusion coefficient $D = l^2/2\tau$.

1.2 Mean values and standard deviations

Let u be a **random variable** that may assume M discrete values, such that u_j occurs with probability $P_j = P(u_j)$, where $0 \leq P_j \leq 1$, for all j. In general, we consider normalized probabilities, so that

$$\sum_{j=1}^{M} P(u_j) = 1.$$

The **average, mean, or expected value** of the variable u is given by

$$\bar{u} = \langle u \rangle = \sum_{j=1}^{M} u_j P(u_j). \tag{1.5}$$

If $f(u)$ is a function of u, the mean value of $f(u)$ is given by

$$\overline{f(u)} = \langle f(u) \rangle = \sum_{j=1}^{M} f(u_j) P(u_j). \tag{1.6}$$

It is easy to show that (i) $\langle f(u) + g(u) \rangle = \langle f(u) \rangle + \langle g(u) \rangle$ and (ii) $\langle cf(u) \rangle = c \langle f(u) \rangle$, where c is a constant, and f and g are (stochastic) functions of the random variable u.

The deviation from the mean value is given by

$$\Delta u = u - \langle u \rangle. \tag{1.7}$$

Thus, we have

$$\langle \Delta u \rangle = \langle (u - \langle u \rangle) \rangle = 0,$$

which indicates that the first moment of u about the mean is of no use.

The quadratic deviation is given by

$$(\Delta u)^2 = (u - \langle u \rangle)^2. \tag{1.8}$$

Therefore, the **dispersion**, or **second moment** with respect to the mean, is given by

$$\left\langle (\Delta u)^2 \right\rangle = \langle (u - \langle u \rangle)^2 \rangle = \langle u^2 \rangle - \langle u \rangle^2. \tag{1.9}$$

It is obvious that $\left\langle (\Delta u)^2 \right\rangle \geq 0$, which leads to $\langle u^2 \rangle \geq \langle u \rangle^2$. The dispersion is also called variance, and the square root of the dispersion is the **standard deviation**. A comparison between the standard deviation and the mean value is interesting to give an idea of the shape of the probability

FIGURE 1.2. Examples of statistical distributions. On the right, $\Delta N_1^* << \langle N_1 \rangle$.

distribution. For a small standard deviation, the probability distribution displays a sharp peak about the mean value.

In general, we can define the nth moment with respect to the mean value,

$$\langle (\Delta u)^n \rangle = \langle (u - \langle u \rangle)^n \rangle. \tag{1.10}$$

From the moments of a distribution, it is possible to reconstruct the probability distribution. However, in many cases of interest, and especially for Gaussian distributions, it is enough to know the mean value and the second moment, $\langle u \rangle$ and $\left\langle (\Delta u)^2 \right\rangle$.

For the random walk problem in one dimension, we have

$$\langle N_1 \rangle = \sum_{N_1=0}^{N} N_1 W_N(N_1) = \sum_{N_1=0}^{N} N_1 \frac{N!}{N_1!N_2!} p^{N_1} q^{N_2},$$

where $N_2 = N - N_1$, and we make $q = 1 - p$ at the end of the calculations. Hence, we can write

$$\langle N_1 \rangle = p\frac{\partial}{\partial p} \sum_{N_1=0}^{N} \frac{N!}{N_1!N_2!} p^{N_1} q^{N_2} = p\frac{\partial}{\partial p}(p+q)^N = pN(p+q)^{N+1} = pN.$$

It is obvious that $\langle N_2 \rangle = \langle N - N_1 \rangle = N - pN = qN$. To summarize, we have

$$\langle N_1 \rangle = pN,$$

$$\langle N_2 \rangle = qN. \tag{1.11}$$

In order to obtain the dispersion with respect to the mean, we note that

$$\langle N_1^2 \rangle = \left(p\frac{\partial}{\partial p} \right) \left\{ \left(p\frac{\partial}{\partial p} \right) \left[\sum_{N_1=0}^{N} \frac{N!}{N_1!N_2!} p^{N_1} q^{N_2} \right] \right\}$$

$$= \left(p\frac{\partial}{\partial p} \right) \left\{ pN(p+q)^{N+1} \right\} = pN + p^2 N(N-1).$$

Thus, we have

$$\left\langle (\Delta N_1)^2 \right\rangle = \langle N_1^2 \rangle - \langle N_1 \rangle^2 = Npq, \qquad (1.12)$$

which leads to the well-known dependence of the standard deviation on \sqrt{N},

$$\Delta N_1^* \equiv \sqrt{\left\langle (\Delta N_1)^2 \right\rangle} = (pq)^{1/2}\sqrt{N}. \qquad (1.13)$$

We then write the relative width,

$$\frac{\Delta N_1^*}{\langle N_1 \rangle} = \left(\frac{p}{q}\right)^{1/2} \frac{1}{\sqrt{N}}, \qquad (1.14)$$

which shows that the binomial distribution becomes very sharp, about the mean value $\langle N_1 \rangle$, for sufficiently large values of N (see figure 1.2).

1.3 Gaussian limit of the binomial distribution

In the limit $N \to \infty$, we have $W_N(0) = q^N \to 0$, and $W_N(N) = p^N \to 0$. Hence, $W_N(N_1)$ as a function of N_1 should be maximum at $N_1 = \tilde{N}_1 = rN$, with $0 < r < 1$. For sufficiently large N, in the neighborhood of this maximum, we assume that $W_N(N_1)$ is a well-behaved function of the random variable N_1. However, instead of working with $W_N(N_1)$, it may be more convenient to work with the slower varying function $\ln W_N(N_1)$. Thus, we write

$$\begin{aligned} f(N_1) &= \ln W_N(N_1) = \ln N! - \ln N_1! - \ln(N - N_1)! \\ &\quad + N_1 \ln p + (N - N_1) \ln q. \end{aligned} \qquad (1.15)$$

Recalling that the logarithm function is monotone increasing, $f(N_1)$ is also maximum at \tilde{N}_1. Close to this maximum, both N_1 and $N - N_1$ are of the order of N. Then, we use the asymptotic expansion of Stirling (see Appendix A.1),

$$\ln N! = N \ln N - N + O(\ln N), \qquad (1.16)$$

to eliminate the factorials. Therefore, we have

$$\begin{aligned} f(N_1) &= N \ln N - N - N_1 \ln N_1 + N_1 - (N - N_1)\ln(N - N_1) \\ &\quad + (N - N_1) + N_1 \ln p + (N - N_1)\ln q + O\left(\ln N, \ln(N - N_1)\right). \end{aligned}$$
$$(1.17)$$

The first derivative with respect to N_1 is given by

$$\frac{\partial f}{\partial N_1} = -\ln N_1 + \ln\left(N - N_1\right) + \ln p - \ln q + O\left(\frac{1}{N}, \frac{1}{N - N_1}\right) = 0.$$

(1.18)

In the limit $N \to \infty$, we have

$$-\ln \tilde{N}_1 + \ln\left(N - \tilde{N}_1\right) + \ln p - \ln q = 0,$$

(1.19)

so that

$$\tilde{N}_1 = Np = \langle N_1 \rangle,$$

(1.20)

which shows that the **most probable** (maximum) is identical to the **mean value** (see equation 1.11). The second derivative is given by

$$\frac{\partial^2 f}{\partial N_1^2} = -\frac{1}{N_1} - \frac{1}{N - N_1} + O\left(\frac{1}{N_1^2}, \frac{1}{(N - N_1)^2}\right).$$

(1.21)

At the maximum, $N_1 = Np$, and $N \to \infty$, we have

$$\left(\frac{\partial^2 f}{\partial N_1^2}\right)_{N_1 = \tilde{N}_1} = -\frac{1}{Npq} = -\frac{1}{\left\langle (\Delta N_1)^2 \right\rangle} < 0.$$

(1.22)

We now perform a Taylor expansion about the maximum,

$$f(N_1) = \ln W_N(N_1) = \ln W_N(\tilde{N}_1) - \frac{1}{2Npq}\left(N_1 - \tilde{N}_1\right)^2 + \dots.$$

(1.23)

To obtain the Gaussian approximation, we discard the higher-order terms of this expansion. This should be reasonable for large N, since $W(N_1)$ is very sharply peaked about the maximum. Noting that $\tilde{N}_1 = \langle N_1 \rangle$ and that $Npq = \left\langle (\Delta N_1)^2 \right\rangle = (\Delta N_1^*)^2$, we write the following Gaussian approximation for the binomial distribution,

$$p_G(N_1) = p_o \exp\left[-\frac{(N_1 - \langle N_1 \rangle)^2}{2\left(\Delta N_1^*\right)^2}\right],$$

(1.24)

where the coefficient p_o comes from the normalization,

$$\int_{-\infty}^{+\infty} p_o \exp\left[-\frac{x^2}{2\left(\Delta N_1^*\right)^2}\right] dx = 1.$$

(1.25)

Using the results of Appendix A.2 for integrals of Gaussian form, we have

$$p_o = \left[2\pi \left(\Delta N_1^*\right)^2\right]^{-1/2}. \tag{1.26}$$

Hence, we write the **normal** or **Gaussian distribution**,

$$p_G(N_1) = \left[2\pi \left(\Delta N_1^*\right)^2\right]^{-1/2} \exp\left[-\frac{(N_1 - \langle N_1 \rangle)^2}{2\left(\Delta N_1^*\right)^2}\right]. \tag{1.27}$$

Using this expression, it is straightforward to see that

$$\langle N_1 \rangle_G = \int\limits_{-\infty}^{+\infty} N_1 p_G(N_1)dN_1 = \langle N_1 \rangle \tag{1.28}$$

and

$$\left\langle (\Delta N_1)^2 \right\rangle_G = \int\limits_{-\infty}^{+\infty} (N_1 - \langle N_1 \rangle_G)^2 \, p_G(N_1)dN_1 = (\Delta N_1^*)^2, \tag{1.29}$$

in full agreement with the results for the original binomial distribution. Higher-order moments, however, as calculated from the Gaussian form, may not be identical to the corresponding moments of the associated binomial distribution.

It is interesting to investigate the limits of validity of the approximation of a binomial by the corresponding Gaussian distribution (with the same values of first and second moments). Consider the third derivative,

$$\frac{\partial^3 f}{\partial N_1^3} = \frac{1}{N_1^2} - \frac{1}{(N - N_1)^2} + O\left(\frac{1}{N_1^3}\right).$$

At the maximum, and for $N \to \infty$, the dominating term of this expression is given by

$$\left(\frac{\partial^3 f}{\partial N_1^3}\right)_{N_1 = \tilde{N}_1} = \frac{q - p}{N^2 p^2 q^2}.$$

The Gaussian approximation should work for

$$\frac{1}{2Npq}\left|N_1 - \tilde{N}_1\right|^2 >> \frac{|q - p|}{6N^2p^2q^2}\left|N_1 - \tilde{N}_1\right|^3,$$

which leads to

$$\left|N_1 - \tilde{N}_1\right| << \frac{3Npq}{|q - p|}.$$

Outside of this interval, that is, for $\left|N_1 - \tilde{N}_1\right| \approx 3Npq/\left|q - p\right|$, we have

$$p_G \sim p_o \exp\left[-\frac{1}{2Npq}\frac{9N^2p^2q^2}{\left|q - p\right|^2}\right] \to 0,$$

as $N \to \infty$. Therefore, in the range of values of N_1 where the approximation does not work, the distribution p_G vanishes exponentially. Because the standard deviation of both binomial and Gaussian distributions is of order \sqrt{N}, these distributions are both exponentially small for $N_1 - \tilde{N}_1$ of the order of N, which justifies the approximation. Now, it is not difficult to calculate higher-order terms to keep checking the validity of this argument.

1.4 Distribution of several random variables. Continuous distributions

We can easily generalize the previous definitions and properties for two or more discrete random variables. For example, consider the random variables u and v. We associate with the pair of values u_j, v_k a joint distribution $P(u_j, v_k)$, such that $0 \leq P(u_j, v_k) \leq 1$, and

$$\sum_{j,k} P(u_j, v_k) = 1. \tag{1.30}$$

We can also define the probability

$$P_u(u_j) = \sum_k P(u_j, v_k) \tag{1.31}$$

that u assumes the value u_j, regardless of the value v_k. It is clear that

$$\sum_j P_u(u_j) = 1. \tag{1.32}$$

Two random variables are **statistically independent**, or non-correlated, if

$$P(u_j, v_k) = P_u(u_j)P_v(v_k), \tag{1.33}$$

as we have implicitly used in the problem of the random walk. It is easy to calculate the mean values of the sum and of the product of several random variables. In particular, the mean value of the product is the product of mean values **only** if the random variables are statistically independent.

Another immediate generalization, which has been implicitly used to construct the Gaussian distribution, consists in the consideration of continuous random variables. Suppose that the random variable u assumes all

real values in the interval between a and b. The differential form $p(u)du$ is then the probability that the random variable u is between the values u and $u + du$, and the function $p(u)$ represents a distribution of a density of probability. The normalization is written as

$$\int_a^b p(u)du = 1. \tag{1.34}$$

It is straightforward to generalize all of the concepts that were developed for discrete distributions. For example, the mean value of the stochastic function $f(u)$ is given by

$$\langle f(u) \rangle = \int_a^b f(u)p(u)du. \tag{1.35}$$

In the continuum limit of the discrete distributions, we emphasize that the interval du is macroscopically small, but microscopically large. In other words, there should be many original discrete values of u within the macroscopically small interval du. The probability vanishes with du, but the density $p(u)$ is independent of du. All of these ideas have been informally used in the last section to construct the Gaussian approximation $p_G(N_1)$ for the binomial distribution.

The problem of the random walk in one dimension may be slightly generalized if we suppose that, in the jth step, the displacement is given by the continuous random length s_j associated with the probability $w(s_j)ds_j$. After performing N steps, we can ask for the probability $p(x; N)dx$ that the particle is in the interval between x and $x + dx$, where $x = \sum_{j=1}^N s_j$. Both s_j, for $j = 1, ..., N$, and x are continuous random variables. The length x is a function of the **independent and identically distributed** random variables $s_1, ..., s_N$. The mean value and the dispersion of the variable x can be easily calculated. Indeed, we have

$$\langle x \rangle = \left\langle \sum_{j=1}^N s_j \right\rangle = \sum_{j=1}^N \langle s_j \rangle = N \langle s \rangle, \tag{1.36}$$

where

$$\langle s \rangle = \int_{-\infty}^{+\infty} sw(s)ds. \tag{1.37}$$

Hence,

$$\Delta x = \sum_{j=1}^N s_j - N \langle s \rangle = \sum_{j=1}^N (s_j - \langle s \rangle) = \sum_{j=1}^N \Delta s_j. \tag{1.38}$$

Thus,

$$\langle \Delta x \rangle = \sum_{j=1}^{N} \langle \Delta s_j \rangle = 0. \tag{1.39}$$

We also have

$$(\Delta x)^2 = \left(\sum_{j=1}^{N} \Delta s_j \right) \left(\sum_{k=1}^{N} \Delta s_k \right) = \sum_{j=1}^{N} (\Delta s_j)^2 + \sum_{j \neq k} (\Delta s_j)(\Delta s_k). \tag{1.40}$$

Thus,

$$\left\langle (\Delta x)^2 \right\rangle = \sum_{j=1}^{N} \left\langle (\Delta s_j)^2 \right\rangle + \sum_{j \neq k} \langle \Delta s_j \Delta s_k \rangle. \tag{1.41}$$

Since the steps are statistically independent, $\langle \Delta s_j \Delta s_k \rangle = \langle \Delta s_j \rangle \langle \Delta s_k \rangle = 0$, for $j \neq k$, we have

$$\left\langle (\Delta x)^2 \right\rangle = \sum_{j=1}^{N} \left\langle (\Delta s_j)^2 \right\rangle = N \left\langle (\Delta s)^2 \right\rangle, \tag{1.42}$$

where

$$\left\langle (\Delta s)^2 \right\rangle = \int_{-\infty}^{+\infty} (\Delta s)^2 \, w(s) ds. \tag{1.43}$$

Finally, we write the relative width,

$$\frac{\sqrt{\left\langle (\Delta x)^2 \right\rangle}}{\langle x \rangle} = \frac{\sqrt{\left\langle (\Delta s)^2 \right\rangle}}{\langle s \rangle} \frac{1}{\sqrt{N}} \sim \frac{1}{\sqrt{N}}. \tag{1.44}$$

Again, for sufficiently large N, and assuming that $w(s)$ is a well-behaved function, which vanishes sufficiently fast for $s \to \pm\infty$, the distribution $p(x; N)$ as a function of x should be sharply peaked in the neighborhood of the mean value. In the following section, we obtain an integral form of $p(x; N)$ to show that it becomes a Gaussian distribution in the limit of large N. This explains the frequent occurrence of Gaussian distributions associated with physical problems involving a very large number of independent events.

1.5 *Probability distribution for the generalized random walk in one dimension. The Gaussian limit

Since the successive steps in the generalized problem of the random walk in one dimension are statistically independent, the probability of occurrence of a given sequence is still given by a simple product, as in the discrete problem. Consider again a sequence of N steps, and suppose that the displacement associated with the jth step has a random length s_j, with a probability $w(s_j)ds_j$. The probability of finding the particle between x and $x + dx$, where $x = \sum_{j=1}^{N} s_j$, is given by

$$p(x; N)dx = \int \cdots \int_{x<s_1+s_2+\ldots+s_N<x+dx} w(s_1)ds_1 \cdots w(s_N)ds_N, \qquad (1.45)$$

where the integrations are performed from $-\infty$ to $+\infty$, with the restrictions that $x < s_1 + s_2 + \ldots + s_N < x + dx$. To remove these restrictions, we use Dirac's δ-function, as introduced in Appendix A.3,

$$\delta(x) = \begin{cases} 1/\gamma, & \text{for} \quad -\gamma/2 < x < \gamma/2, \\ 0, & \text{for} \quad x > \gamma/2, \end{cases} \qquad (1.46)$$

with $\gamma \to 0$. Therefore, we have

$$p(x; N)dx = \int_{-\infty}^{+\infty} \cdots \int_{-\infty}^{+\infty} w(s_1)ds_1 \cdots w(s_N)ds_N dx \delta\left(x - \sum_{j=1}^{N} s_j\right). \qquad (1.47)$$

Using an integral representation of the δ-function (see Appendix A.3),

$$\delta(x) = \frac{1}{2\pi} \int_{-\infty}^{+\infty} \exp(-ikx) \, dk, \qquad (1.48)$$

we can write

$$p(x; N) = \frac{1}{2\pi} \int_{-\infty}^{+\infty} \cdots \int_{-\infty}^{+\infty} w(s_1)ds_1 \cdots w(s_N)ds_N$$

$$\times \int_{-\infty}^{+\infty} \exp\left[-ik\left(x - \sum_{j=1}^{N} s_j\right)\right] dk. \qquad (1.49)$$

We now see that the factorization of the integrals over the variables $s_1, ..., s_N$ leads to the expression

$$p(x; N) = \frac{1}{2\pi} \int\limits_{-\infty}^{+\infty} \exp{(-ikx)} \left[\hat{w}(k)\right]^N dk, \qquad (1.50)$$

where the **characteristic function** $\hat{w}(k)$ is the Fourier transform of $w(s)$,

$$\hat{w}(k) = \int\limits_{-\infty}^{+\infty} \exp{(iks)} \, w(s) ds. \qquad (1.51)$$

As an exercise, let us check that this formalism reproduces the binomial distribution for the discrete random walk in one dimension with steps of the same length l. The function $w(s)$ is then given by

$$w(s) = p\delta{(s - l)} + q\delta{(s + l)},$$

which leads to

$$\hat{w}(k) = p\exp{(ikl)} + q\exp{(-ikl)},$$

from which we have

$$[\hat{w}(k)]^N = \sum_{n=0}^{N} \frac{N!}{n!\,(N - n)!} \left(pe^{ikl}\right)^n \left(qe^{-ikl}\right)^{N-n}.$$

Hence, we can write

$$
\begin{aligned}
p(x; N) &= \frac{1}{2\pi} \int\limits_{-\infty}^{+\infty} \exp{(-ikx)} \sum_{n=0}^{N} \frac{N!}{n!\,(N - n)!} p^n q^{N-n} \\
&\quad \times \exp{[iknl - ikl\,(N - n)]}\, dk \\
&= \sum_{n=0}^{N} \frac{N!}{n!\,(N - n)!} p^n q^{N-n} \delta\,(x - 2nl + Nl).
\end{aligned}
$$

Therefore, $p(x)$ vanishes except at $x = (2n - N)\,l$, with $n = 0, 1, ..., N$. In order to obtain the discrete binomial distribution, we integrate $p(x)$ over an infinitesimal interval around $x = (2n - N)\,l$. Thus, we finally have

$$P_N(n) = \int\limits_{(2n-N)l-\epsilon}^{(2n-N)l+\epsilon} p(x; N) dx = \frac{N!}{n!\,(N - n)!} p^n q^{N-n}.$$

The probability density $p(x; N)$ can also be obtained, in a more direct and somewhat easier way, from the associated stochastic equation, as it has been done in Section 1.1 for the random walk with steps of equal length. From equation (1.3), it is intuitive to write the slightly more general recurrence relation

$$p(x; N+1) = \int_{-\infty}^{+\infty} p(x-s; N)w(s)ds, \qquad (1.52)$$

whose right-hand side is a convolution integral. Introducing the Fourier transformation,

$$\hat{p}(k; N) = \int_{-\infty}^{+\infty} \exp(ikx)\,p(x; N)dx, \qquad (1.53)$$

and using $\hat{w}(k)$, given by equation (1.51), we have

$$\hat{p}(k; N+1) = \hat{p}(k; N)\hat{w}(k). \qquad (1.54)$$

From this equation, and taking into account that at the initial time (that is, for $N = 0$) the particle is at the origin, in other words, that $p(x; 0) = \delta(x)$, we have

$$\hat{p}(k; N) = [\hat{w}(k)]^N. \qquad (1.55)$$

The distribution $p(x; N)$ comes from the inverse Fourier transform,

$$p(x; N) = \frac{1}{2\pi} \int_{-\infty}^{+\infty} \exp(-ikx) [\hat{w}(k)]^N dk, \qquad (1.56)$$

which is the same expression as equation (1.50).

We now obtain an asymptotic expression for $p(x; N)$ in the limit of large N. Owing to the oscillating factor $\exp(iks)$, the function $\hat{w}(k)$ is only appreciable in the immediate vicinity of $k = 0$. This is further enhanced for large values of N. Thus, we write the expansion

$$\hat{w}(k) = \int_{-\infty}^{+\infty} \exp(iks)\,w(s)ds = \int_{-\infty}^{+\infty} w(s) \left\{ 1 + iks - \frac{1}{2}k^2 s^2 + ... \right\} ds$$

$$= 1 + ik\langle s \rangle - \frac{1}{2}k^2 \langle s^2 \rangle + \qquad (1.57)$$

Hence, we have

$$[\hat{w}(k)]^N = \exp[N \ln \hat{w}(k)]$$

$$= \exp\left\{ N \left[ik\langle s \rangle + \frac{1}{2}k^2 \left(\langle s \rangle^2 - \langle s^2 \rangle \right) + O(k^3) \right] \right\} \qquad (1.58)$$

If we discard terms of higher order than k^2, we have the integral

$$p(x) = \frac{1}{2\pi} \int\limits_{-\infty}^{+\infty} \exp\left[-ikx + Nik\langle s\rangle - \frac{1}{2}N\left\langle (\Delta s)^2\right\rangle k^2\right] dk, \qquad (1.59)$$

which leads to the Gaussian form

$$p(x) = \left(2\pi\sigma^2\right)^{1/2} \exp\left[-\frac{(x - N\langle s\rangle)^2}{2\sigma^2}\right], \qquad (1.60)$$

where $\sigma^2 = N\left\langle (\Delta s)^2\right\rangle$. Again, we have a Gaussian distribution, with the same mean value and second moment as the original distribution. It should be pointed out that, for $\langle s\rangle = 0$, with $N = t\tau$, $D = \left\langle (\Delta s)^2\right\rangle / (2\tau)$, in the continuum limit ($t \to \infty$, $\left\langle (\Delta s)^2\right\rangle \to 0$, and $\tau \to 0$), the Gaussian form of equation (1.60) is a solution of the diffusion equation

$$\frac{\partial p}{\partial t} = D\frac{\partial^2 p}{\partial x^2}. \qquad (1.61)$$

The existence of a Gaussian limit comes from the well-known **central limit theorem** of the theory of probabilities. This theorem holds if (i) the steps are statistically independent; (ii) the function $w(s)$ vanishes sufficiently fast as $s \to \pm\infty$; and (iii) N is sufficiently large. These conditions are met by a huge variety of physically interesting phenomena, which is the basis for the widespread use and relevance of Gaussian distributions.

Exercises

1. Obtain the probability of adding up six points if we toss three distinct dice.

2. Consider a binomial distribution for a one-dimensional random walk, with $N = 6$, $p = 2/3$, and $q = 1 - p = 1/3$.

 (a) Draw a graph of $P_N(N_1)$ versus N_1/N.

 (b) Use the values of $\langle N_1\rangle$ and $\langle N_1^2\rangle$ to obtain the corresponding Gaussian distribution, $p_G(N_1)$. Draw a graph of $p_G(N_1)$ versus N_1/N to compare with the previous result.

 (c) Repeat items (a) and (b) for $N = 12$ and $N = 36$. Are the new answers too different?

3. Obtain an expression for the third moment of a binomial distribution. What is the behavior of this moment for large N?

4. Consider an event of probability p. The probability of n occurrences of this event out of N trials is given by the binomial distribution

$$W_N(n) = \frac{N!}{n!(N-n)!} p^n (1-p)^{N-n}.$$

If p is small ($p << 1$), $W_N(n)$ is very small, except for $n << N$. In this limit, show that we obtain the Poisson distribution,

$$W_N(n) \to P(n) = \frac{\lambda^n}{n!} \exp(-\lambda),$$

where $\lambda = np$ is the mean number of events. Check that $P(n)$ is normalized. Calculate $\langle n \rangle$ and $\langle (\Delta n)^2 \rangle$ for this Poisson distribution. Formulate a statistical problem to be solved in terms of this distribution.

5. Consider an experiment with N equally likely outcomes, involving two events A and B. Let N_1 be the number of events in which A occurs, but not B; N_2 be the number of events in which B occurs, but not A; N_3 be the number of events in which both A and B occur; and N_4 be the number of events in which neither A nor B occur.

 (a) Check that $N_1 + N_2 + N_3 + N_4 = N$.

 (b) Check that

 $$P(A) = \frac{N_1 + N_3}{N} \text{ and } P(B) = \frac{N_2 + N_3}{N},$$

 where $P(A)$ and $P(B)$ are the probabilities of occurrence of A and B, respectively.

 (c) Calculate the probability $P(A+B)$ of occurrence of either A or B.

 (d) Calculate the probability $P(AB)$ of occurrence of both A and B.

 (e) Calculate the conditional probability $P(A \mid B)$ that A occurs given that B occurs.

 (f) Calculate the conditional probability $P(B \mid A)$ that B occurs given that A occurs.

 (g) Show that

 $$P(A+B) = P(A) + P(B) - P(AB)$$

 and

 $$P(AB) = P(B) P(A \mid B) = P(A) P(B \mid A).$$

(h) Considering a third event C, show that

$$\frac{P(B \mid A)}{P(C \mid A)} = \frac{P(B)}{P(C)} \frac{P(A \mid B)}{P(A \mid C)},$$

which is an expression of Bayes' theorem.

6. A random variable x is associated with the probability density

$$p(x) = \exp(-x),$$

for $0 < x < \infty$.

(a) Find the mean value $\langle x \rangle$.

(b) Two values x_1 and x_2 are chosen independently. Find $\langle x_1 + x_2 \rangle$ and $\langle x_1 x_2 \rangle$.

(c) What is the probability distribution of the random variable $y = x_1 + x_2$?

7. Consider a random walk in one dimension. After N steps from the origin, the position is given by $x = \sum_{j=1}^{N} s_j$, where $\{s_j\}$ is a set of independent, identically distributed, random variables, given by the probability distribution

$$w(s) = \left(2\pi\sigma^2\right)^{-1/2} \exp\left[-\frac{(s-l)^2}{2\sigma^2}\right],$$

where σ and l are positive constants. After N steps, what is the average displacement from the origin? What is the standard deviation of the random variable x? In the large N limit, what is the form of the Gaussian distribution associated with this problem?

8. Consider again problem 7, with a distribution $w(s)$ of the Lorentzian form

$$w(s) = \frac{1}{\pi} \frac{a}{s^2 + a^2},$$

with $a > 0$. Obtain an expression for the probability distribution associated with the random variable x. Is it possible to write a Gaussian approximation for large N? Why?

(b) Considering a third event C, show that

$$\frac{P(B|A)}{P(C|A)} = \frac{P(B)}{P(C)} \frac{P(A|B)}{P(A|C)}$$

which is an expression of Bayes' theorem.

6. A random variable x is associated with the probability density

$$p(x) = \exp(-x),$$

for $0 < x < \infty$.

(a) Find the mean value $\langle x \rangle$.

(b) Two values x_1 and x_2 are chosen independently. Find $\langle x_1 + x_2 \rangle$ and $\langle x_1 x_2 \rangle$.

(c) What is the probability distribution of the random variable $y = x_1 + x_2$?

7. Consider a random walk in one dimension. After N steps from the origin, the position is given by $x = \sum_{j=1}^{N} s_j$, where $\{s_j\}$ is a set of independent, identically distributed random variables, given by the probability distribution

$$w(s) = (2\pi\sigma^2)^{-1/2} \exp\left[-\frac{(s-l)^2}{2\sigma^2} \right]$$

where σ and l are positive constants. After N steps, what is the average displacement from the origin? What is the standard deviation of the random variable x? In the large N limit, what is the form of the Gaussian distribution associated with this problem?

8. Consider again problem 7, with a distribution $w(s)$ of the Lorentzian form

$$w(s) = \frac{1}{\pi} \frac{a}{s^2 + a^2}$$

with $a > 0$. Obtain an expression for the probability distribution associated with the random variable x. Is it possible to write a Gaussian approximation for large N? Why?

2
Statistical Description of a Physical System

The mechanico-statistical analysis of a physical system in **equilibrium** comprises the following steps: (i) The specification of the microscopic states of the system (which form a set called **statistical ensemble**); (ii) the statement of a statistical postulate in order to use the theory of probabilities – for a system with fixed total energy, we invoke the simplifying assumption of **equal a priori probabilities**, which leads to the **microcanonical ensemble** – and (iii) the establishment of a connection with the visible variables of the macroscopic world (that is, **with thermodynamics**).

A physical system of particles is governed by the laws of mechanics (either classical or quantum, depending on the level of description and on the particular interests of the analysis), which provide the means for the specification of the microscopic states. Depending on the phenomenon under consideration, we can construct specific models, sometimes of a semiclassical character, with a drastically simplified microscopic dynamics. We then take into account the basic mechanisms associated with the physical manifestations of interest. For instance, to analyze the magnetic properties of an ionic insulating crystal, it is convenient to consider a rigid crystalline lattice and to neglect vibrational motions of the magnetic ions. In magnetic models of this type, we are usually interested in the magnetic moments (that is, the spins) of the electronic shell, disregarding the effects of the additional degrees of freedom (including the nuclear spins). In some cases, owing to technical reasons, it may be interesting to introduce a discrete model of a gas of N particles within a container of volume V. This volume is divided into V discrete cells that may be either empty or occupied by

only one particle, so that we simulate the effects of impenetrability of the hard core of interatomic potentials. In this "lattice gas" model, the microscopic specification of the states of the system consists in the identification of the configurations of N particles distributed in V cells. The statistical ensemble is the set of all these microscopic states, to each one of which we assign a well-determined probabilistic weight.

In this chapter, we use several examples to illustrate the specification of the microscopic states of statistical models. We also state the fundamental postulate of statistical mechanics, introduce the microcanonical ensemble, and present a brief discussion of the basis of the theory. The connection with the macroscopic world will be treated later, after discussing the basic ideas of Gibbsian thermodynamics.

2.1 Specification of the microscopic states of a quantum system

In quantum mechanics a stationary system is characterized by the wave function $\Psi(q_1, q_2, ...)$. In general, this wave function may be written in terms of a complete orthonormal basis of eigenfunctions of an operator as the Hamiltonian of the system. We thus write

$$\Psi = \sum_n c_n \phi_n, \tag{2.1}$$

with

$$\mathcal{H}\phi_n = E_n \phi_n, \tag{2.2}$$

where \mathcal{H} is the Hamiltonian operator. The eigenstates ϕ_n, characterized by the set of n quantum numbers, give rise to a simple way to count the "microscopic states" of the system. Later, we will return to this question, to argue that quantum mechanics itself displays an intrinsic statistical character, distinct from the statistics that is needed to describe the distribution of the microscopic states of the system.

Example 1 *Localized particle of spin* 1/2.

There are two eigenstates,

$$\alpha = \begin{pmatrix} 1 \\ 0 \end{pmatrix} \text{ and } \beta = \begin{pmatrix} 0 \\ 1 \end{pmatrix}, \tag{2.3}$$

associated with "spin up, or ↑," and "spin down, or ↓," respectively. In the presence of an external applied magnetic field \vec{H}, the energy (Hamiltonian) is given by

$$\mathcal{H} = -\vec{\mu} \cdot \vec{H} = -\mu_z H = \begin{cases} +\mu_o H, & \text{for spin } \downarrow, \\ -\mu_o H, & \text{for spin } \uparrow, \end{cases} \tag{2.4}$$

where $\vec{\mu}$ is the magnetic moment, with projection μ_z along the direction of the applied field, so that either $\mu_z = +\mu_o$ or $\mu_z = -\mu_o$.

Example 2 *Three localized, noninteracting (maybe very weakly interacting), particles of spin 1/2, in the presence of an external field \vec{H}.*

In this case the Hamiltonian is given by he sum

$$\mathcal{H} = -\vec{\mu}_1 \cdot \vec{H} - \vec{\mu}_2 \cdot \vec{H} - \vec{\mu}_3 \cdot \vec{H}.$$

We have eight eigenstates that are given by the following products: (i) $\uparrow\uparrow\uparrow$ or $\alpha_1\alpha_2\alpha_3$, with energy $-3\mu_o H$; (ii) $\uparrow\uparrow\downarrow$, (iii) $\uparrow\downarrow\uparrow$, and (iv) $\downarrow\uparrow\uparrow$, with energy $-\mu_o H$; (v) $\uparrow\downarrow\downarrow$, (vi) $\downarrow\uparrow\downarrow$, and (vii) $\downarrow\downarrow\uparrow$, with energy $+\mu_o H$; and (viii) $\downarrow\downarrow\downarrow$, with energy $+3\mu_o H$. The eigenstates with energies $\pm\mu_o H$ are triply degenerate.

Example 3 *N localized and noninteracting particles of spin 1/2, in the presence of an external field \vec{H}.*

Now the Hamiltonian is given by

$$\mathcal{H} = -\sum_{j=1}^{N} \vec{\mu}_j \cdot \vec{H} = -\mu_o H \sum_{j=1}^{N} \sigma_j, \qquad (2.5)$$

where the set of "spin variables", $\{\sigma_j; j = 1, ..., N\}$, with σ_j assuming the values ± 1 for all j, label each one of the accessible microstates of the system. Given the energy E, we have a huge degeneracy in this system. Indeed, the energy can be written in terms of the number of spins up, N_1, and the number of spins down, $N_2 = N - N_1$. We then have

$$E = -\mu_o H N_1 + \mu_o H (N - N_1). \qquad (2.6)$$

Therefore,

$$N_1 = \frac{1}{2}\left(N - \frac{E}{\mu_o H}\right); \text{ and } N_2 = N - N_1 = \frac{1}{2}\left(N + \frac{E}{\mu_o H}\right). \qquad (2.7)$$

As the energy depends on N_1 and N only, we can use the same combinatorial notions as used in the problem of the random walk in one dimension (see Chapter 1) to obtain the number of eigenstates associated with the system for a given energy E,

$$\Omega(E, N) = \frac{N!}{N_1! N_2!} = \frac{N!}{\left[\frac{1}{2}\left(N - \frac{E}{\mu_o H}\right)\right]! \left[\frac{1}{2}\left(N + \frac{E}{\mu_o H}\right)\right]!}. \qquad (2.8)$$

In the following chapters, we shall see that, given the energy E, the fundamental postulate of statistical mechanics states that all of these microstates

are equally probable. The connection with thermodynamics is provided by the entropy function, which is given by the natural logarithm of $\Omega\left(E,N\right)$, in the so-called thermodynamic limit, where $E,N \to \infty$, for a fixed value of the ratio E/N. This model of noninteracting spins is a good representation of the thermal behavior of an ideal paramagnet. The introduction of interactions between spins, which makes the statistical problem extremely complicated, gives rise to a model for describing the phenomena of magnetic ordering, as ferromagnetism.

Example 4 *One-dimensional harmonic oscillator of frequency* ω.

Now the eigenstates are given by the Hermite polynomials, associated with the energy eigenvalues

$$E_n = \left(n + \frac{1}{2}\right)\hbar\omega, \tag{2.9}$$

with $n = 0, 1, 2,$

Example 5 *Two one-dimensional localized and independent harmonic oscillators, with the same fundamental frequency* ω.

As in the example of localized and noninteracting spins, in these quantum problems the Hamiltonian is a sum, $\mathcal{H} = \mathcal{H}_1 + \mathcal{H}_2$, the eigenstates are a product, $\phi = \phi_1 \cdot \phi_2$, and the corresponding energy eigenvalues are also a sum, $E = E_1 + E_2$. Therefore, the eigenenergies of the system are given by

$$E_{n_1,n_2} = \left(n_1 + \frac{1}{2}\right)\hbar\omega + \left(n_2 + \frac{1}{2}\right)\hbar\omega = (n_1 + n_2 + 1)\hbar\omega, \tag{2.10}$$

where the pair (n_1, n_2) labels a quantum eigenstate. The eigenstate $(0,0)$ has energy $\hbar\omega$; the eigenstates $(0,1)$ and $(1,0)$ have the same energy, $2\hbar\omega$; the eigenstates $(0,2)$, $(2,0)$, and $(1,1)$, have the same energy, $3\hbar\omega$, and so on.

Example 6 *Set of* N *localized and noninteracting one-dimensional harmonic oscillators, with the same fundamental frequency* ω.

This a mere generalization of the last example. It gives rise to a slightly more sophisticated combinatorial problem, associated with a famous model proposed by Einstein in 1905 to explain the thermal behavior of the specific heat of solids. The eigenenergies are given by

$$\begin{aligned} E_{n_1,...,n_N} &= \left(n_1 + \frac{1}{2}\right)\hbar\omega + ... + \left(n_N + \frac{1}{2}\right)\hbar\omega \\ &= \left(n_1 + ... + n_N + \frac{N}{2}\right)\hbar\omega, \end{aligned} \tag{2.11}$$

where the set of quantum numbers $(n_1, ..., n_N)$, with $n_j = 0, 1, 2,,$ for all j, labels the corresponding eigenstate. We can write this energy in the form

$$E_{n_1,...,n_N} = M\hbar\omega + \frac{N}{2}\hbar\omega, \tag{2.12}$$

where the integer $M = n_1 + ... + n_N$ represents the total number of "quanta" of energy in this system. To find the degeneracy of the corresponding eigenstates, it is sufficient to obtain the number of ways to distribute $M = E/\hbar\omega - N/2$ "quanta" of energy among N localized oscillators. The combinatorial problem is analogous to the calculation of the distribution of M identical balls along a one-dimensional arrangement of N boxes. Consider the following illustration.

$$\bullet\,\bullet\,\bullet\,|\bullet\,\bullet\,\bullet\,\bullet|\,\bullet\,|............|\,\bullet\,\bullet$$

In the first box, there are 3 balls; in the second, there are 4 balls; in the third, there is just 1 ball, and so on, up to the last box, where there are two balls. To find out the total number of possible configurations, we have to calculate all the permutations of $M + N - 1$ elements (that is, of the number of balls plus the number of divisions that are defining the boxes), and divide this number by $M!$ (as the balls are identical) and by $(N - 1)!$ (as the separating walls are also identical). We then have the number of eigenstates associated with the system with energy E,

$$\Omega(E, N) = \frac{(M + N - 1)!}{M!\,(N - 1)!} = \frac{\left(\frac{E}{\hbar\omega} + \frac{N}{2} - 1\right)!}{\left(\frac{E}{\hbar\omega} - \frac{N}{2}\right)!\,(N - 1)!}. \tag{2.13}$$

In Chapter 4, we use this expression of $\Omega(E, N)$ to establish a connection with thermodynamics for obtaining the well-known Einstein law for the dependence of the specific heat of solids on temperature.

Example 7 *Free particle of mass m, in a one-dimensional space, within a "box of potential" of side L (that is, such that $0 \leq x \leq L$, with zero potential inside the box, and infinite potential outside).*

The Hamiltonian of this system is given by

$$\mathcal{H} = \frac{1}{2m}p_x^2 = -\frac{\hbar^2}{2m}\frac{d^2}{dx^2}. \tag{2.14}$$

Therefore, we have the Schrödinger equation

$$-\frac{\hbar^2}{2m}\frac{d^2\phi(x)}{dx^2} = E\phi(x), \tag{2.15}$$

whose solution may be written in the form

$$\phi(x) = A\sin kx + B\cos kx, \tag{2.16}$$

with real constants A and B, and energy

$$E = \frac{\hbar^2 k^2}{2m}.$$

(2.17)

From the boundary conditions, $\phi(0) = \phi(L) = 0$, we obtain the discrete spectrum of the energy eigenvalues and associated eigenstates of this system,

$$\phi_n(x) = A \operatorname{sen} k_n x \; ; \; E_n = \frac{\hbar^2 k_n^2}{2m},$$

(2.18)

with

$$k_n = \frac{n\pi}{L}, \text{ where } n = 1, 2, 3, \ldots.$$

(2.19)

In the following chapters, we will prefer writing this single-particle wave function in the complex form,

$$\phi_k(x) = C \exp(ikx),$$

(2.20)

and use periodic boundary conditions, $\phi(0) = \phi(L)$, such that

$$k = 0, \pm \frac{2\pi}{L}, \pm 2\frac{2\pi}{L}, \pm 3\frac{2\pi}{L}, \ldots$$

(2.21)

It should be pointed out that, in the thermodynamic limit, $L \to \infty$, distinct boundary conditions lead to the same thermodynamical results.

Example 8 *System of N noninteracting free particles of mass m, in one dimension, inside a "potential box" of side L (that is, such that $0 \leq x_1 \leq L$, $0 \leq x_2 \leq L$, ..., $0 \leq x_N \leq L$, with zero potential inside the box, and infinite potential outside).*

The Hamiltonian of this system is given by

$$\mathcal{H} = \frac{1}{2m} \sum_{j=1}^{N} p_j^2 = -\frac{\hbar^2}{2m} \sum_{j=1}^{N} \frac{d^2}{dx_j^2}.$$

(2.22)

We then have the Schrödinger equation,

$$-\frac{\hbar^2}{2m} \frac{d^2 \Phi(x_1, \ldots, x_N)}{dx^2} = E\Phi(x_1, \ldots, x_N),$$

(2.23)

whose solution is given by the product

$$\Phi(x_1, \ldots, x_N) = \phi_{k_1}(x_1) \ldots \phi_{k_N}(x_N),$$

(2.24)

where the single-particle wave functions may be written in the form $\phi_k(x) = C\exp(ikx)$, as in equation (2.20). The energy is given by

$$E = E_{k_1,...,k_N} = \frac{\hbar^2}{2m}\left(k_1^2 + \cdots + k_N^2\right). \tag{2.25}$$

The requirements of periodic boundary conditions lead to the quantization of the wave numbers,

$$k_1 = n_1\frac{2\pi}{L}, ..., k_N = n_N\frac{2\pi}{L}, \tag{2.26}$$

where $n_1, ..., n_N = 0, \pm 1, \pm 2, \pm 3,$ A particular microscopic state of the system will be labeled by the set of quantum numbers $\{k_1, ..., k_N\}$. However, the wave functions of identical particles are required to be either symmetric (for bosons) or antisymmetric (for fermions) under the permutation of two position variables. The analysis of the microstates of the quantum system of N identical particles is much more complicated. It will be discussed in detail in one of the more advanced chapters of this text.

2.2 Specification of the microscopic state of a classical system of particles

In classical mechanics, the state of a system of n degrees of freedom is fully specified if we give n generalized coordinates of position, $q_1, ..., q_n$, and n generalized coordinates of momentum, $p_1, ..., p_n$. For example, in the case of N free particles in the Euclidean space, we need $3N$ coordinates of position and $3N$ momenta, that is, $x_1, y_1, z_1, ..., x_N, y_N, z_N, p_{x1}, p_{y2}, p_{z3}, ..., p_{xN}, p_{yN}, p_{zN}$. It is then convenient to introduce the phase space, with $2n$ dimensions, such that each microscopic state of the system of n degrees of freedom is represented by one single point of this space. In figure 2.1, the pair (q, p) represents the set of variables $q_1, ..., q_n, p_1, ..., p_n$. In this phase space, we can represent all the points (microscopic states) that are compatible with the macroscopic conditions (energy, volume) of a given system. In contrast to the examples of discrete quantum models, now we have to count the number of microscopic states in a continuum space. Therefore, it is convenient to introduce the density function $\rho(q, p)$, such that $\rho dq dp$ gives the number of microscopic states with generalized coordinates within the limits of the cell $dq dp$ in phase space.

Example 9 *Free particle of mass m, in one dimension, with energy E, inside a box of length L (that is, such that $0 \leq x \leq L$).*

As the energy has the form $E = p^2/2m$, the momentum will be given by $p = \pm\sqrt{2mE}$. In figure 2.2, we represent the phase space associated with

FIGURE 2.1. Two-dimensional representation of a classical phase space.

FIGURE 2.2. The line segments indicate the accessible regions of phase space for a particle with energy E and position $0 \leq x \leq L$.

FIGURE 2.3. Accessible regions of phase space for a free particle in one dimension, with energy between E and $E + \delta E$, and position $0 \leq x \leq L$. The line segments of figure 2.2 are replaced by the thin hatched surfaces.

this system. All the points belonging to the two segments of the sketch can be reached by the particle with energy E.

In this case, the phase space is two dimensional, but the accessible region of points representing the system is one dimensional. As this reduction of dimensionality introduces some technical difficulties, it is interesting to devise a scheme such that the accessible region has the same dimensionality of the phase space (that is, an area in two dimensions, a volume in three dimensions, a hypervolume of dimension d on a d-dimensional phase space). To deal with this problem, instead of considering a fixed value E for the energy, let us say that the energy may assume all values between E and $E + \delta E$, where δE is a macroscopically small quantity, but of fixed value (in the following, we will see that, in the thermodynamic limit, this trick simplifies the calculations, and does not affect the connections with thermodynamics). In figure 2.3, we illustrate the accessible regions of phase space associated with this system (the hatched regions, with $\delta p = \sqrt{m/2E}\delta E$). Therefore, the accessible volume of phase space is given by

$$\Omega\left(E, L; \delta E\right) = 2L\delta p = \left(\frac{2m}{E}\right)^{1/2} L\delta E. \qquad (2.27)$$

Later, we will see that, according to the fundamental postulate of statistical physics, the density of points is assumed to be a constant inside the hatched region (with a normalized value given by $\rho = 1/\Omega$), and zero otherwise. The classical entropy (although it does not make any sense to refer to an entropy of a system of a single particle!) would be given by the natural logarithm of Ω.

Example 10 *One-dimensional harmonic oscillator with energy between E and $E + \delta E$.*

FIGURE 2.4. Accessible region of phase space for a one-dimensional harmonic oscillator with energy between E and $E + \delta E$.

The classical Hamiltonian of this system is given by

$$\mathcal{H} = \frac{p^2}{2m} + \frac{1}{2}kq^2, \qquad (2.28)$$

where m is the mass and $k > 0$ is the spring constant. Therefore, given the energy E, the accessible region of points in phase space is given by the ellipse

$$\frac{p^2}{2mE} + \frac{q^2}{2E/k} = 1. \qquad (2.29)$$

With energy between E and $E + \delta E$, the accessible region is an elliptical shell (see figure 2.4), whose area is given by

$$\Omega(E; \delta E) = 2\pi \left(\frac{m}{k}\right)^{1/2} \delta E. \qquad (2.30)$$

In this very simple case, the accessible volume of phase space (that is, the area Ω) does not even depend on energy!

We can now propose a number of additional examples. However, beyond two dimensions it becomes difficult to draw the accessible regions in phase space. Also, it may not be easy to calculate the hypervolumes of limited regions of phase space. So, let us present just one additional example, of considerable physical interest.

Example 11 *Classical ideal gas of N monatomic and noninteracting particles of mass m, in a container of volume V, with energy between E and $E + \delta E$.*

The Hamiltonian of the system is given by

$$\mathcal{H} = \sum_{j=1}^{N} \frac{1}{2m}\vec{p}_j^2. \qquad (2.31)$$

The position coordinates, $\{\vec{r}_j; j = 1, ..., N\}$, may assume values inside the boundaries of the container of volume V. Each component of the momentum coordinates may assume values between $-\infty$ and $+\infty$, with the restriction that the total energy is between E and $E + \delta E$. Therefore, the accessible volume in phase space, $\Omega = \Omega(E, V, N; \delta E)$, will be written as

$$\Omega = \int \cdots \int_V d^3\vec{r}_1 \cdots d^3\vec{r}_N \int_{2mE \leq \vec{p}_1^2 + ... + \vec{p}_N^2 \leq 2m(E+\delta E)} \cdots \int d^3\vec{p}_1 \cdots d^3\vec{p}_N$$

$$= V^N \int_{2mE \leq \vec{p}_1^2 + ... + \vec{p}_N^2 \leq 2m(E+\delta E)} \cdots \int d^3\vec{p}_1 \cdots d^3\vec{p}_N,$$

(2.32)

where the integrations over the position coordinates give a trivial contribution. To calculate the multiple integral over momentum coordinates, at first order in δE, let us recall the formula of the hypervolume of a spherical shell of radius R and thickness δR on an n-dimensional space (see Appendix A.4),

$$\Omega_n(R; \delta R) = C_n R^{n-1} \delta R, \tag{2.33}$$

where C_n depends on the dimension n (but does not depend either on R or on δR). In the present example, $n = 3N$ and $R = (2mE)^{1/2}$. So, we have

$$\Omega(E, V, N; \delta E) = \left(\frac{m}{2}\right)^{1/2} C_{3N} (2m)^{\frac{3N}{2} - \frac{1}{2}} V^N E^{\frac{3N}{2} - 1} \delta E, \tag{2.34}$$

where C_{3N} may be obtained from Appendix A.4. Systems of this kind, where volume and energy, in the limit of large N, show up in the expression for Ω as powers of a fraction of N, are important examples of ideal fluids, sometimes called "normal systems".

2.3 *Ergodic hypothesis and fundamental postulate of statistical mechanics

The set of eigenstates of a quantum model, or the set points in classical phase space, which are accessible to a certain physical system (that is, which are compatible with given macroscopic constraints of the system) define a statistical ensemble.

To illustrate the meaning of the ergodic hypothesis, which provides the basis for the establishment of the fundamental postulate of statistical physics, let us consider a classical system of n degrees of freedom (see figure 2.5), and look at the trajectory of a point in phase space, beginning at an initial

FIGURE 2.5. Trajectory of a point on classical phase space.

time t_o. In the Hamiltonian formulation of classical mechanics, where the Hamiltonian \mathcal{H} is a function of the independent variables q, p, and t (and where q and p are parametrized by the time t), the trajectory in phase space is governed by the Hamilton equations,

$$\frac{\partial \mathcal{H}}{\partial p} = \frac{dq}{dt} = \dot{q}; \quad -\frac{\partial \mathcal{H}}{\partial p} = \frac{dp}{dt} = \dot{p}. \tag{2.35}$$

Given the initial conditions, the solutions of these equations are unique. Therefore, although very complicated, the trajectories never cross in phase space. Now consider a (macroscopically dense) set of points in a certain region of phase space. The number of points in this region, at time t, is characterized by the density $\rho = \rho(q, p, t)$. Then, $\rho(q, p, t) \, dqdp$ represents the number of points, at time t, with coordinates between q and $q + dq$, and p and $p + dp$. Now it is easy to establish an equation for the time evolution of the density,

$$\frac{d\rho}{dt} = \frac{\partial \rho}{\partial q}\dot{q} + \frac{\partial \rho}{\partial p}\dot{p} + \frac{\partial \rho}{\partial t} = \frac{\partial \rho}{\partial q}\frac{\partial \mathcal{H}}{\partial p} - \frac{\partial \rho}{\partial p}\frac{\partial \mathcal{H}}{\partial q} + \frac{\partial \rho}{\partial t}. \tag{2.36}$$

Introducing the parenthesis of Poisson,

$$\{\rho, \mathcal{H}\} = \frac{\partial \rho}{\partial q}\frac{\partial \mathcal{H}}{\partial p} - \frac{\partial \rho}{\partial p}\frac{\partial \mathcal{H}}{\partial q}, \tag{2.37}$$

we can write the differential equation

$$\frac{d\rho}{dt} = \{\rho, \mathcal{H}\} + \frac{\partial \rho}{\partial t}. \tag{2.38}$$

As the number of points in phase space is a constant, that is, the representative points of a physical system cannot be either created or destroyed with time, we can write an equation of continuity (as in electromagnetism,

for the conservation of electric charge). Then we have,

$$\oint_S \vec{J} \cdot d\vec{S} = -\frac{d}{dt} \int_{V(S)} \rho dV, \qquad (2.39)$$

where S is a closed hypersurface in phase space defined by the hypervolume V, and the flux is given by $\vec{J} = \rho \vec{v}$, where the vector $\vec{v} = (\dot{q}, \dot{p})$ is a generalized velocity. Using the Gauss theorem of vector calculus, the integral equation (2.39) is transformed into the differential form

$$\vec{\nabla} \cdot (\rho \vec{v}) = -\frac{\partial \rho}{\partial t}, \qquad (2.40)$$

where $\vec{\nabla} = (\partial/\partial q, \partial/\partial p)$ is a generalized gradient operator in phase space. To make clear the form of the generalized divergent, we write

$$\frac{\partial}{\partial q}(\rho \dot{q}) + \frac{\partial}{\partial p}(\rho \dot{p}) = -\frac{\partial \rho}{\partial t}. \qquad (2.41)$$

Now, using Hamilton's equations, we have

$$\frac{\partial}{\partial q}\dot{q} + \frac{\partial}{\partial p}\dot{p} = \frac{\partial}{\partial q}\frac{\partial \mathcal{H}}{\partial p} - \frac{\partial}{\partial p}\frac{\partial \mathcal{H}}{\partial q} = 0. \qquad (2.42)$$

Therefore, equation (2.41) can be written as

$$\frac{\partial \rho}{\partial q}\dot{q} + \dot{p}\frac{\partial \rho}{\partial p} = \{\rho, \mathcal{H}\} = -\frac{\partial \rho}{\partial t}. \qquad (2.43)$$

Comparing equations (2.38) and (2.43), we obtain the well-known **theorem of Liouville,**

$$\frac{d\rho}{dt} = 0 \Longrightarrow \rho = \text{constant}, \qquad (2.44)$$

that is, a departing point for various formulations of the statistical physics of systems out of equilibrium (as will be seen in Chapter 15). From the theorem of Liouville, ρ is a constant with time, which means that the points in phase space move as an incompressible fluid. In **equilibrium,** that is, in a **stationary** situation, when the density ρ cannot be an explicit function of time, we have

$$\frac{\partial \rho}{\partial t} = 0 \Longrightarrow \{\rho, \mathcal{H}\} = 0, \qquad (2.45)$$

that is, the function $\rho = \rho(q, p)$ depends on the generalized coordinates q and p through a function of the Hamiltonian of the system, $\mathcal{H} = \mathcal{H}(q, p)$. It should be noted that this result already justifies the establishment of the fundamental postulate of equilibrium statistical mechanics.

Now let us consider the ergodic hypothesis. To calculate the time average of a certain quantity f in the laboratory, we usually take the average of several sample values of f along a reasonably large time interval τ. From the mathematical point of view, we write

$$\langle f \rangle_{lab} = \lim_{\tau \to \infty} \frac{1}{\tau} \int_0^\tau f(t)\,dt. \qquad (2.46)$$

The ergodic hypothesis consists in assuming that, in equilibrium, the same value can be obtained from an average in phase space,

$$\langle f \rangle_{est} = \frac{\int f(q,p)\,\rho(q,p)\,dqdp}{\int \rho(q,p)\,dqdp}, \qquad (2.47)$$

where the ensemble comprises all the accessible microstates of the system. The temporal average is substituted by an average over the statistical ensemble. We are supposing that all points of the ensemble are faithful copies of the macroscopic system under considerations, and that, as time goes by, the trajectory of the physical system in phase space visits all points of the ensemble. Therefore, it becomes fully justified to replace the temporal average by the instantaneous statistical average over the ensemble. We will implicitly use the ergodic hypothesis as a tool to obtain results that will be justified by their very consequences. However, at least since the early days of the twentieth century (the ergodic hypothesis was proposed and used by Boltzmann), there are many discussions about the validity of this hypothesis, whose study has given rise to an active branch of pure mathematics. In its stronger form, the ergodic hypothesis can only be verified in some extremely simple cases (as the classical one-dimensional harmonic oscillator). However, there are weak forms of the ergodic hypothesis that are valid for "almost" all points in phase space (that is, except for a set of points of zero measure), which have been extensively analyzed in the literature. The interested reader should check a paper by Joel Lebowitz and Oliver Penrose in *Physics Today*, February of 1973, page 23.

Now we are prepared to state the **fundamental postulate of statistical mechanics** (postulate of equal *a priori* probabilities) that will be justified *a posteriori* from its many consequences. **In a closed statistical system, with fixed energy, all the accessible microstates are equally probable.** In some respects, this assumption of equal *a priori* probabilities represents our own ignorance about the true microstate of the system. In the classical phase space, the density ρ should be a constant in the accessible region to the system, and should vanish otherwise. A duly

normalized density is given by the definition

$$\rho = \begin{cases} 1/\Omega; & \text{for } E \leq H(q,p) \leq E + \delta E, \\ 0; & \text{otherwise.} \end{cases} \tag{2.48}$$

For quantum systems, as well as discrete models, the probability is just the inverse of the number Ω of accessible microstates of the system. Given the definition of the distribution of probabilities associated with the ensemble (which is called a **microcanonical ensemble** in this situation of fixed energy), we can apply the standard methods of the theory of probabilities.

2.4 *Formulation of statistical mechanics for quantum systems

In quantum mechanics a system is characterized by the wave function Ψ, that may be expanded in terms of a complete set $\{\phi_n\}$ of orthonormalized eigenfunctions of a certain operator. The process of measurement of a certain quantity may be described by the so-called reduction of the wave packet. So, from the more traditional point of view of quantum mechanics, with each physical quantity o we associate an operator \hat{O} such that

$$\hat{O}\psi_n = o_n\psi_n, \tag{2.49}$$

where $\{\psi_n\}$ is a complete and orthonormalized set of eigenfunctions, and o_n is the eigenvalue corresponding to ψ_n. The wave function of the system may be written as

$$\Psi = \sum_n c_n\psi_n. \tag{2.50}$$

The process of measurement of the physical quantity o, which requires the interaction with a classical object, may produce a particular value o_k, which reduces the quantum system to the state ψ_k. However, a priori, before the measurement, we can only say that there is a probability, given by $|c_k|^2$, of obtaining the value o_k. If we have a large number of identical systems, duly prepared in the same state Ψ (for example, from a selection by means of previous measurements of another physical quantity), we obtain a distribution of values of the physical quantity o, with an average or expected value given by

$$\langle\Psi|\hat{O}|\Psi\rangle = \sum_{m,n} c_m^* c_n \langle\psi_m|\hat{O}|\psi_n\rangle = \sum_{m,n} c_m^* c_n o_n \langle\psi_m|\psi_n\rangle = \sum_n |c_n|^2 o_n.$$
$$\tag{2.51}$$

Up to this point, the process involves the intrinsic statistical character of quantum mechanics. There are no references to the ignorance of the microscopic state of the system. What happens if we do not know the exact expression of the wave function Ψ? In other words, what happens if the system can be found in a set of wave functions, similar to Ψ, all of them compatible with the macroscopic conditions? Under these circumstances, we have to perform an extrinsic statistical average, directly related to our ignorance about the system, and very similar to the averages of classical statistical mechanics.

Consider again the expected value of the operator \hat{O}, and perform an expansion of Ψ in terms of the eigenfunctions ϕ_n of the Hamiltonian operator,

$$\Psi = \sum_n a_n \phi_n, \tag{2.52}$$

where we know the set of coefficients $\{a_n\}$. The intrinsic average value of quantum mechanics is given by

$$\langle \Psi | \hat{O} | \Psi \rangle = \langle \hat{O} \rangle = \sum_{m,n} a_m^* a_n \langle \phi_m | \hat{O} | \phi_n \rangle. \tag{2.53}$$

If we do not know the exact wave function Ψ, we have to perform a statistical average over all functions Ψ, in other words, over all values of the coefficients a_n. Let us call $\langle ... \rangle_{est}$ this second average. Thus we have

$$\left\langle \langle \hat{O} \rangle \right\rangle_{est} = \sum_{m,n} \langle a_m^* a_n \rangle_{est} \langle \phi_m | \hat{O} | \phi_n \rangle = \sum_{m,n} \langle a_m^* a_n \rangle_{est} \hat{O}_{mn}. \tag{2.54}$$

In general, the statistical average $\langle a_m^* a_n \rangle_{est}$ may be very difficult to calculate. Let us simply define the **density matrix**,

$$\rho_{nm} = \langle a_m^* a_n \rangle_{est}. \tag{2.55}$$

We then write

$$\left\langle \langle \hat{O} \rangle \right\rangle_{est} = \sum_{m,n} \rho_{nm} \hat{O}_{mn} = \sum_n (\rho O)_{nn} = \mathrm{Tr}\, \rho O = \mathrm{Tr}\, O \rho. \tag{2.56}$$

If \hat{O}_{mn} is diagonal in the representation $\{\phi_n\}$, that is, if $\hat{O}_{mn} = o_m \delta_{m,n}$, we have

$$\left\langle \langle \hat{O} \rangle \right\rangle_{est} = \sum_n \rho_{nn} o_n, \tag{2.57}$$

where the sum is over all quantum eigenstates defined by the set $\{\phi_n\}$.

These expressions should be compared to the average in the classical phase space with a normalized density,

$$\langle f(q,p) \rangle = \int f(q,p) \, \rho(q,p) \, dqdp. \tag{2.58}$$

The matrix ρ plays the same role as the classical normalized density in phase space, $\rho(q,p)$. In the quantum case, the fundamental postulate requires that the diagonal elements of the density matrix, ρ_{nn}, should be equal to the same constant for all eigenstates with the same energy E (if it is more convenient, with energy between E and $E + \delta E$).

Exercises

1. Neglect the complexities of classical phase space, and consider a system of N distinguishable and noninteracting particles, which may be found in two states of energy, with $\epsilon = 0$ and $\epsilon > 0$, respectively. Given the total energy U of this system, obtain an expression for the associated number of microscopic states.

2. Calculate the number of accessible microscopic states of a system of two localized and independent quantum oscillators, with fundamental frequencies ω_o and $3\omega_o$, respectively, and total energy $E = 10\hbar\omega_o$.

3. Consider a classical one-dimensional system of two noninteracting particles of the same mass m. The motion of the particles is restricted to a region of the x axis between $x = 0$ and $x = L > 0$. Let x_1 and x_2 be the position coordinates of the particles, and p_1 and p_2 be the canonically conjugated momenta. The total energy of the system is between E and $E + \delta E$. Draw the projection of phase space in a plane defined by position coordinates. Indicate the region of this plane that is accessible to the system. Draw similar graphs in the plane defined by the momentum coordinates.

4. The position of a one-dimensional harmonic oscillator is given by

$$x = A \cos(\omega t + \varphi),$$

where A, ω, and φ are positive constants. Obtain $p(x) \, dx$, that is, the probability of finding the oscillator with position between x and $x + dx$. Note that it is enough to calculate dT/T, where T is a period of oscillation, and dT is an interval of time, within a period, in which the amplitude remains between x and $x + dx$. Draw a graph of $p(x)$ versus x.

Now consider the classical phase space of an ensemble of identical one-dimensional oscillators with energy between E and $E + \delta E$. Given the

energy E, we have an ellipse in phase space. So, the accessible region in phase space is a thin elliptical shell bounded by the ellipses associated with energies E and $E + \delta E$, respectively. Obtain an expression for the small area δA of this elliptical shell between x and $x + dx$. Show that the probability $p(x)dx$ may also be given by $\delta A/A$, where A is the total area of the elliptical shell. This is one of the few examples where we can check the validity of the ergodic hypothesis and the postulate of equal *a priori* probabilities.

5. Consider a classical system of N localized and weakly interacting one-dimensional harmonic oscillators, whose Hamiltonian is written as

$$\mathcal{H} = \sum_{j=1}^{N} \left(\frac{1}{2m} p_j^2 + \frac{1}{2} k x_j^2 \right),$$

where m is the mass and k is an elastic constant. Obtain the accessible volume of phase space for $E \leq \mathcal{H} \leq E + \delta E$, with $\delta E << E$. This classical model for the elastic vibrations of a solid leads to a constant specific heat with temperature (law of Dulong and Petit). The solid of Einstein is a quantum version of this model. The specific heat of Einstein's model decreases with temperature, in qualitative agreement with experimental data.

6. The spin Hamiltonian of a system of N localized magnetic ions is given by

$$\mathcal{H} = D \sum_{j=1}^{N} S_j^2,$$

where $D > 0$ and the spin variable S_j may assume the values ± 1 or 0, for all $j = 1, 2, 3....$ This spin Hamiltonian describes the effects of the electrostatic environment on spin-1 ions. An ion in states ± 1 has energy $D > 0$, and an ion in state 0 has zero energy. Show that the number of accessible microscopic states of this system with total energy U can be written as

$$\Omega(U, N) = \frac{N!}{(N - \frac{U}{D})!} \sum_{N_-} \left[\left(\frac{U}{D} - N_- \right)! N_-! \right]^{-1},$$

for N_- ranging from 0 to N, with $N > U/D$ and $N_- < U/D$. Thus, we have

$$\Omega(U, N) = N! 2^{U/D} \left[\left(N - \frac{U}{D} \right)! \left(\frac{U}{D} \right)! \right]^{-1}.$$

Using Stirling's asymptotic series, show that

$$\frac{1}{N}\ln\Omega \to \frac{u}{D}\ln 2 - \left(1 - \frac{u}{D}\right)\ln\left(1 - \frac{u}{D}\right) - \frac{u}{D}\ln\frac{u}{D},$$

for $N, U \to \infty$, with $U/N = u$ fixed. This last expression is the entropy per particle in units of Boltzmann's constant, k_B.

7. In a simplified model of a gas of particles, the system is divided into V cells of unit volume. Find the number of ways to distribute N distinguishable particles (with $0 \leq N \leq V$) within V cells, such that each cell may be either empty or filled up by only one particle. How would your answer be modified for indistinguishable particles?

8. The atoms of a crystalline solid may occupy either a position of equilibrium, with zero energy, or a displaced position, with energy $\epsilon > 0$. To each equilibrium position, there corresponds a unique displaced position. Given the number N of atoms, and the total energy U, calculate the number of accessible microscopic states of this system.

Using Stirling's asymptotic series, show that

$$\frac{S}{N} = \frac{u}{2}\ln 3 - \left(1 - \frac{u}{2}\right)\ln\left(1 - \frac{u}{2}\right) - \frac{u}{2}\ln\frac{u}{2}$$

for $N, U \to \infty$, with $U/N = u$ fixed. This last expression is the entropy per particle in units of Boltzmann's constant, k_B.

7. In a simplified model of a gas of particles, the system is divided into V cells of unit volume. Find the number of ways to distribute N distinguishable particles (with $0 \leq N \leq V$) within V cells, such that each cell may be either empty or filled up by only one particle. How would your answer be modified for indistinguishable particles?

8. The atoms of a crystalline solid may occupy either a position of equilibrium, with zero energy, or a displaced position, with energy $\varepsilon > 0$. To each equilibrium position, there corresponds a unique displaced position. Given the number N of atoms, and the total energy U, calculate the number of accessible microscopic states of this system.

3

Overview of Classical Thermodynamics

Thermodynamics is the branch of physics that organizes in a systematic way the empirical laws referring to the thermal behavior of macroscopic matter. In contrast to statistical mechanics, it does not require any assumptions about the underlying microscopic elements of the material bodies. Equilibrium thermodynamics, which is the subject of this chapter, offers a complete description of the thermal properties of physical systems whose macroscopic parameters are not changing with time.

In this review of equilibrium thermodynamics, we assume that some basic notions, as the definitions of the macroscopic extensive parameters (energy, volume, and number of moles, for example, which are proportional to the size of the system) are well-known concepts. Also, we assume the validity of the "law of conservation of energy," and all of its consequences (heat is a form of energy that may be transformed into mechanical work).

3.1 Postulates of equilibrium thermodynamics

To state the postulates of equilibrium thermodynamics, let us define a **simple system**. The simple systems are macroscopically homogeneous, isotropic, chemically inert, without any charges, and sufficiently big. In this chapter we mainly consider a **pure fluid**, that is, a simple system with one single component and in the absence of external (electric, magnetic, or gravitational) fields. The thermodynamic (macroscopic) state of this pure fluid is fully characterized by a very small number of macroscopic

FIGURE 3.1. Composite thermodynamic system of three simple fluids separated by adiabatic, fixed, and impermeable walls.

variables (additional variables may be easily introduced to describe a more-complicated physical system).

First postulate: "The macroscopic state of a pure fluid is completely characterized by the internal energy U, the volume V, and the amount of matter (that may be given by the number of moles n)." To make the connection with statistical mechanics, instead of using the number of moles, we express the amount of matter by the total number N of particles. In the more general case of a fluid with r distinct components, we give the set $\{N_j; j = 1, ..., r\}$, where N_j is the number of particles of the jth chemical component.

A **composite system** is a set of simple systems that are separated by walls or constraints. The walls are ideal separations that may restrict the change of certain variables. Adiabatic walls do not allow the flux of energy in the form of heat (otherwise, the wall is diathermal). Fixed walls do not allow the change of volume. Impermeable walls do not allow the flux of particles (of either one or more components of a fluid).

The postulates will provide an answer to the fundamental problem of equilibrium thermodynamics, which consists in the determination of the final state of equilibrium of a composite system if we remove some internal constraints. For example, what is the final state of equilibrium if an adiabatic wall becomes diathermal? And if a fixed wall is allowed to move?

Second postulate: "There is a function of the extensive parameters of a composite system, called entropy, $S = S(U_1, V_1, N_1, U_2, V_2, N_2, \ldots)$, that is defined for all states of equilibrium. If we remove an internal constraint, in the new state of equilibrium the extensive parameters of the system assume a set of values that maximize the entropy." The entropy as a function of the extensive parameters is a **fundamental equation** of the system. It contains all the thermodynamic information about this system.

Third postulate: "The entropy of a composite system is additive over each one of the individual components. The entropy is a continuous, differentiable, and monotonically increasing function of the energy."

According to this third postulate, for a composite system of two pure fluids, we can write

$$S(U_1, V_1, N_1, U_2, V_2, N_2) = S_1(U_1, V_1, N_1) + S_2(U_2, V_2, N_2). \qquad (3.1)$$

Given $S = S(U, V, N)$, from the third postulate we have $(\partial S/\partial U) > 0$ (we will see that this inequality is related to the assumption of a positive temperature). Therefore, we can invert the functional form of S to write $U = U(S, V, N)$, which is also a fundamental equation, with the complete thermodynamic information about the system, and the same status of the entropy $S = S(U, V, N)$.

The additivity of entropy means that $S = S(U, V, N)$ is a homogeneous function of first degree of all of its variables. For all values of λ, we then write

$$S(\lambda U, \lambda V, \lambda N) = \lambda S(U, V, N). \qquad (3.2)$$

In particular, making $\lambda = 2$, we see that, if we double energy, volume, and number of particles, the entropy will double too. Taking $\lambda = 1/N$, we have

$$\frac{1}{N} S(U, V, N) = S\left(\frac{U}{N}, \frac{V}{N}, 1\right) = s(u, v), \qquad (3.3)$$

which leads to the definitions of the **densities** $u = U/N$, $v = V/N$, and $s = S/N$.

Forth postulate: "The entropy vanishes in a state where $(\partial U/\partial S)_{V,N} = 0$." We will later see that this is the statement of Nernst law, also called the "third law of thermodynamics." From this postulate, we shall see that the entropy vanishes at the absolute zero of temperature (although this may be violated by classical statistical mechanics!).

3.2 Intensive parameters of thermodynamics

In the **energy representation** the fundamental equation of a pure fluid is given by the relation $U = U(S, V, N)$. Hence, we can write the differential form

$$dU = \left(\frac{\partial U}{\partial S}\right)_{V,N} dS + \left(\frac{\partial U}{\partial V}\right)_{S,N} dV + \left(\frac{\partial U}{\partial N}\right)_{S,V} dN, \qquad (3.4)$$

which describes a thermodynamic or quasistatic process (that is, an infinitesimal succession of equilibrium thermodynamic states). From a comparison with the usual expression for the conservation of energy,

$$\Delta U = \Delta Q + \Delta W_{mechanical} + \Delta W_{chemical}$$

$$= T\Delta S - p\Delta V + \mu\Delta N, \tag{3.5}$$

we have the following definitions of the **intensive parameters** or thermodynamic **fields**,

$$\textbf{temperature:} \qquad T = \left(\frac{\partial U}{\partial S}\right)_{V,N};$$

$$\textbf{pressure:} \qquad p = -\left(\frac{\partial U}{\partial V}\right)_{S,N}; \tag{3.6}$$

$$\textbf{chemical potential:} \qquad \mu = \left(\frac{\partial U}{\partial N}\right)_{S,V}.$$

The functions $T = T(S,V,N)$, $p = p(S,V,N)$, and $\mu = \mu(S,V,N)$, are the **equations of state** in the energy representation. The knowledge of a single equation of state is not enough to obtain a fundamental equation. However, two equations of state are already enough, since $T = T(S,V,N)$, $p = p(S,V,N)$, and $\mu = \mu(S,V,N)$, are homogeneous functions of zero degree of their variables [that is, $T(\lambda S, \lambda V, \lambda N) = T(S,V,N)$, and so on]. Note that we are purposely using a complicated notation for the partial derivatives. We are making a point to keep the subscripts to indicate the variables that are supposed to remain fixed during a differentiation. This special care is particularly useful in thermodynamics, where it may be convenient to work with different representations, which are characterized by different sets of independent variables.

In the entropy representation, we have

$$dS = \left(\frac{\partial S}{\partial U}\right)_{V,N} dU + \left(\frac{\partial S}{\partial V}\right)_{U,N} dV + \left(\frac{\partial S}{\partial N}\right)_{U,V} dN = \sum_{k=1}^{3} F_k dX_k, \tag{3.7}$$

where $X_1 = U$, $X_2 = V$, and $X_3 = N$. Therefore, from equation (3.5), we have the equations of state in the entropy representation,

$$F_1 = \left(\frac{\partial S}{\partial U}\right)_{V,N} = \frac{1}{T}, \tag{3.8}$$

$$F_2 = \left(\frac{\partial S}{\partial V}\right)_{U,N} = \frac{p}{T}, \tag{3.9}$$

$$F_3 = \left(\frac{\partial S}{\partial N}\right)_{U,V} = -\frac{\mu}{T}. \tag{3.10}$$

As $U(S, V, N) = Nu(s, v)$, we can write a differential form in terms of the densities,

$$du = \left(\frac{\partial u}{\partial s}\right)_v ds + \left(\frac{\partial u}{\partial v}\right)_s dv = Tds - pdv, \tag{3.11}$$

since

$$\left(\frac{\partial U}{\partial S}\right)_{V,N} = \left(\frac{\partial u}{\partial s}\right)_v = T,$$

$$\left(\frac{\partial U}{\partial V}\right)_{S,N} = \left(\frac{\partial u}{\partial v}\right)_s = -p. \tag{3.12}$$

Also, we can write

$$ds = \frac{1}{T}du + \frac{p}{T}dv. \tag{3.13}$$

For example, let us consider a classical monatomic gas, given by the equations of state $pV = Nk_BT$ (Boyle law) and $U = (3/2)Nk_BT$ (which means that the energy per particle is a function of temperature only, that is, the specific heat is a constant). These equations may be easily written in the entropic form,

$$\frac{p}{T} = \frac{Nk_B}{V} \rightarrow \left(\frac{\partial s}{\partial v}\right)_u = \frac{k_B}{v} \tag{3.14}$$

and

$$\frac{1}{T} = \frac{3Nk_B}{2U} \rightarrow \left(\frac{\partial s}{\partial u}\right)_v = \frac{3k_B}{2u}. \tag{3.15}$$

Integrating equations (3.14) and (3.15), we have

$$s(u, v) = k_B \ln v + f(u) \tag{3.16}$$

and

$$s(u, v) = \frac{3}{2}k_B \ln u + g(v), \tag{3.17}$$

where $f(u)$ and $g(v)$ are arbitrary functions of u and v, respectively. Comparing these expressions, we finally have

$$s(u, v) = \frac{3}{2}k_B \ln u + k_B \ln v + k_B c, \tag{3.18}$$

FIGURE 3.2. Composite system of two simple fluids separated by adiabatic, fixed, and impermeable walls.

where c is a constant. Therefore, except for a constant, the classical entropy of the monatomic gas is given by

$$S(U, V, N) = Ns\left(\frac{U}{N}, \frac{V}{N}\right) = \frac{3}{2}Nk_B \ln \frac{U}{N} + Nk_B \ln \frac{V}{N} + Nk_B c. \quad (3.19)$$

3.3 Equilibrium between two thermodynamic systems

To give an example of **thermal equilibrium**, let us consider a closed composite system of two pure fluids. In the initial state, the fluids are separated by an adiabatic, fixed, and impermeable wall. Then, there is a change in one of the internal constraints. The separating wall remains fixed and impermeable, but becomes diathermal (see figure 3.2). After a certain time, the system reaches another state of thermodynamic equilibrium. This new state of equilibrium will be given by the maximization of the total entropy,

$$S = S_1(U_1, V_1, N_1) + S_2(U_2, V_2, N_2), \quad (3.20)$$

where V_1, V_2, N_1, and N_2 are fixed parameters, and

$$U_1 + U_2 = U_o = \text{constant}, \quad (3.21)$$

where U_o is the total energy of the composite system (that remains fixed for the closed global system). Hence, we have

$$\frac{\partial S}{\partial U_1} = \frac{\partial S_1}{\partial U_1} + \frac{\partial S_2}{\partial U_1} = \frac{\partial S_1}{\partial U_1} - \frac{\partial S_2}{\partial U_2} = \frac{1}{T_1} - \frac{1}{T_2} = 0. \quad (3.22)$$

Therefore, in equilibrium, $T_1 = T_2$, which corresponds to the intuitive expectations about the final equilibrium state of two systems in thermal contact.

We now look at the second derivative,

$$\frac{\partial^2 S}{\partial U_1^2} = \frac{\partial^2 S_1}{\partial U_1^2} - \frac{\partial}{\partial U_1}\left(\frac{\partial S_2}{\partial U_2}\right) = \frac{\partial^2 S_1}{\partial U_1^2} + \frac{\partial^2 S_2}{\partial U_2^2}. \tag{3.23}$$

Using the equation of state for the inverse of the temperature, we write

$$\frac{\partial^2 S}{\partial U_1^2} = \frac{\partial}{\partial U_1}\left(\frac{1}{T_1}\right) + \frac{\partial}{\partial U_2}\left(\frac{1}{T_2}\right) = -\frac{1}{T_1^2}\frac{\partial T_1}{\partial U_1} - \frac{1}{T_2^2}\frac{\partial T_2}{\partial U_2}. \tag{3.24}$$

As this second derivative is required to be negative, we have

$$\left(\frac{\partial T}{\partial U}\right)_{V,N} > 0 \tag{3.25}$$

for a pure fluid. Introducing the definition of the specific heat at constant volume,

$$c_V \equiv \frac{T}{N}\left(\frac{\partial S}{\partial T}\right)_{V,N} = \frac{1}{N}\left(\frac{\partial U}{\partial T}\right)_{V,N}, \tag{3.26}$$

and taking the inverse of the derivative $(\partial T/\partial U)_{V,N}$, we have

$$\left(\frac{\partial T}{\partial U}\right)_{V,N} = \frac{1}{\left(\frac{\partial U}{\partial T}\right)_{V,N}} = \frac{1}{N c_V} > 0. \tag{3.27}$$

Therefore, the maximization of entropy is directly related to a fundamental property of thermal stability of matter. In a thermally stable system, the specific heat (proportional to the ratio between the quantity of heat injected into the system and the consequent change of temperature) cannot be negative.

Consider again equation (3.23). We may rewrite the condition of stability in the form

$$\left(\frac{\partial}{\partial U}\frac{1}{T}\right)_{V,N} = \left(\frac{\partial^2 S}{\partial U^2}\right)_{V,N} < 0, \tag{3.28}$$

that is, the entropy should be a **concave** function of energy. At this point, it is interesting to make some considerations about **convex** (or concave) functions. A differentiable function $f(x)$ is convex if the second derivative is positive for all values of x. If the second derivative is negative, the function is concave. For example, the function $f(x) = \exp(x)$ is convex, and the function $f(x) = \ln x$ is concave. Indeed, we could have used a more-general definition, to include nondifferentiable functions, at least at a few points (as it may be the case for thermodynamic functions at a phase transition). We shall see that, to guarantee the thermal and mechanical stability of a

physical system, the density of entropy, $s = s(u, v)$, has to be a concave function of both independent variables, u and v. On the other hand, the density of energy, $u = u(s, v)$, is required to be a convex function of the independent variables s and v.

Now we give an example of thermal and mechanical equilibrium. Consider the same problem as before, but with a separating wall that may become diathermal and free to move after a certain initial time. The global entropy is still given by equation (3.20),

$$S = S_1(U_1, V_1, N_1) + S_2(U_2, V_2, N_2), \tag{3.29}$$

where N_1 and N_2 are still fixed parameters, but

$$U_1 + U_2 = U_o = \text{constant} \tag{3.30}$$

and

$$V_1 + V_2 = V_o = \text{constant}, \tag{3.31}$$

where V_o is the (fixed) volume of the closed composite system. In the new equilibrium state, we have

$$\frac{\partial S}{\partial U_1} = \frac{\partial S_1}{\partial U_1} - \frac{\partial S_2}{\partial U_2} = \frac{1}{T_1} - \frac{1}{T_2} = 0 \tag{3.32}$$

and

$$\frac{\partial S}{\partial V_1} = \frac{\partial S_1}{\partial V_1} - \frac{\partial S_2}{\partial V_2} = \frac{p_1}{T_1} - \frac{p_2}{T_2} = 0. \tag{3.33}$$

From these equations, we have the intuitive conditions for thermal and mechanical equilibrium,

$$T_1 = T_2, \text{ and } p_1 = p_2, \tag{3.34}$$

that is, an equalization of temperatures and pressures. To proceed with the analysis, we have to consider the second derivatives of entropy with respect to energy and volume. As the number of moles is fixed, it is enough to analyze the sign of the quadratic form

$$d^2s = \frac{1}{2}\frac{\partial^2 s}{\partial u^2}(du)^2 + \frac{\partial^2 s}{\partial u \partial v}du\,dv + \frac{1}{2}\frac{\partial^2 s}{\partial v^2}(dv)^2. \tag{3.35}$$

If the entropy function, $s = s(u, v)$, is concave with respect to both independent variables, u and v, this quadratic form is always negative. However, as it demands a bit of algebraic manipulations with partial derivatives, we will not show the concavity of entropy at this point. We remark that the conditions for thermal and mechanical stability are related to the positive signs of the specific heats and the compressibility of a pure fluid [the second derivative $(\partial p / \partial v)_T$ is always negative for all physical systems!].

3.4 The Euler and Gibbs–Duhem relations

We have already seen that the additivity may be expressed in the form

$$U(\lambda S, \lambda V, \lambda N) = \lambda U(S, V, N). \tag{3.36}$$

Taking the derivative of both sides with respect to λ, we have

$$\frac{\partial U(\lambda S, \lambda V, \lambda N)}{\partial(\lambda S)}S + \frac{\partial U(\lambda S, \lambda V, \lambda N)}{\partial(\lambda V)}V + \frac{\partial U(\lambda S, \lambda V, \lambda N)}{\partial(\lambda N)}N = U(S, V, N). \tag{3.37}$$

For $\lambda = 1$, we obtain the **Euler equation** of thermodynamics,

$$TS - pV + \mu N = U. \tag{3.38}$$

From the differential form of this relation, we have

$$TdS + SdT - pdV - Vdp + \mu dN + Nd\mu = dU, \tag{3.39}$$

that is,

$$SdT - Vdp + Nd\mu = 0, \tag{3.40}$$

which can also be written as

$$d\mu = vdp - sdT, \tag{3.41}$$

which is known as the **Gibbs–Duhem relation**. From this differential form we see that the chemical potential as a function of temperature and pressure is a fundamental equation of the pure fluid. On the other hand, the densities v and s, as a function of temperature and pressure, are just equations of state.

3.5 Thermodynamic derivatives of physical interest

The relatively easy experimental access to some thermodynamic derivatives makes them particularly interesting. In many cases, it is convenient to express the thermodynamic quantities under consideration in terms of these derivatives, whose values for different compounds may be given in tables of thermodynamic data.

For a pure fluid, the following derivatives are of special interest:

(a) The coefficient of thermal expansion,

$$\alpha = \lim_{\Delta T \to 0} \frac{1}{V}\left(\frac{\Delta V}{\Delta T}\right)_{p,N} = \frac{1}{V}\left(\frac{\partial V}{\partial T}\right)_{p,N}, \tag{3.42}$$

that measures the relative thermal expansion of the volume of a system at constant pressure.

(b) The isothermal compressibility,

$$\kappa_T = \lim_{\Delta p \to 0} -\frac{1}{V}\left(\frac{\Delta V}{\Delta p}\right)_{T,N} = -\frac{1}{V}\left(\frac{\partial V}{\partial p}\right)_{T,N}, \tag{3.43}$$

that measures the relative change of volume with increasing pressure at fixed temperature. It may also be interesting to measure the adiabatic compressibility, κ_S, that is defined at fixed entropy (instead of constant temperature).

(c) The specific heat at constant pressure,

$$c_p = \lim_{\Delta T \to 0} \frac{1}{N}\left(\frac{\Delta Q}{\Delta T}\right)_{p,N} = \frac{T}{N}\left(\frac{\partial S}{\partial T}\right)_{p,N}, \tag{3.44}$$

that corresponds to an experiment where we measure the ratio between the injected heat, into a closed system under constant pressure, and the consequent change of temperature.

(d) The specific heat at constant volume,

$$c_V = \lim_{\Delta T \to 0} \frac{1}{N}\left(\frac{\Delta Q}{\Delta T}\right)_{V,N} = \frac{T}{N}\left(\frac{\partial S}{\partial T}\right)_{V,N}. \tag{3.45}$$

To guarantee thermal stability, we must have $c_p, c_V \geq 0$. Mechanical stability will be guaranteed by $\kappa_T, \kappa_S \geq 0$ (note the sign in the definition of the compressibility). On the other hand, the coefficient α does not have a well-defined sign (the phenomenon of volume contraction of water with increasing temperature in the neighborhood of $4°C$ is a standard example of negative α). From a well-known relation for fluids (it will be proved later),

$$c_p = c_V + \frac{TV\alpha^2}{N\kappa_T}, \tag{3.46}$$

we can see that $c_p \geq c_V$.

It should be noted that α, κ_T, and c_p are functions of the variables T, p, and N. For fluid systems, there are no doubts that these variables are much more practical and convenient than S, V, and N. This suggests the usefulness of constructing alternative representations of thermodynamics, as it will be done in the following section by using the formalism of the Legendre transformations.

3.6 Thermodynamic potentials

Instead of working either in the energy representation, where S, V, and N are the independent variables, or in the entropy representation, where U,

FIGURE 3.3. Legendre transformation of the function $y = y(x)$. We indicate the slope p at the point (x, y) and the intersection $\psi(p)$ between the tangent line and the y axis.

V, and N are independent, it may be much more convenient to work with independent variables of easier experimental access, such as temperature T or pressure p (note that the temperature T, for example, is a partial derivative of U with respect to S). To deal with this problem, let us consider a function $y = y(x)$, with derivative $p = dy/dx$. We wish to find another function, of the form $\psi = \psi(p)$, that is equivalent to $y = y(x)$. This may be achieved by a Legendre transformation.

A function $y = y(x)$ may be constructed from a table of pairs of values (y, x). Let us consider whether a table of pairs (y, p) is also useful to define this function. Each pair of this type corresponds to a family of straight lines on the $x - y$ plane, so that there is no chance of defining the function $y = y(x)$. Indeed, the relation $y = y(p)$, that can be written in the form $y = y(dy/dx)$, is just a differential equation, whose solution is given except for an additive constant. However, we may construct a different table, involving the value of the slope (p) and the intersection with the y axis (which will be called ψ) of the straight line tangent to the curve $y = y(x)$ at the point x. See, for example, the graph of figure 3.3. We then construct a family of tangents to the curve $y = y(x)$. For a function with a well-defined convexity, as the fundamental equations of thermodynamics, the "convex envelope" of the curve $y = y(x)$ will determine it completely. Looking at figure 3.3, the Legendre transformation of the function $y = y(x)$ is given by the function $\psi = \psi(p)$, such that

$$\psi = \psi(p) = y(x) - px, \qquad (3.47)$$

where the variable x is eliminated from the equation

$$p = \frac{dy}{dx}. \qquad (3.48)$$

It should be noted that we can also use a more-general definition, including cases where the derivative p is not defined,

$$\psi\left(p\right) = \inf_{x}\left\{y\left(x\right) - px\right\}, \tag{3.49}$$

where inf means the smallest value of the relation $y\left(x\right) - px$ with respect to x.

Example 1 *Legendre transformation of a quadratic function.*

Let us calculate the Legendre transformation of the function $y = y(x) = ax^2 + bx + c$. Using equation (3.47), the Legendre transformation can be written as

$$\psi\left(p\right) = ax^2 + bx + c - px,$$

where

$$p = \frac{dy}{dx} = 2ax + b.$$

Therefore,

$$x = \frac{p - b}{2a}$$

and

$$\psi\left(p\right) = -\frac{1}{4a}p^2 + \frac{b}{2a}p - \frac{b^2}{4a}.$$

Note that if the function $y(x)$ is convex, then the function $\psi\left(p\right)$ will be concave, and vice versa (that is, the Legendre transformation is an operation that reverses the convexity of a function).

Example 2 *Hamiltonian formulation of classical mechanics.*

In the Lagrangian formulation of classical mechanics, the Lagrange function, given by $\mathcal{L} = \mathcal{L}\left(q, \dot{q}, t\right)$, in terms of the independent variables q, \dot{q}, and t, is a fundamental equation that contains all the mechanical information about the system. The generalized momentum is defined by the relation

$$p = \frac{\partial \mathcal{L}}{\partial \dot{q}}.$$

In the Hamiltonian formulation that contains the same physical information, the function of Hamilton is given by the Legendre transformation,

$$-\mathcal{H}\left(q, p, t\right) = \mathcal{L}\left(q, \dot{q}, t\right) - p\dot{q}.$$

For a one-dimensional harmonic oscillator, we have

$$\mathcal{L} = \frac{1}{2}m\dot{x}^2 - \frac{1}{2}kx^2.$$

Hence,

$$-\mathcal{H} = \mathcal{L} - \dot{x}p_x,$$

where

$$p_x = \frac{\partial \mathcal{L}}{\partial \dot{x}} = m\dot{x}.$$

Therefore,

$$\mathcal{H} = \frac{1}{2m}p_x^2 + \frac{1}{2}kx^2.$$

Now, let us define the Legendre transformations of the energy, $U = U(S, V, N)$. Recalling that

$$T = \left(\frac{\partial U}{\partial S}\right)_{V,N} ; \quad p = -\left(\frac{\partial U}{\partial V}\right)_{S,N} ; \quad \mu = \left(\frac{\partial U}{\partial N}\right)_{S,V} ; \qquad (3.50)$$

we have the following **thermodynamic potentials:**

(i) The **Helmholtz free energy,**

$$U[T] = F(T, V, N) = U - TS, \qquad (3.51)$$

where the variable S has been replaced by the temperature T.

(ii) The **enthalpy,**

$$U[p] = H(S, p, N) = U + pV, \qquad (3.52)$$

where the volume V has been replaced by the pressure p.

(iii) The function

$$U[\mu] = f_1(S, V, \mu) = U - \mu N, \qquad (3.53)$$

where N has been replaced by the chemical potential μ.

(iv) The **Gibbs free energy,**

$$U[T, p] = G(T, p, N) = U - TS + pV, \qquad (3.54)$$

where we have performed a double Legendre transformation, with respect to the variables S and V.

(v) The **grand thermodynamic potential,**

$$U[T, \mu] = \Phi(T, V, \mu) = U - TS - \mu N, \tag{3.55}$$

where the transformation has been done with respect to the variables S and N.

(vi) The function

$$U[p, \mu] = f_2(S, p, \mu) = U + pV - \mu N, \tag{3.56}$$

which has no special name.

Since $U = U(S, V, N)$ is a homogeneous function of first degree of its variables, taking the Legendre transformation with respect to all three variables leads to an identically null function,

$$U[T, p, \mu] = U - TS + pV - \mu N = 0, \tag{3.57}$$

which gives again the Euler relation of thermodynamics. Hence, we can also write

$$G(T, p, N) = N\mu = N\mu(T, p), \tag{3.58}$$

which shows that the chemical potential as a function of temperature and pressure is the Gibbs free energy per particle, with the same status of a fundamental equation, as we have already seen from the Gibbs–Duhem relation. Furthermore, we have

$$\Phi(T, V, \mu) = -Vp = -Vp(T, \mu), \tag{3.59}$$

which means that the pressure (as a function of temperature and chemical potential) times the volume also leads to a fundamental equation in the representation of the grand thermodynamic potential.

We can also build a set of analogous thermodynamic potentials from the Legendre transformations of the entropy function, $S = S(U, V, N)$, with respect to its variables. These Massieu potentials are less well-known, and will not be considered here.

3.7 The Maxwell relations

In the Helmholtz representation, the free energy is given by $F = U - TS$. Hence, we have the differential form

$$dF = dU - TdS - SdT = -SdT - pdV + \mu dN, \tag{3.60}$$

where

$$-S = \left(\frac{\partial F}{\partial T}\right)_{V,N} ; \quad -p = \left(\frac{\partial F}{\partial V}\right)_{T,N} ; \quad \mu = \left(\frac{\partial F}{\partial N}\right)_{T,V} . \tag{3.61}$$

Therefore, $S = S(T, V, N)$, $p = p(T, V, N)$, and $\mu = \mu(T, V, N)$ are equations of state in the representation of Helmholtz (note that the entropy is a fundamental equation for a fluid if it is expressed in terms of energy, volume, and number of particles, only). From equation (3.61), and taking the crossed derivatives, we have three Maxwell relations,

$$\left(\frac{\partial S}{\partial V}\right)_{T,N} = \left(\frac{\partial p}{\partial T}\right)_{V,N} ; \tag{3.62}$$

$$-\left(\frac{\partial S}{\partial N}\right)_{T,V} = \left(\frac{\partial \mu}{\partial T}\right)_{V,N} ; \tag{3.63}$$

$$-\left(\frac{\partial p}{\partial N}\right)_{T,V} = \left(\frac{\partial \mu}{\partial V}\right)_{T,N} . \tag{3.64}$$

In each representation of thermodynamics we obtain a set of Maxwell's relations. Although there is a mnemonic scheme (the diagrams of Born) to recall them, it may be easier to make the proper deductions for each case of interest. As it can be seen from the exercises, the Maxwell relations are very useful to connect thermodynamic quantities of interest that might otherwise seem completely unrelated.

In the Gibbs representation, using the free energy $G = U - TS + pV$, we have

$$dG = -SdT + Vdp + \mu dN. \tag{3.65}$$

Hence,

$$-S = \left(\frac{\partial G}{\partial T}\right)_{p,N} ; \quad V = \left(\frac{\partial G}{\partial p}\right)_{T,N} ; \quad \mu = \left(\frac{\partial G}{\partial N}\right)_{T,p} , \tag{3.66}$$

from which we have some additional Maxwell relations,

$$-\left(\frac{\partial S}{\partial p}\right)_{T,N} = \left(\frac{\partial V}{\partial T}\right)_{p,N} ; \tag{3.67}$$

$$-\left(\frac{\partial S}{\partial N}\right)_{T,p} = \left(\frac{\partial \mu}{\partial T}\right)_{p,N} ; \tag{3.68}$$

and

$$\left(\frac{\partial V}{\partial N}\right)_{T,p} = \left(\frac{\partial \mu}{\partial p}\right)_{T,N} . \tag{3.69}$$

Example 3 *Ideal monatomic classical gas of particles.*

According to equation (3.19), in the entropy representation the fundamental equation of the classical ideal monatomic gas is given by

$$S = \frac{3}{2}Nk_B \ln \frac{U}{N} + Nk_B \ln \frac{V}{N} + Nk_Bc,$$

where c is a constant. In the energy representation, we have

$$U = N\left(\frac{N}{V}\right)^{2/3} \exp\left[\frac{2}{3}\left(\frac{S}{Nk_B} - c\right)\right].$$

Hence, the Helmholtz free energy is given by

$$F = U - TS = N\left(\frac{N}{V}\right)^{2/3} \exp\left[\frac{2}{3}\left(\frac{S}{Nk_B} - c\right)\right] - TS,$$

where the entropy S can be eliminated from the equation

$$T = \left(\frac{\partial U}{\partial S}\right)_{V,N} = \frac{2}{3k_B}\left(\frac{N}{V}\right)^{2/3} \exp\left[\frac{2}{3}\left(\frac{S}{Nk_B} - c\right)\right].$$

Therefore, we have

$$F = -k_BTN \ln \frac{V}{N} - \frac{3}{2}k_BTN \ln \frac{3k_BT}{2} + Nk_BT\left(\frac{3}{2} - c\right).$$

In the Helmholtz representation, we can write the following equations of state:

$$\text{(i) } S = -\left(\frac{\partial F}{\partial T}\right)_{V,N} = k_BN \ln \frac{V}{N} + \frac{3}{2}k_BN \ln \frac{3k_BT}{2} + Nk_Bc,$$

which is a well-known expression for the entropy of the classical gas, where the constant c may be found from the classical limit of the entropy of an ideal monatomic quantum gas;

$$\text{(ii) } p = -\left(\frac{\partial F}{\partial V}\right)_{T,N} = \frac{k_BTN}{V},$$

which is the well-known Boyle law of a perfect gas;

$$\text{(iii) } \mu = \left(\frac{\partial F}{\partial N}\right)_{T,V} = -k_BT \ln \frac{V}{N} - \frac{3}{2}k_BT \ln \frac{3k_BT}{2} + Nk_Bc,$$

which indicates that the chemical potential increases monotonically with density $\rho = N/V$.

Example 4 **Concavity of the Gibbs free energy.*

We have already mentioned that the density of entropy, $s = s(u, v)$, is a concave function of its variables. On the other hand, the density of energy, $u = u(s, v)$, is a convex function of the variables s and v. The Gibbs free energy per particle, $g = g(T, p) = G/N$, may be obtained from a double Legendre transformation of the density of energy, $g = u - Ts + pv$. Therefore, the density of Gibbs free energy should be a concave function of temperature and pressure. Indeed, we can show that the quadratic form

$$d^2 g = \frac{1}{2} \left(\frac{\partial^2 g}{\partial T^2} \right) (dT)^2 + \left(\frac{\partial^2 g}{\partial T \partial p} \right) dT dp + \frac{1}{2} \left(\frac{\partial^2 g}{\partial p^2} \right) (dp)^2$$

is always negative. Since $dg = -sdT + vdp$, we have

$$s = - \left(\frac{\partial g}{\partial T} \right)_p \quad \text{and} \quad v = \left(\frac{\partial g}{\partial p} \right)_T .$$

Hence, we write

$$\frac{\partial^2 g}{\partial T^2} = - \left(\frac{\partial s}{\partial T} \right)_p = -\frac{1}{T} c_p < 0$$

and

$$\frac{\partial^2 g}{\partial p^2} = \left(\frac{\partial v}{\partial p} \right)_T = -v\kappa_T < 0,$$

where we used the definitions of the specific heat at constant pressure, c_p, and of the isothermal compressibility, κ_T, respectively. We also have

$$\frac{\partial^2 g}{\partial T \partial p} = \left(\frac{\partial v}{\partial T} \right)_p = v\alpha,$$

where we used the definition of the coefficient of thermal expansion. To show that the quadratic form $d^2 g$ is always negative, and thus complete the proof that the density of Gibbs free energy is a concave function, it is not enough to obtain the signs of the second derivatives of g with respect to T and p. We still have to show that

$$\frac{\partial^2 g}{\partial T^2} \frac{\partial^2 g}{\partial p^2} - \left(\frac{\partial^2 g}{\partial T \partial p} \right)^2 \geq 0,$$

where

$$\frac{\partial^2 g}{\partial T^2} \frac{\partial^2 g}{\partial p^2} - \left(\frac{\partial^2 g}{\partial T \partial p} \right)^2 = \frac{1}{T} c_p v\kappa_T - v^2 \alpha^2 = \frac{v\kappa_T}{T} \left[c_p - \frac{T v \alpha^2}{\kappa_T} \right].$$

To find the meaning of the term inside the square brackets [see equation (3.46)], let us obtain an expression for the specific heat at constant volume. The technique of Jacobians (see Appendix A.5) is particularly useful for this problem. Indeed, we can write

$$\frac{1}{T}c_V = \left(\frac{\partial s}{\partial T}\right)_v = \frac{\partial(s,v)}{\partial(T,v)} = \frac{\partial(s,v)}{\partial(T,p)}\frac{\partial(T,p)}{\partial(T,v)}$$

$$= \left[\left(\frac{\partial s}{\partial T}\right)_p\left(\frac{\partial v}{\partial p}\right)_T - \left(\frac{\partial s}{\partial p}\right)_T\left(\frac{\partial v}{\partial T}\right)_p\right]\left(\frac{\partial p}{\partial v}\right)_T.$$

Using one of Maxwell's relations in the Gibbs representation, we can also write

$$\left(\frac{\partial s}{\partial p}\right)_T = -\left(\frac{\partial v}{\partial T}\right)_p = -v\alpha.$$

Therefore, we have

$$\frac{1}{T}c_V = \left[-\frac{1}{T}vc_p\kappa_T + v^2\alpha^2\right]\frac{-1}{v\kappa_T},$$

from which we write

$$c_V = c_p - \frac{Tv\alpha^2}{\kappa_T},$$

that is exactly the same expression that had been used before (see equation 3.46). Hence, we finally have

$$\frac{\partial^2 g}{\partial T^2}\frac{\partial^2 g}{\partial p^2} - \left(\frac{\partial^2 g}{\partial T\partial p}\right)^2 = \frac{v}{T}\kappa_T c_V \geq 0,$$

which completes the proof of the concavity of the Gibbs free energy. Furthermore, as $(Tv\alpha^2)/\kappa_T \geq 0$, where the equality holds is some extreme cases, we also show that $c_p \geq c_V$.

3.8 Variational principles of thermodynamics

The second postulate of thermodynamics establishes a variational principle that gives rise to several consequences. According to the postulate, upon the removal of an internal constraint, the extensive parameters assume a set of equilibrium values that maximize the entropy of a closed composite system. In other words, at fixed total energy, the entropy reaches a maximum value. As the entropy is a monotonically increasing function of energy (we are assuming positive temperatures!), we can show that, for total fixed entropy,

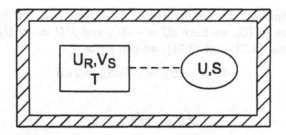

FIGURE 3.4. System with energy U and entropy S in contact with a thermal reservoir (with energy U_R, entropy S_R, and temperature T). The composite system is closed.

the same equilibrium state corresponds to a minimum value of energy. This is an alternative formulation, in the energy representation, of the second postulate of thermodynamics. There is a famous problem of plane geometry that can also be expressed by alternative variational principles: Given a perimeter, the circle is the figure on the plane with maximum area; given an area, the circle is the planar geometric figure with minimum perimeter.

We can use the second postulate to establish variational principles of usefulness in alternative representations of thermodynamics. For example, in the case of a system in contact with a heat reservoir (that is, at fixed temperature), the removal of an internal constraint leads to an equilibrium state where the Helmholtz free energy is minimum. In figure 3.4 we sketch a certain system, with energy U and entropy S, and a heat reservoir, with energy U_R and entropy S_R (at fixed temperature T). The composite system is closed, but there can be transport of energy, in the form of heat, between the system under consideration and the heat reservoir. Therefore, in the equilibrium state, we have

$$U + U_R = U_o, \qquad (3.70)$$

with

$$d(S + S_R) = 0 \qquad (3.71)$$

and

$$d^2(S + S_R) < 0, \qquad (3.72)$$

where U_o is a constant (total energy of the closed composite system) and all additional macroscopic parameters (volume, number of particles) that characterize the state of the reservoir are fixed. The heat reservoir is described by the equations

$$dU_R = TdS_R \; ; \; d^2U_R = 0, \text{ and } d^2S_R = 0, \qquad (3.73)$$

since, by definition, the temperature T of a heat reservoir is a constant. From equation (3.70), we have $dU = -dU_R$ and $d^2U = -d^2U_R$. Therefore, using equations (3.73) and (3.71), we can write

$$dU = -dU_R = -TdS_R = TdS \tag{3.74}$$

and

$$d^2U = -d^2U_R = 0. \tag{3.75}$$

Using these equations, we have

$$d(U - TS) = dU - TdS = 0 \tag{3.76}$$

and

$$d^2(U - TS) = d^2U - Td^2S = -Td^2S. \tag{3.77}$$

However, as $d^2S_R = 0$, from equation (3.72), we have

$$d^2(U - TS) = -Td^2S > 0. \tag{3.78}$$

The Helmholtz free energy of this system, $F = U - TS$, is then a minimum in equilibrium.

For a system in contact with heat (fixed temperature) and mechanical work (fixed pressure) reservoirs, we can show that the Gibbs free energy is minimized in equilibrium. At fixed pressure, as in several chemical reactions of interest, the enthalpy is minimized. At fixed temperature and chemical potential, the grand thermodynamic potential should assume a minimum value.

Now it is important to make some comments about systems with additional degrees of freedom. Although we have restricted most of the considerations to a pure fluid, all of the thermodynamic principles so far discussed can be generalized for more complex systems. We just introduce additional extensive variables in the definition of the entropy function (or the corresponding energy function). More convenient fundamental equations will be produced by the usual mechanism of the Legendre transformations. For example, in the case of a simple fluid with several components, the differential form in the energy representation is written as

$$dU = TdS - pdV + \sum_j \mu_j dN_j, \tag{3.79}$$

where μ_j is the chemical potential of the jth component. To describe a solid body in the presence of anisotropic tensions, the elements of the deformation tensor play the role of the additional extensive variables. In the presence of electric or magnetic fields, we may introduce the (electric or magnetic) polarization as the additional extensive variable. In this case, there are alternative forms to establish the thermodynamic equations of the system, depending on the way of treating the energy associated with the field (we refer to this question in the exercises).

Exercises

1. The chemical potential of a simple fluid of a single component is given by the expression

$$\mu = \mu_o(T) + k_B T \ln \frac{p}{p_o(T)},$$

where T is the temperature, p is the pressure, k_B is the Boltzmann constant, and the functions $\mu_o(T)$ and $p_o(T)$ are well behaved. Show that this system obeys Boyle's law, $pV = N k_B T$. Obtain an expression for the specific heat at constant pressure. What are the expressions for the thermal compressibility, the specific heat at constant volume, and the thermal expansion coefficient? Obtain the density of Helmholtz free energy, $f = f(T, v)$.

2. Consider a pure fluid of one component. Show that

$$\left(\frac{\partial c_V}{\partial v} \right)_T = T \left(\frac{\partial^2 p}{\partial T^2} \right)_v.$$

Use this result to show that the specific heat of an ideal gas does not depend on volume. Show that

$$\left(\frac{\partial \alpha}{\partial p} \right)_{T,N} = - \left(\frac{\partial \kappa_T}{\partial T} \right)_{p,N}.$$

3. Consider a pure fluid characterized by the grand thermodynamic potential

$$\Phi = V f_o(T) \exp \left(\frac{\mu}{k_B T} \right),$$

where $f_o(T)$ is a well-behaved function. Write the equations of state in this thermodynamic representation. Obtain an expression for the internal energy as a function of T, V, and N. Obtain an expression for the Helmholtz free energy of this system. Calculate the thermodynamic derivatives κ_T and α as a function of temperature and pressure.

4. Obtain an expression for the Helmholtz free energy per particle, $f = f(T, v)$, of a pure system given by the equations of state

$$u = \frac{3}{2} p v \quad \text{and} \quad p = a v T^4,$$

where a is a constant.

5. Obtain an expression for the Gibbs free energy per particle, $g = g(T, p)$, for a pure system given by the fundamental equation

$$\left(\frac{S}{N} - c\right)^4 = a\frac{VU^2}{N^3},$$

where a and c are constants.

6. Consider an elastic ribbon of length L under a tension f. In a quasi-static process, we can write

$$dU = TdS + fdL + \mu dN.$$

Suppose that the tension is increased very quickly, from f to $f + \Delta f$, keeping the temperature T fixed. Obtain an expression for the change of entropy just after reaching equilibrium. What is the change of entropy per mole for an elastic ribbon that behaves according to the equation of state $L/N = cf/T$, where c is a constant?

7. A magnetic compound behaves according to the Curie law, $m = CH/T$, where C is a constant, H is the applied magnetic field, m is the magnetization per particle (with corrections due to presumed surface effects), and T is temperature. In a quasi-static process, we have

$$du = Tds + Hdm,$$

where $u = u(s, m)$ plays the role of an internal energy. For an infinitesimal adiabatic process, show that we can write

$$\Delta T = \frac{CH}{c_H T}\Delta H,$$

where c_H is the specific heat at constant magnetic field.

8. From stability arguments, show that the enthalpy of a pure fluid is a convex function of entropy and a concave function of pressure.

9. *Show that the entropy per mole of a pure fluid, $s = s(u, v)$, is a concave function of its variables. Note that we have to analyze the sign of the quadratic form

$$d^2s = \frac{1}{2}\frac{\partial^2 s}{\partial u^2}(du)^2 + \frac{\partial^2 s}{\partial u \partial v}dudv + \frac{1}{2}\frac{\partial^2 s}{\partial v^2}(dv)^2.$$

4

Microcanonical Ensemble

According to the fundamental postulate of statistical mechanics, the **accessible microscopic states of a closed system in equilibrium are equally probable.** The number of microscopic states of a thermodynamic fluid, with energy E, volume V, and number of particles N, in the presence of a set $\{X_i\}$ of internal constraints, is given by $\Omega = \Omega(E, V, N; \{X_i\})$. For example, in the previous chapter we looked at a system of two simple fluids subjected to an internal constraint given by an adiabatic, fixed, and impermeable wall (see figure 4.1). The probability $P(\{X_i\})$ of finding a system subjected to the set of constraints $\{X_i\}$ is proportional to $\Omega(E, V, N; \{X_i\})$, that is,

$$P(\{X_i\}) \propto \Omega(E, V, N; \{X_i\}). \tag{4.1}$$

We now use the distribution of probabilities $P(\{X_i\})$ to perform a statistical treatment of the thermodynamic process that comes from removing a certain subset of internal constraints. In this chapter, we give some examples of the use of statistical techniques and of the statistical interpretation of thermodynamics. Upon removing a set of internal constraints, the most probable values of the new parameters of the system are identified with the corresponding thermodynamic quantities in equilibrium. Of course, there are fluctuations about these most probable values. However, except in some very special cases, the relative widths are extremely small in the **thermodynamic limit** of a large enough physical system. We will give several arguments to justify the definition of entropy as the logarithm of the number of accessible microscopic states of the system (second postulate of statistical mechanics in equilibrium). Also, the formalism of the **microcanonical**

FIGURE 4.1. Two simple fluids separated by an adiabatic, fixed, and impermeable wall.

ensemble will be used to calculate the thermodynamic properties of some statistical models, including an ideal classical gas of particles.

4.1 Thermal interaction between two microscopic systems

We will use the statistical language to repeat the same sort of analysis that has already been performed in Chapter 3, in the context of classical thermodynamics, for the process of equilibrium between two simple fluids in thermal contact. Initially, fluids (1) and (2) are inside closed reservoirs, separated by an adiabatic, fixed, and impermeable wall (see figure 4.1).

The number of accessible microscopic states of subsystem (1) is given by $\Omega_1(E_1, V_1, N_1)$. The number of accessible microscopic states of subsystem (2) is given by $\Omega_2(E_2, V_2, N_2)$. Therefore, the number of accessible microscopic states of the composite system of independent subsystems is given by the product

$$\Omega = \Omega_1(E_1, V_1, N_1)\,\Omega_2(E_2, V_2, N_2). \tag{4.2}$$

Suppose that, at a given time, the separating wall becomes diathermal. The total energy remains fixed, but the energies of each subsystem, E_1 and E_2, may freely fluctuate, under the condition $E_1 + E_2 = E_o$, where E_o is the total energy of the system. Note that the additional macroscopic parameters of the two subsystems (V_1, V_2, N_1, N_2) remain fixed. Note also that we are still using the concept of an ideal wall, whose internal energy is disregarded. The total number of microscopic states of the composite system, such that subsystem (1) has energy E_1 and subsystem (2) has energy $E_2 = E_o - E_1$, may be written as

$$\Omega(E_1; E_o) = \Omega_1(E_1)\,\Omega_2(E_o - E_1), \tag{4.3}$$

where we simplify the notation by omitting the fixed parameters. Hence, using the fundamental postulate of statistical mechanics, the probability of

finding the composite system in a microscopic state such that the energy of subsystem (1) is E_1 [with the energy of subsystem (2) given by $E_2 = E_o - E_1$] is written in the form

$$P(E_1) = c\,\Omega(E_1; E_o) = c\,\Omega_1(E_1)\,\Omega_2(E_o - E_1),\qquad(4.4)$$

where the inverse of the constant c is the total number of accessible microscopic states of the global composite system,

$$\frac{1}{c} = \Omega_C = \sum_{E_1=0}^{E_o} \Omega_1(E_1)\,\Omega_2(E_o - E_1).\qquad(4.5)$$

These heuristic arguments refer to the simple case of a model with discretized energy states. It should be pointed out that, although a bit more awkward, it is not difficult to rewrite all of these steps for a system with continuous parameters.

In general, $\Omega(E)$ increases with increasing values of the energy E (since there should be more and more available microscopic states as the energy increases). Therefore, $\Omega_1(E_1)$ increases while $\Omega_2(E_o - E_1)$ decreases with the energy E_1, thus indicating that $P(E_1)$ displays a maximum as a function of E_1. It is convenient to take the logarithm of the probability to write

$$f(E_1) = \ln P(E_1) = \ln c + \ln \Omega_1(E_1) + \ln \Omega_2(E_o - E_1).\qquad(4.6)$$

At the maximum value, we have

$$\frac{\partial \ln P(E_1)}{\partial E_1} = \frac{\partial \ln \Omega_1(E_1)}{\partial E_1} - \frac{\partial \ln \Omega_2(E_2)}{\partial E_2} = 0,\qquad(4.7)$$

where $E_2 = E_o - E_1$. Introducing the definition of entropy,

$$S(E) = k_B \ln \Omega(E),\qquad(4.8)$$

where k_B is Boltzmann's constant, we recover the thermodynamic condition of thermal equilibrium, $T_1 = T_2$. Therefore, assuming the definition (4.8), **the maximization of the probability corresponds to the maximization of the thermodynamic entropy!**

Example 1 *Thermal contact between two classical ideal gases.*

In Chapter 2, we have obtained an expression for the volume in phase space of an ideal monatomic classical gas [see example (3) of Section 2.2]. For large enough values of N, we may use the expression of Chapter 2 to write

$$\Omega(E) = CE^{\frac{3}{2}N},$$

where the constant C includes the dependence on the fixed parameters N and V. Therefore, we have

$$\ln P\left(E_1\right) = \text{constant} + \frac{3}{2} N_1 \ln E_1 + \frac{3}{2} N_2 \ln E_2,$$

where $E_2 = E_o - E_1$. Calculating the first derivative with respect to E_1,

$$\frac{\partial \ln P\left(E_1\right)}{\partial E_1} = \frac{3}{2} \frac{N_1}{E_1} - \frac{3}{2} \frac{N_2}{E_2},$$

we have the condition

$$\frac{\tilde{E}_1}{N_1} = \frac{\tilde{E}_2}{N_2}.$$

Using the definition of entropy, given by equation (4.8), and identifying the most probable value of the energy with the thermodynamic internal energy of the system, that is, taking $U = \tilde{E}$, we write

$$\frac{1}{k_B T_1} = \frac{3N_1}{2U_1} = \frac{1}{k_B T_2} = \frac{3N_2}{2U_2},$$

from which we have the equalization of temperatures, $T_1 = T_2 = T$, and the equation of state for an ideal monatomic classical gas of particles, $U = \frac{3}{2} N k T$. Now we consider the second derivative,

$$\frac{\partial^2 \ln P\left(E_1\right)}{\partial E_1^2} = -\frac{3}{2} \frac{N_1}{E_1^2} - \frac{3}{2} \frac{N_2}{E_2^2}.$$

If we write

$$E_1 = U_1 = \frac{3}{2} N_1 k_B T, \quad \text{and} \quad E_2 = U_2 = \frac{3}{2} N_2 k_B T,$$

we have

$$\left(\frac{\partial^2 \ln P\left(E_1\right)}{\partial E_1^2} \right)_{\max} = -\frac{2}{3} \frac{N_1 + N_2}{k_B^2 T^2 N_1 N_2} < 0,$$

which shows that the probability is a maximum indeed. In the neighborhood of this maximum, we can write the Taylor expansion

$$\ln P\left(E_1\right) = \text{constant} - \frac{1}{3} \frac{N_1 + N_2}{k_B^2 T^2 N_1 N_2} \left(E_1 - \frac{3}{2} N_1 k_B T \right)^2 + \dots.$$

Hence, we have the Gaussian approximation (see Chapter 1),

$$p_G\left(E_1\right) = A \exp\left[-\frac{1}{3} \frac{N_1 + N_2}{k_B^2 T^2 N_1 N_2} \left(E_1 - \frac{3}{2} N_1 k_B T \right)^2 \right],$$

where A is a normalization constant. Using this Gaussian distribution, we calculate the average value,

$$\langle E_1 \rangle_G = \frac{3}{2} N_1 k_B T,$$

and the quadratic deviation,

$$\left\langle (\Delta E_1)^2 \right\rangle_G = \frac{3}{2} \frac{k_B^2 T^2 N_1 N_2}{N_1 + N_2}.$$

Hence, we can write the relative width,

$$\frac{\sqrt{\left\langle (\Delta E_1)^2 \right\rangle_G}}{\langle E_1 \rangle_G} = \left[\frac{2}{3} \frac{N_2}{N_1 (N_1 + N_2)} \right]^{1/2} \to 0,$$

which becomes very small for a sufficiently large system (that is, such that $N_1, N_2 \to \infty$). In this thermodynamic limit the fluctuations about the average value (which is identical to the most probable value) are extremely small, and thus justify the identification between the probabilistic average of the energy and the internal thermodynamic energy itself.

4.2 Thermal and mechanical interaction between two systems

Consider again the system sketched in figure 4.1. Initially, the internal separating wall does not allow any contact between the two subsystems (the wall is adiabatic, fixed, and impermeable). At a certain time, it becomes diathermal and free to move. Therefore, although N_1 and N_2 remain fixed, the energy and the volume of each subsystem may fluctuate, with the global constraints

$$E_1 + E_2 = E_o, \quad \text{and} \quad V_1 + V_2 = V_o, \tag{4.9}$$

where E_o and V_o are constant values. In equilibrium, the number of accessible microscopic states of the composite system, such that the simple system (1) has energy E_1 and volume V_1, will be given by

$$\Omega (E_1, V_1; E_o, V_o) = \Omega_1 (E_1, V_1) \Omega_2 (E_o - E_1, V_o - V_1), \tag{4.10}$$

where we are not showing the dependence on the fixed parameters N_1 and N_2. The probability of finding the composite system in a state such that subsystem (1) is characterized by the macroscopic parameters E_1, V_1, and N_1, will be given by

$$P(E_1, V_1) = c \, \Omega_1 (E_1, V_1) \Omega_2 (E_o - E_1, V_o - V_1), \tag{4.11}$$

where the constant prefactor comes from the normalization,

$$\frac{1}{c} = \sum_{E_1=0}^{E_o} \sum_{V_1=0}^{V_o} \Omega_1 \left(E_1, V_1\right) \Omega_2 \left(E_o - E_1, V_o - V_1\right). \tag{4.12}$$

Maximizing the logarithm of the probability $P\left(E_1, V_1\right)$, we have

$$\frac{\partial \ln P\left(E_1, V_1\right)}{\partial E_1} = \frac{\partial \ln \Omega_1 \left(E_1, V_1\right)}{\partial E_1} - \frac{\partial \ln \Omega_2 \left(E_2, V_2\right)}{\partial E_2} = 0 \tag{4.13}$$

and

$$\frac{\partial \ln P\left(E_1, V_1\right)}{\partial V_1} = \frac{\partial \ln \Omega_1 \left(E_1, V_1\right)}{\partial V_1} - \frac{\partial \ln \Omega_2 \left(E_2, V_2\right)}{\partial V_2} = 0. \tag{4.14}$$

Using the definition of entropy, given by equation (4.8), we have the usual conditions of thermodynamic equilibrium,

$$\frac{1}{T_1} = \frac{1}{T_2}, \quad \text{and} \quad \frac{p_1}{T_1} = \frac{p_2}{T_2}, \tag{4.15}$$

that is, $T_1 = T_2$, and $p_1 = p_2$, which are the standard results of equilibrium thermodynamics. Maximizing probabilities turns out to be identical to maximizing entropies. In the limit of a very large system, we hope that fluctuations about the equilibrium state are very small, so that we can assume the identification between the average (or more probable) values and the values of the macroscopic thermodynamic quantities.

To emphasize the role of the **thermodynamic limit**, let us check the conditions under which the statistical entropy of a composite system equals the sum of the statistical entropies of the simple components. To keep the notation as simple as possible, consider again the example of the previous section, where two subsystems in thermal contact are separated by a fixed and impermeable wall. Using the definition (4.8), the entropy of the composite system may be written as

$$S_C = k_B \ln \Omega_C = k_B \ln \left\{ \sum_{E_1=0}^{E_o} \Omega_1 \left(E_1\right) \Omega_2 \left(E_o - E_1\right) \right\}. \tag{4.16}$$

Consider now the sum over the energy values. This sum of positive numbers is larger than its largest term. Also, it is smaller than the maximum term times the number of terms. If we keep working with discretized values of energy, the total number of terms is of the order of $E_o/\delta E$, where δE is a small but fixed value. Therefore, we can write the trivial inequality

$$\Omega_1 \left(U_1\right) \Omega_2 \left(E_o - U_1\right) \leq \sum_{E_1=0}^{E_o} \Omega_1 \left(E_1\right) \Omega_2 \left(E_o - E_1\right)$$

$$\leq \frac{E_o}{\delta E} \Omega_1 \left(U_1\right) \Omega_2 \left(E_o - U_1\right). \tag{4.17}$$

Since all terms are positive, this inequality still works for the logarithms,

$$\ln \Omega_1 (U_1) + \ln \Omega_2 (E_o - U_1) \leq \ln \left\{ \sum_{E_1=0}^{E_o} \Omega_1 (E_1) \Omega_2 (E_o - E_1) \right\}$$

$$\leq \ln \Omega_1 (U_1) + \ln \Omega_2 (E_o - U_1) + \ln E_o - \ln \delta E. \qquad (4.18)$$

Therefore, we have

$$S_1 (U_1) + S_2 (U_2) \leq S_C \leq S_1 (U_1) + S_2 (U_2) + k_B \ln E_o - k_B \ln \delta E. \qquad (4.19)$$

In the thermodynamic limit, we take $N_1, N_2 \to \infty$, and $U_1, U_2 \to \infty$, with fixed values of the ratios N_1/N, N_2/N, U_1/N, and U_2/N (this limit can be obtained from an increasing collection of identical systems, such that the extensive parameters increase while the densities remain fixed). Hence, the entropies S_1, S_2, and S_C are of the order of N, while $\ln E_o$ is of the order of $\ln N$, and $\ln \delta E$ does not depend on N. Therefore, in the thermodynamic limit, we have the strict additivity of entropy,

$$S_C \to S_1 (U_1) + S_2 (U_2). \qquad (4.20)$$

In the 1960s, there was an effort to base these heuristic arguments on much firmer mathematical grounds. We shall later return to this question, in the context of the canonical ensemble. Nowadays, it is possible to prove the existence of the entropy function, in the thermodynamic limit, with the correct properties of convexity, for all systems of classical and quantum particles with interactions of a physical character.

4.3 Connection between the microcanonical ensemble and thermodynamics

From the first postulate of equilibrium statistical mechanics, we know that all the microscopic states of a closed system (with fixed energy) in equilibrium are equally probable. The second postulate is the definition of entropy, which should be given by the logarithm of the number of accessible microscopic states. For a pure fluid, where the thermodynamic state is characterized by the energy E, the volume V, and the quantity of matter represented by the number of particles N, the entropy is given by

$$S (E, V, N) = k_B \ln \Omega (E, V, N), \qquad (4.21)$$

where k_B is the Boltzmann constant and $\Omega (E, V, N)$ is the number of microstates (that is, the volume of phase space, for a classical model).

This connection with thermodynamics, however, works in the thermodynamic limit only, that is, for $E, V, N \to \infty$, with fixed densities, $E/N = u$, $V/N = v$, where u and v are kept constant. In this limit, we eliminate the effects of the boundary conditions of a mathematical model, and the entropy, given by equation (4.21), becomes a homogeneous function of first degree of the extensive variables (as required by classical thermodynamics). From the physical point of view, there should be no doubts about the relevance of this limit. The thermodynamic derivatives of experimental interest, as the specific heat, the coefficients of expansion, and the compressibility, are all of them measured and given by mole (or particle), without any reference to the size of the system (as far as it is large enough). The thermodynamic properties always refer to the bulk of the materials, and implicitly assume that we are taking the thermodynamic limit. From the mathematical point of view, we have already presented some illustrations to show that the thermodynamical limit is essential to allow the connection between the average values of statistical mechanics and the macroscopic values of the thermodynamic quantities. In the following chapters, we will see that the thermodynamic limit is required to prove the equivalence among the different ensembles of statistical mechanics. Therefore, the second postulate, associated with the definition (4.21), should be reformulated as

$$s(u, v) = \lim \frac{1}{N} S\left(E, V, N\right) = \lim \frac{1}{N} k_B \ln \Omega\left(E, V, N\right), \qquad (4.22)$$

where the limit is taken for $E, V, N \to \infty$, with $E/N = u$, $V/N = v$, where u and v are fixed. Now we revisit the same examples of Chapter 2.

Example 2 *Ideal paramagnet of spin 1/2.*

Let us revisit example (3) of section 2.1. We consider N localized particles, of spin 1/2, in the presence of a magnetic field. As we have already seen, the Hamiltonian of this system is given by

$$\mathcal{H} = -\mu_o H \sum_{j=1}^{N} \sigma_j, \qquad (4.23)$$

where the set of spin variables, $\{\sigma_j\}$, with $\sigma_j = \pm 1$, for $j = 1, ..., N$, characterizes the microscopic states of the system. In Chapter 2, we calculated the number of accessible microscopic states, with fixed energy E,

$$\Omega\left(E, N\right) = \frac{N!}{\left[\frac{1}{2}\left(N - \frac{E}{\mu_o H}\right)\right]! \left[\frac{1}{2}\left(N + \frac{E}{\mu_o H}\right)\right]!}. \qquad (4.24)$$

Therefore,

$$\ln \Omega\left(E, N\right) = \ln N! - \ln \left[\frac{1}{2}\left(N - \frac{E}{\mu_o H}\right)\right]! - \ln \left[\frac{1}{2}\left(N + \frac{E}{\mu_o H}\right)\right]!.$$

Since we are interested in the entropy in the thermodynamic limit (that is, for $E, N \to \infty$, with $E/N = u$ fixed), we use Stirling's expansion to write

$$
\ln \Omega (E, N) = N \ln N - \frac{1}{2} \left(N - \frac{E}{\mu_o H} \right) \ln \left[\frac{1}{2} \left(N - \frac{E}{\mu_o H} \right) \right]
$$
$$
- \frac{1}{2} \left(N + \frac{E}{\mu_o H} \right) \ln \left[\frac{1}{2} \left(N + \frac{E}{\mu_o H} \right) \right] + O(\ln N, \ln E).
$$

Hence, we have

$$
\lim_{E, N \to \infty;\, \frac{E}{N} = u} \frac{1}{N} \ln \Omega (E, N) = \ln 2 - \frac{1}{2} \left(1 - \frac{u}{\mu_o H} \right) \ln \left(1 - \frac{u}{\mu_o H} \right)
$$
$$
- \frac{1}{2} \left(1 + \frac{u}{\mu_o H} \right) \ln \left(1 + \frac{u}{\mu_o H} \right),
$$

where the density of energy is taken as a continuum variable. The entropy per particle of this ideal paramagnet will be given by

$$
s(u) = k_B \ln 2 - \frac{1}{2} k_B \left(1 - \frac{u}{\mu_o H} \right) \ln \left(1 - \frac{u}{\mu_o H} \right)
$$
$$
- \frac{1}{2} k_B \left(1 + \frac{u}{\mu_o H} \right) \ln \left(1 + \frac{u}{\mu_o H} \right).
$$

$$(4.25)$$

From this equation, we obtain all the thermodynamic properties of the ideal paramagnet of spin 1/2. For example, the temperature is given by the equation of state

$$
\frac{1}{T} = \frac{\partial s}{\partial u} = \frac{k_B}{2\mu_o H} \ln \left(1 - \frac{u}{\mu_o H} \right) - \frac{k_B}{2\mu_o H} \ln \left(1 + \frac{u}{\mu_o H} \right). \tag{4.26}
$$

In figure 4.2, we sketch a graph of $s(u)$ versus energy per particle u. Note that the density of entropy is a concave function of energy. Note also that the physical region, associated with positive values of temperature, corresponds to $u < 0$. For $u = -\mu_o H$, all spins are aligned with the magnetic field, and there is a unique accessible microscopic state, with zero entropy. For $u = 0$, the entropy is maximum (in this situation, half of the spins are pointing up, and the other half are pointing down). Later, we shall see a clever trick to reach the regions of negative temperatures, which correspond to larger values of energy.

From the equation of state in the entropy representation, we obtain an expression for the energy per particle of the ideal paramagnet,

$$
u = -\mu_o H \tanh \left(\frac{\mu_o H}{k_B T} \right). \tag{4.27}
$$

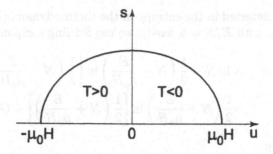

FIGURE 4.2. Entropy versus energy per particle for the ideal paramagnet of localized spins.

As the energy may be written in the form

$$E = -\mu_o H N_1 + \mu_o H N_2,$$

where N_1 and N_2 are the numbers of up and down spins, respectively, and defining the magnetization per particle,

$$m = \frac{M}{N} = \frac{\mu_o N_1 - \mu_o N_2}{N},$$

we obtain the relation

$$u = -Hm,$$

from which comes the famous equation of state for the magnetization of an ideal paramagnet,

$$m = \mu_o \tanh\left(\frac{\mu_o H}{k_B T}\right). \tag{4.28}$$

In figure 4.3, we sketch a graph of the dimensionless magnetization per particle versus the magnetic field in units of $k_B T/\mu_o$. For $\mu_o H \gg k_B T$, that is, for large enough fields and low enough temperatures, the system saturates ($m/\mu_o \rightarrow \pm 1$, depending on the direction of the field H). According to the characteristic behavior of paramagnetic crystals, for $\mu_o H \ll k_B T$, the magnetization changes linearly with the applied field.

From the equation of state for the magnetization, $m = m(T, H)$, we have the magnetic susceptibility (which is the analog of the isothermal compressibility of a fluid),

$$\chi(T, H) = \left(\frac{\partial m}{\partial H}\right)_T = \frac{\mu_o^2}{k_B T} \cosh^{-2}\left(\frac{\mu_o H}{k_B T}\right). \tag{4.29}$$

In zero field, we have,

$$\chi_o(T) = \chi(T, H = 0) = \frac{C}{T}, \tag{4.30}$$

FIGURE 4.3. Dimensionless magnetization per particle versus applied magnetic field for the ideal paramagnet of localized spins.

where $C = \mu_o^2/k_B > 0$. This is the well-known expression of the Curie law of paramagnetism. It has been experimentally confirmed for a wide variety of paramagnetic compounds (being sometimes used as a thermometer at low temperatures). In general, the experimental data are plotted in a graph where the coordinates are the inverse of the susceptibility in zero field and the temperature. The linear coefficient of this straight line provides information about the moments of the magnetic ions of the system. We will return to this problem in the more convenient context of the canonical ensemble.

Example 3 *Solid of Einstein.*

In example (6) of section 2.1, we considered a system of N localized and noninteracting one-dimensional quantum harmonic oscillators, with the same fundamental frequency ω. To mimic the elastic vibrations of a three-dimensional crystal, we locate the oscillators on the sites of a crystalline lattice, and consider independent oscillations along three directions. This is still a rather crude model, with a single frequency of oscillation, which does not take into account the interactions among the oscillators. In one of his pioneering works of 1905, using the ideas of energy quantization as proposed by Planck, Einstein has shown that this simple model of quantized oscillators is capable of predicting the dependence of the specific heat of solids on temperature.

In Chapter 2, we have already shown that the microscopic states of this model of N independent oscillators is characterized by the set of quantum numbers $(n_1, ..., n_N)$, where $n_j = 0, 1, 2, ...$ labels the number of quanta of energy of the jth oscillator. Hence, the total energy is given by

$$E_{n_1,...,n_N} = M\hbar\omega + \frac{1}{2}N\hbar\omega, \tag{4.31}$$

and the statistical problem is reduced to the calculation of the number of possibilities of distributing M quanta of energy among N oscillators. In the microcanonical ensemble, this counting of microscopic states, which cannot be done exactly except in some very simple cases, is the more difficult step of the problem. In Chapter 2, given the total energy E and the number of oscillators N, we had already obtained the number of accessible eigenstates of the system,

$$\Omega(E, N) = \frac{\left(\frac{E}{\hbar\omega} + \frac{N}{2} - 1\right)!}{\left(\frac{E}{\hbar\omega} - \frac{N}{2}\right)!(N-1)!}. \tag{4.32}$$

Therefore, using the Stirling expansion, we can write

$$
\begin{aligned}
\ln\Omega(E, N) &= -N\ln N + \left(\frac{E}{\hbar\omega} + \frac{N}{2} - 1\right)\ln\left(\frac{E}{\hbar\omega} + \frac{N}{2} - 1\right) \\
&\quad - \left(\frac{E}{\hbar\omega} - \frac{N}{2}\right)\ln\left(\frac{E}{\hbar\omega} - \frac{N}{2}\right) + O(\ln N, \ln E).
\end{aligned}
$$

In the thermodynamic limit, we have the entropy per oscillator,

$$
\begin{aligned}
s(u) &= \lim_{E, N\to\infty; \frac{E}{N}=u} \frac{1}{N} k_B \ln\Omega(E, N) \\
&= k_B\left(\frac{u}{\hbar\omega} + \frac{1}{2}\right)\ln\left(\frac{u}{\hbar\omega} + \frac{1}{2}\right) \\
&\quad - k_B\left(\frac{u}{\hbar\omega} - \frac{1}{2}\right)\ln\left(\frac{u}{\hbar\omega} - \frac{1}{2}\right), \tag{4.33}
\end{aligned}
$$

from which we fully describe the thermodynamic behavior of the solid of Einstein.

The equation of state for the temperature is given by

$$\frac{1}{T} = \frac{\partial s}{\partial u} = \frac{k_B}{\hbar\omega}\left[\ln\left(\frac{u}{\hbar\omega} + \frac{1}{2}\right) - \ln\left(\frac{u}{\hbar\omega} - \frac{1}{2}\right)\right].$$

From this expression, we write the energy per oscillator in terms of temperature,

$$u = \frac{1}{2}\hbar\omega + \frac{\hbar\omega}{\exp\left(\frac{\hbar\omega}{k_B T}\right) - 1}. \tag{4.34}$$

In the limit of low temperatures (that is, for $k_B T \ll \hbar\omega$), we obtain the quantum energy of point zero, $u \to \hbar\omega/2$. In the limit of high temperatures ($k_B T \gg \hbar\omega$), we recover the well-known classical result $u \to k_B T$. The specific heat of this model is given by

$$c = \frac{\partial u}{\partial T} = k_B\left(\frac{\hbar\omega}{kT}\right)^2 \frac{\exp\left(\frac{\hbar\omega}{k_B T}\right)}{\left[\exp\left(\frac{\hbar\omega}{k_B T}\right) - 1\right]^2}, \tag{4.35}$$

FIGURE 4.4. Specific heat versus temperature for the Einstein solid.

with the form sketched in figure 4.4. In the limit of high temperatures, for $T >> T_E = \hbar\omega/k_B$, where the Einstein temperature T_E is of the order 100 K for most crystals, we recover the classical result, $c \to k_B$, known as the Dulong and Petit law. At low temperatures, for $T << T_E = \hbar\omega/k_B$, the specific heat vanishes exponentially with temperature,

$$c \to k_B \left(\frac{\hbar\omega}{kT}\right)^2 \exp\left(-\frac{\hbar\omega}{kT}\right).$$

Several years after the pioneering work of Einstein, there appeared more precise measurements for the dependence of the specific heat of solids on temperature. Instead of an exponential decay, the specific heat of most solids behaves asymptotically as T^3, for $T \to 0$. This behavior has been accounted for by a model proposed by Debye, including a treatment of the interactions between the quantum oscillators of the Einstein model.

Example 4 *System of particles with two levels of energy.*

Before considering the more realistic (and much more difficult) problem of a gas of particles in classical phase space, let us discuss another drastically simplified model of N particles that may be found in one of two states, with energies 0 and $\epsilon > 0$, respectively. The specification of the microscopic states of this system requires the knowledge of the energy of each particle. We face the same combinatorial problem of localized ions of spin 1/2, which has already been treated in this section. Consider a situation with N_1 particles in the state of zero energy and $N_2 = N - N_1$ particles in the state with energy $\epsilon > 0$. The number of accessible microscopic states of this system is given by the combinatorial factor

$$\Omega = \frac{N!}{N_1! \, (N - N_1)!}.$$

Given the energy $E = \epsilon (N - N_1)$, we determine the values of N_1 and N_2. Therefore,

$$\Omega(E, N) = \frac{N!}{(N - \frac{E}{\epsilon})! \, (\frac{E}{\epsilon})!}.$$

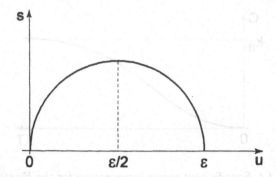

FIGURE 4.5. Entropy versus energy per particle for the system of particles with two energy levels.

Using Stirling's expansion, we write

$$\ln \Omega (E, N) = N \ln N - \left(N - \frac{E}{\epsilon} \right) \ln \left(N - \frac{E}{\epsilon} \right)$$

$$- \frac{E}{\epsilon} \ln \left(\frac{E}{\epsilon} \right) + O \left(\ln N, \ln E \right).$$

In the thermodynamic limit, $N, E \to \infty$, with $E/N = u$ fixed, we have the entropy per particle,

$$s \left(u \right) = \lim \frac{1}{N} k_B \ln \Omega \left(E, N \right) = -k_B \left(1 - \frac{u}{\epsilon} \right) \ln \left(1 - \frac{u}{\epsilon} \right) - k_B \frac{u}{\epsilon} \ln \left(\frac{u}{\epsilon} \right).$$

$$(4.36)$$

In figure 4.5, we sketch a graph of the density of entropy s versus the density of energy u. Note that the entropy is a concave function of energy. For $u = 0$, the entropy vanishes, since all particles are in the ground state, with zero energy (the slope of the curve is infinite, since the temperature vanishes). The entropy is maximum at $u = \epsilon/2$, which corresponds to infinite temperature, where the two energy levels of the particles are practically identical $(k_B T \gg \epsilon)$, with half of the particles in each level. For $u > \epsilon/2$, we have the non-physical region, associated with negative temperatures.

The equation of state in the entropy representation is given by

$$\frac{1}{T} = \frac{\partial s}{\partial u} = \frac{k_B}{\epsilon} \left[\ln \left(1 - \frac{u}{\epsilon} \right) - \ln \frac{u}{\epsilon} \right],$$

from which we have

$$\frac{\partial^2 s}{\partial u^2} = - \frac{k_B}{u \left(\epsilon - u \right)},$$

indicating that the entropy function, $s = s \left(u \right)$, is concave for $0 \leq u \leq \epsilon$. From this equation of state in the entropy representation, we obtain an

expression for the energy as a function of temperature,

$$u = \frac{\epsilon e^{-\beta\epsilon}}{1 + e^{-\beta\epsilon}}, \tag{4.37}$$

where we have used the standard notation $\beta = 1/(k_B T)$. This formula leads to a simple probabilistic interpretation that will become more transparent in the framework of the canonical ensemble. For a single particle, we assign the probabilities

$$P_1 = \frac{1}{1 + e^{-\beta\epsilon}} \quad \text{and} \quad P_2 = \frac{e^{-\beta\epsilon}}{1 + e^{-\beta\epsilon}} \tag{4.38}$$

for the occupancies of the levels with zero and $\epsilon > 0$ energies, respectively. These are the famous **Boltzmann factors**, which define the probabilities of occupancy of the single-particle states of a system of noninteracting particles at a certain temperature T. The probabilistic average, given by equation (4.37), may be obtained from the standard expression for the mean value, $u = 0 \times P_1 + \epsilon \times P_2$. In figure 4.6, we draw u versus the temperature T. As $P_1 \geq P_2$, with the equality holding for $T \to \infty$ only, the energy increases from zero, at the ground state, to the maximum value $\epsilon/2$, where the two levels become equally populated. We are also including the branch of negative temperatures, which is associated with a more energetic situation, with larger energies than any case of positive temperature. In terms of the Boltzmann factors, P_1 and P_2, in a case of negative temperatures, we should have $P_1 < P_2$, that is, we should reverse the population of the two levels. This can be accomplished by reverting the sign of ϵ. For example, in the case of the paramagnetic model of ions of spin 1/2, the splitting between energy levels is given by $\epsilon = 2\mu_o H$, and may be reversed through a change of sign (direction) of the applied magnetic field. Just after reversing the field, the magnetic lattice, with a very high energy, is out of thermal equilibrium with the crystalline lattice of the environment. Indeed, there are some experiments of this nature, for magnetic compounds where the spin-lattice relaxation time is long enough. The interested reader may check the pioneering work of N. F. Ramsey, in Phys. Rev. **103**, 20 (1956).

From equation (4.37), for the energy as a function of temperature, we obtain an expression for the specific heat,

$$c = \frac{\partial u}{\partial T} = k_B (\beta\epsilon)^2 \frac{e^{-\beta\epsilon}}{(1 + e^{-\beta\epsilon})^2}. \tag{4.39}$$

It is easy to see that the specific heat vanishes in the limits of very high, $\beta\epsilon \to 0$, and low, $\beta\epsilon \to \infty$, temperatures. In figure 4.7, we sketch a graph of the specific heat versus temperature, which displays a broad maximum at $\beta\epsilon \sim 1$, that is, at a temperature of order ϵ/k. This is the well-known **Schottky effect**, a brand mark of systems where single particles may occupy a few discrete energy levels.

FIGURE 4.6. Sketch of the energy per particle versus temperature for the system of particles with two energy levels. Note that the region associated with negative temperatures corresponds to larger values of energy.

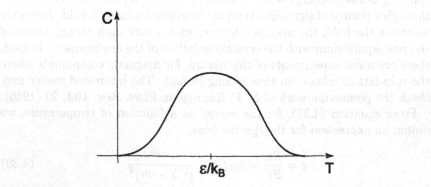

FIGURE 4.7. Sketch of the specific heat versus temperature for the system of particles of two energy levels.

Example 5 *The Boltzmann gas.*

We now introduce a generalization of the model of particles occupying two energy levels. To emphasize the role of probabilistic arguments in the definition of entropy, Boltzmann proposed a model where the particles of the gas may be found in a discrete set of energy values, $\{\epsilon_j;\ j = 1, 2, 3, ...\}$. The two-level model is a particular case, with $\epsilon_1 = 0$ and $\epsilon_2 = \epsilon > 0$. The microscopic state of the Boltzmann gas is specified by giving the energy of each one of the particles. Let us then define the set of "occupation numbers," $\{N_j\}$, where N_1 labels the number of particles with energy ϵ_1, N_2 labels the number of particles with energy ϵ_2, and so on. Therefore, given the set of numbers of occupation $\{N_j\}$, the total energy E, and the total number of particles N, we write the number of accessible microstates of the gas,

$$\Omega\left(\{N_j\}; E, N\right) = \frac{N!}{N_1! N_2! N_3! \cdots}, \tag{4.40}$$

with the restrictions

$$N = \sum_j N_j \quad \text{and} \quad E = \sum_j \epsilon_j N_j. \tag{4.41}$$

The probability of finding the system under these conditions is proportional to $\Omega\left(\{N_j\}; E, N\right)$, with the restrictions imposed by equations (4.41). To find the occupation numbers in equilibrium, we maximize $\Omega\left(\{N_j\}; E, N\right)$ with respect to the set $\{N_j\}$, taking into account conditions (4.41). Using the method of the Lagrange multipliers, we define the function

$$f\left(\{N_j\}, \lambda_1, \lambda_2\right) = \ln \Omega(\{N_j\}; E, N) + \lambda_1 \left(N - \sum_j N_j\right)$$

$$+ \lambda_2 \left(E - \sum_j \epsilon_j N_j\right). \tag{4.42}$$

The extremization with respect to the Lagrange multipliers λ_1 and λ_2 gives equations (4.41). Now, using the Stirling expansion, we have

$$\frac{\partial f}{\partial N_k} = -\ln N_k - \lambda_1 - \lambda_2 \epsilon_k = 0. \tag{4.43}$$

Eliminating the Lagrange multiplier λ_1, we have the equilibrium distribution,

$$\frac{N_k}{N} = \frac{\exp\left(-\lambda_2 \epsilon_k\right)}{Z_1}, \tag{4.44}$$

where the normalization factor is given by

$$Z_1 = \sum_j \exp\left(-\lambda_2 \epsilon_j\right). \tag{4.45}$$

The expression for N_k/N already has the form of a Boltzmann factor, and may be interpreted as the probability of occupation of the energy level ϵ_k. The Lagrange multiplier λ_2, which should be inversely proportional to temperature, may be found from the condition that the total energy E is given by the classical expression, $(3/2)NkT$. Hence, we have

$$E = \sum_k \epsilon_k N_k = N \frac{\sum_k \epsilon_k \exp\left(-\lambda_2 \epsilon_k\right)}{\sum_j \exp\left(-\lambda_2 \epsilon_j\right)} = \frac{3}{2} N k_B T, \tag{4.46}$$

which may be written in the more compact form

$$-\frac{\partial}{\partial \lambda_2} \ln Z_1 = \frac{3}{2} k_B T. \tag{4.47}$$

In fact, in the limit of the continuum, given by

$$\epsilon_k \to \frac{1}{2} m \vec{v}^2, \tag{4.48}$$

we may write

$$Z_1 \to \int d^3\vec{v} \exp\left(-\lambda_2 \frac{1}{2} m \vec{v}^2\right) = \left(\frac{2\pi}{\lambda_2 m}\right)^{3/2}. \tag{4.49}$$

Using equations (4.49) and (4.47), we finally establish that

$$\lambda_2 = \frac{1}{k_B T}. \tag{4.50}$$

In this continuum limit, we also recover the famous Maxwell–Boltzmann distribution of the molecular velocities,

$$\frac{N_k}{N} \to p\left(\vec{v}\right) d^3\vec{v}, \tag{4.51}$$

where

$$p\left(\vec{v}\right) = \left(\frac{2\pi k_B T}{m}\right)^{-3/2} \exp\left(-\frac{m\vec{v}^2}{2k_B T}\right). \tag{4.52}$$

This proposal by Boltzmann is a pioneering attempt, previous to the work of Planck, at introducing a model with discrete energy levels. However,

although being useful to illustrate the role of the theory of probabilities, and reproducing some physical results, the Boltzmann gas is just a toy model. A more adequate discussion of the classical gas of particles has to be carried out in the phase space of classical mechanics. Also, the discretization of the energy levels of a gas of particles may be correctly dealt with only in the framework of quantum mechanics. Later, we shall see that Boltzmann's model of a gas of particles, with the inclusion of some corrections and clarifications, may indeed be obtained in the classical limit of the quantum theory of an ideal monatomic gases.

4.4 Classical monatomic ideal gas

The Hamiltonian of a classical gas of N particles may be written as

$$\mathcal{H} = \sum_{i=1}^{N} \frac{1}{2m} \vec{p}_i^2 + \sum_{i<j} V\left(|\vec{r}_i - \vec{r}_j|\right), \tag{4.53}$$

where the pair potential $V(r)$ should have a hard core, which reflects the impenetrability of matter, and a small attractive part that vanishes (fast enough) for $r \to \infty$. A realistic form of $V(r)$, for which we may demonstrate the existence of the thermodynamic limit, is given by the well-known potential of Lennard-Jones. More realistic models, however, cannot be treated without approximations, and will not be considered at this point. In this section, in order to restrict the considerations to the analysis of an ideal gas of particles, we disregard the interaction terms in equation (4.53).

In example (3) of Section 2.2, we have already analyzed a classical ideal gas, of N monatomic particles of mass m, inside a reservoir of volume V, with total energy between E and $E+\delta E$. The accessible volume of classical phase space is given by

$$\Omega(E, V, N; \delta E) = \left(\frac{m}{2}\right)^{1/2} C_{3N} \left(2m\right)^{\frac{3N}{2}-1} V^N E^{\frac{3N}{2}-1} \delta E, \tag{4.54}$$

where the prefactor C_{3N} depends only on the number of particles N (an expression for C_{3N} is given in Appendix A.4). To take the thermodynamic limit, we write $E/N = u$ and $V/N = v$. Therefore, we have

$$\begin{aligned}
\frac{1}{N} \ln \Omega(E, V, N; \delta E) &= \left(\frac{3}{2} - \frac{1}{N}\right) \ln u + \ln v + \left(\frac{3}{2} - \frac{1}{N}\right) \ln(2m) \\
&\quad + \frac{1}{N} \ln C_{3N} + \ln N + \left(\frac{3}{2} - \frac{1}{N}\right) \ln N \\
&\quad + \frac{1}{2N} \ln\left(\frac{m}{2}\right) + \frac{1}{N} \ln \delta E.
\end{aligned} \tag{4.55}$$

It should be mentioned that

$$\lim_{N\to\infty}\frac{1}{N}\ln\delta E = 0,$$

since δE is a small but fixed quantity (which is a justification for the convenience of choosing an ensemble with the energy between E and $E + \delta E$). Therefore, in the thermodynamic limit, $E, V, N \to \infty$, with $E/N = u$ and $V/N = v$ fixed, we recover the known results for the entropy of the ideal monatomic gas,

$$s\left(u,v\right) = \lim\frac{1}{N}k\ln\Omega\left(E,V,N;\delta E\right) = \frac{3}{2}k_B\ln u + k_B\ln v + s_o, \quad (4.56)$$

where s_o is a constant. To write this limit, however, as it can be seen from a mere inspection of equation (4.55), we have to suppose that the prefactor C_{3N}, for large N, behaves asymptotically as

$$C_{3N} \to N^{-\frac{5}{2}N}a^N, \quad (4.57)$$

where a is a constant. From the fundamental equation (4.56), we obtain the equations of state in the entropy representation,

$$\frac{1}{T} = \frac{\partial s}{\partial u} = \frac{3k_B}{2u}$$

and

$$\frac{p}{T} = \frac{\partial s}{\partial v} = \frac{k_B}{v},$$

which may be rewritten in the more usual forms

$$u = \frac{3}{2}k_BT \quad \text{and} \quad pv = k_BT. \quad (4.58)$$

From the classical point of view, there is no possibility of obtaining the constant s_o, since the very choice of generalized coordinates in classical phase space is not unique (and the volume V is a dimensional quantity).

We now make some comments on these results. The existence of an entropy function, $s = s\left(u,v\right)$, in the thermodynamic limit, has been established under the supposition of a well-defined dependence of the prefactor C_{3N} on the number N of particles [see again equation (4.57)]. As a matter of fact, it is not even necessary to invoke the thermodynamic limit to obtain the equations of state of the ideal gas. In fact, from equation (4.54), we may write

$$S\left(E,V,N;\delta E\right) = k_B\ln\Omega\left(E,V,N;\delta E\right) = \frac{3}{2}k_B\ln E + k_B\ln V + f\left(N;\delta E\right),$$

where $f(N; \delta E)$ is a function of N and δE. Therefore, we have the expressions

$$\frac{1}{T} = \frac{\partial S}{\partial E} = \frac{3k_B}{2E} \quad \text{and} \quad \frac{p}{T} = \frac{\partial S}{\partial V} = \frac{k_B}{V}$$

that are identical to the previously obtained equations of state. However, either to obtain the chemical potential or to work with the entropy of a mixture of gases, we have to analyze the correct dependence on the number of particles. There is then a serious problem that had already been detected by Boltzmann and Gibbs. Looking more carefully at the equations of Appendix A.4, we see that the volume of a hyperspherical shell of radius R and thickness δR, in a space of n dimensions, is given by

$$\Omega_n(R; \delta R) = C_n R^{n-1} \delta R = 2 \frac{\pi^{n/2}}{\Gamma\left(\frac{n}{2}\right)} R^{n-1} \delta R. \tag{4.59}$$

Hence,

$$C_{3N} = 2 \frac{\pi^{3N/2}}{\Gamma\left(\frac{3N}{2}\right)} \sim N^{-\frac{3}{2}N} b^N, \tag{4.60}$$

in disagreement with the exponent of N in equation (4.57). A long time ago, it was recognized the need of introducing an *ad-hoc* correction factor in the formulas obtained from the classical phase space. The accessible volume of phase space, Ω, has to be divided by $N!$, to guarantee the existence of the thermodynamic limit. In fact, using the correct form of C_{3N}, given by equation (4.60), we have

$$\frac{1}{N!} C_{3N} \sim N^{-\frac{5}{2}N} a^N,$$

with the asymptotic behavior assumed in equation (4.57). For classical particles on the phase space, we will always introduce the correction

$$\Omega \rightarrow \frac{1}{N!} \Omega \tag{4.61}$$

to guarantee the existence of the thermodynamic limit. This **correct counting factor of Boltzmann** will be present in all the manipulations of the classical statistical mechanics of a gas of particles. Although the division by $N!$ reflects the distinguishable character of classical particles, the introduction of this factor is a mere artifact of classical statistical physics to make sure that the definition of entropy does indeed produce an additive function, with the homogeneity properties of the thermodynamic entropy. We will see that this artifact ends up being very well justified by taking the classical limit of the entropy of a quantum gas of particles. In this classical limit, we regain the correct counting factor, as well as the constant terms that lead to a unique expression for the entropy of a gas of particles.

Exercises

1. Consider a model of N localized magnetic ions, given by the spin Hamiltonian

$$\mathcal{H} = D \sum_{j=1}^{N} S_j^2,$$

where the spin variable S_j may assume the values -1, 0, or $+1$, for all j (see exercise 1 of Chapter 2). Given the total energy E, use the expression for the number of accessible microstates, $\Omega(E, N)$, to obtain the entropy per particle, $s = s(u)$, where $u = E/N$. Obtain an expression for the specific heat c in terms of the temperature T. Sketch a graph of c versus T. Check the existence of a broad maximum associated with the Schottky effect. Write an expression for the entropy as a function of temperature. What are the limiting values of the entropy for $T \to 0$ and $T \to \infty$?

2. In the solid of Einstein, we may introduce a volume coordinate if we make the phenomenological assumption that the fundamental frequency ω as a function of $v = V/N$ is given by

$$\omega = \omega(v) = \omega_o - A \ln\left(\frac{v}{v_o}\right),$$

where ω_o, A, and v_o are positive constants. Obtain expressions for the expansion coefficient and the isothermal compressibility of this model system.

3. Consider the semiclassical model of N particles with two energy levels (0 and $\epsilon > 0$). As in the previous exercise, suppose that the volume of the gas may be introduced by the assumption that the energy of the excited level depends on $v = V/N$,

$$\epsilon = \epsilon(v) = \frac{a}{v^\gamma},$$

where a and γ are positive constants. Obtain an equation of state for the pressure, $p = p(T, v)$, and an expression for the isothermal compressibility (note that the constant γ plays the role of the Grüneisen parameter of the solid).

4. The total number of the accessible microscopic states of the Boltzmann gas, with energy E and number of particles N, may be written as

$$\Omega(E, N) = \sum_{N_1, N_2, \ldots} \frac{N!}{N_1! N_2! \cdots},$$

with the restrictions

$$\sum_j N_j = N \quad \text{and} \quad \sum_j \epsilon_j N_j = E.$$

Except for an additive constant, show that the entropy per particle is given by

$$s = -k_B \sum_j \left(\frac{\tilde{N}_j}{N}\right) \ln \left(\frac{\tilde{N}_j}{N}\right),$$

where $\left\{\tilde{N}_j\right\}$ is the set of occupation numbers at equilibrium. Using the continuum limit of the Boltzmann gas, show that the entropy depends on temperature according to a term of the form $-k_B T \ln T$.

5. Consider a lattice gas of N particles distributed among V cells (with $N \leq V$). Suppose that each cell may be either empty or occupied by a single particle. The number of microscopic states of this system will be given by

$$\Omega(V, N) = \frac{V!}{N!(V - N)!}.$$

Obtain an expression for the entropy per particle, $s = s(v)$, where $v = V/N$. From this fundamental equation, obtain an expression for the equation of state p/T. Write an expansion of p/T in terms of the density $\rho = 1/v$. Show that the first term of this expansion gives the Boyle law of the ideal gases. Sketch a graph of μ/T, where μ is the chemical potential, in terms of the density ρ. What is the behavior of the chemical potential in the limits $\rho \to 0$ and $\rho \to 1$?

with the restrictions

$$\sum_j N_j = N \quad \text{and} \quad \sum_j \epsilon_j N_j = E.$$

Except for an additive constant, show that the entropy per particle is given by

$$s = -k_B \sum_j \left(\frac{N_j}{N}\right) \ln\left(\frac{N_j}{N}\right)$$

where $\{N_j\}$ is the set of occupation numbers at equilibrium. Using the continuum limit of the Boltzmann gas, show that the entropy depends on temperature according to a term of the form $-k_B T \ln T$.

5. Consider a lattice gas of N particles distributed among V cells (with $M \geq N$). Suppose that each cell may be either empty or occupied by a single particle. The number of microscopic states of this system will be given by

$$\Omega(V, N) = \frac{V!}{N!(V - N)!}.$$

Obtain an expression for the entropy per particle, $s = s(v)$, where $v = V/N$. From this fundamental equation, obtain an expression for the equation of state p/T. Write an expansion of p/T in terms of the density $\rho = 1/v$. Show that the first term of this expansion gives the Boyle law of the ideal gases. Sketch a graph of μ/T, where μ is the chemical potential, in terms of the density ρ. What is the behavior of the chemical potential in the limits $\rho \to 0$ and $\rho \to 1$?

5
Canonical Ensemble

As in the formalism of thermodynamics, in which alternative representations may be more convenient to use depending on the circumstances, in statistical mechanics we also use other ensembles, besides the microcanonical ensemble.

Consider, for example, a simple thermodynamic system in contact with a thermal reservoir through a diathermal but fixed and impermeable wall. In figure 5.1, the reservoir R is very large with respect to the system S of interest (that is, R is characterized by a very large number of degrees of freedom as compared with S). The isolated composite system, S+R, with fixed total E_o, can be analyzed by applying the fundamental postulates of equilibrium statistical mechanics. Therefore, the probability P_j of finding

FIGURE 5.1. System S in contact with a thermal reservoir R (at temperature T).

system S in a particular microscopic state j is given by

$$P_j = c \, \Omega_R \, (E_o - E_j) \, , \tag{5.1}$$

where c is a normalization constant, E_j is the energy of system S in the particular microscopic state j, and $\Omega_R \, (E)$ is the number of accessible microscopic states of the thermal reservoir R with energy $E = E_o - E_j$. Since the reservoir is very large, energy E_j is much smaller than the total energy E_o for all j. Hence, we can write the expansion

$$\ln P_j = \ln c + \ln \Omega_R \, (E_o) + \left[\frac{\partial \ln \Omega_R \, (E)}{\partial E} \right]_{E=E_o} (-E_j)$$
$$+ \frac{1}{2} \left[\frac{\partial^2 \ln \Omega_R \, (E)}{\partial E^2} \right]_{E=E_o} (-E_j)^2 + \dots . \tag{5.2}$$

Using the statistical definition of entropy, given by the second postulate of statistical mechanics, we have

$$\frac{\partial \ln \Omega_R \, (E)}{\partial E} = \frac{1}{k_B T}, \tag{5.3}$$

where T is the temperature of the reservoir. We also have

$$\frac{\partial^2 \ln \Omega_R \, (E)}{\partial E^2} = \frac{1}{k_B} \frac{\partial}{\partial E} \left(\frac{1}{T} \right) \to 0, \tag{5.4}$$

in the limit of a true thermal reservoir, in which the temperature is practically fixed. Therefore, expansion (5.2) is reduced to the form

$$\ln P_j = \text{constant} - \frac{1}{k_B T} E_j, \tag{5.5}$$

that is,

$$P_j = \frac{\exp \left(-\beta E_j \right)}{\sum_k \exp \left(-\beta E_k \right)}, \tag{5.6}$$

where we are using the standard notation of statistical physics, $\beta = 1/ \left(k_B T \right)$. The **canonical ensemble is defined by the set of microstates** $\{j\}$, **associated with a distribution of probabilities given by equation (5.6), which are accessible to a system S in contact with a thermal reservoir at temperature** T.

A. *Connection with thermodynamics*

The **partition function** (also called sum over states, from the German Zustandsumme) is given by the sum

$$Z = \sum_j \exp \left(-\beta E_j \right), \tag{5.7}$$

which is related to the normalization of the probability P_j. We should note that this sum is over all microscopic states. Therefore, given a certain value of energy, there may be several identical terms, corresponding to microscopic states with the same particular value of energy. Taking into account this degeneracy factor, we can write

$$Z = \sum_j \exp\left(-\beta E_j\right) = \sum_E \Omega\left(E\right) \exp\left(-\beta E\right), \qquad (5.8)$$

where $\Omega\left(E\right)$ is the number of microscopic states of system S with energy E. Using the same heuristic arguments of the last chapter, in the limit of a large system we can substitute the sum by its maximum term. Thus,

$$Z = \sum_E \exp\left[\ln \Omega\left(E\right) - \beta E\right] \sim \exp\left[-\beta \min_E \left\{E - TS\left(E\right)\right\}\right], \qquad (5.9)$$

where we use again the definition of entropy, $S\left(E\right) = k_B \ln \Omega\left(E\right)$. Since the minimization with respect to energy corresponds to a Legendre transformation, this heuristic argument already suggests a connection between the canonical ensemble and thermodynamics, through the correspondence

$$Z \rightarrow \exp\left(-\beta F\right), \qquad (5.10)$$

where F is the Helmholtz free energy. For a pure fluid, the energy levels E_j depend on the volume V and the number of particles N. Therefore, the partition function Z depends on T, V, and N, that is, we write $Z = Z\left(T, V, N\right)$. Then, we have

$$F = F\left(T, V, N\right) \rightarrow -\frac{1}{\beta} \ln Z\left(T, V, N\right). \qquad (5.11)$$

To be more precise, we should explicitly refer to the thermodynamic limit to define the connection with thermodynamics,

$$f\left(T, v\right) = -\frac{1}{\beta} \lim_{V, N \to \infty; \frac{V}{N} = v} \frac{1}{N} \ln Z\left(T, V, N\right). \qquad (5.12)$$

B. Canonical ensemble in classical phase space

With slight changes of language, we can use similar heuristic arguments in the context of classical phase space. Indeed, in classical statistical mechanics we define the **canonical distribution of Gibbs** as the density of probability in phase space,

$$\rho\left(q, p\right) = \frac{1}{Z} \exp\left\{-\beta \mathcal{H}_S\left(q, p\right)\right\}, \qquad (5.13)$$

where

$$Z = \int \exp\left\{-\beta\mathcal{H}_S\left(q,p\right)\right\} dqdp, \tag{5.14}$$

and $\mathcal{H}_S\left(q,p\right)$ is the Hamiltonian of system S, in thermal equilibrium with an infinitely large reservoir R, at absolute temperature $T = 1/\left(k_B\beta\right)$. The connection between the canonical ensemble and thermodynamics is carried out in the thermodynamic limit, expressed by the Helmholtz free energy per particle, given by equation (5.12). In some cases, assuming a convenient form for the thermodynamic functions of the reservoir R, it is possible to rigorously deduce the canonical distribution (5.13) from the fundamental postulate of statistical mechanics. The interested reader may check Chapter 2 of the text by C. J. Thompson, *Classical Equilibrium Statistical Mechanics*, Clarendon Press, Oxford, 1988.

C. Fluctuations of energy

Of course, there are fluctuations of the energy values of the system S in the canonical ensemble. Using the distribution of probabilities P_j, given by equation (5.6), we may obtain the average (probabilistic expected) value of the energy of system S,

$$\langle E_j \rangle = \frac{\sum_j E_j \exp\left(-\beta E_j\right)}{\sum_j \exp\left(-\beta E_j\right)} = -\frac{\partial}{\partial\beta}\ln Z, \tag{5.15}$$

where the partition function Z is given by equation (5.7). In the thermodynamic limit, from the connection with thermodynamics, $F \to -k_BT\ln Z$, it is easy to see that the expected value of energy, $\langle E_j \rangle$, should be identified with the thermodynamic internal energy U of system S, that is, $U \to \langle E_j \rangle$.

Now we write an expression for the quadratic deviation,

$$\left\langle \left(E_j - \langle E_j \rangle\right)^2 \right\rangle = \left\langle E_j^2 \right\rangle - \left\langle E_j \right\rangle^2$$

$$= Z^{-1}\sum_j E_j^2 \exp\left(-\beta E_j\right) - \left[Z^{-1}\sum_j E_j \exp\left(-\beta E_j\right)\right]^2, \tag{5.16}$$

that is,

$$\left\langle \left(E_j - \langle E_j \rangle\right)^2 \right\rangle = \frac{\partial}{\partial\beta}\left[\frac{1}{Z}\frac{\partial Z}{\partial\beta}\right] = -\frac{\partial}{\partial\beta}\langle E_j \rangle. \tag{5.17}$$

Making the identification between the expected value of energy and the internal thermodynamic energy, we have

$$\left\langle (E_j - \langle E_j \rangle)^2 \right\rangle = -\frac{\partial}{\partial \beta} U = k_B T^2 \frac{\partial U}{\partial T} = N k_B T^2 c_V \geq 0, \qquad (5.18)$$

where it becomes clear that the specific heat at constant volume should be positive. Finally, we write the relative width,

$$\frac{\left\langle (E_j - \langle E_j \rangle)^2 \right\rangle^{1/2}}{\langle E_j \rangle} = \frac{\sqrt{N k_B T^2 c_V}}{N u} \sim \frac{1}{\sqrt{N}}. \qquad (5.19)$$

Therefore, according to expectations, the relative width vanishes as $1/\sqrt{N}$, in the limit $N \to \infty$, which guarantees the connection with thermodynamics. However, in special situations, the specific heat may be very large, which brings about enormous fluctuations of energy in the canonical ensemble. We shall see that this may happen in the neighborhood of phase transitions of second order, giving rise to the peculiar critical behavior of matter.

D. Alternative derivation of the canonical distribution

We may deduce the canonical distribution from another line of argument, still of a heuristic nature, that has been particularly useful in certain applications in information theory and optimization problems.

If we write the Helmholtz free energy in the form

$$F = -\frac{1}{\beta} \ln Z = -\frac{1}{\beta} \ln \sum_j \exp(-\beta E_j), \qquad (5.20)$$

the entropy is given by

$$S = -\frac{\partial F}{\partial T} = k_B \beta^2 \frac{\partial F}{\partial \beta} = k_B \ln Z + k_B \beta \frac{1}{Z} \sum_j E_j \exp(-\beta E_j). \qquad (5.21)$$

However, from the expression for P_j, given by equation (5.6), we have

$$-\beta E_j = \ln(Z P_j). \qquad (5.22)$$

Inserting into equation (5.21), we obtain an expression for the entropy in terms of the distribution of probabilities,

$$S = -k_B \sum_j P_j \ln P_j, \qquad (5.23)$$

that is sometimes called the entropy of Shannon. Indeed, this same expression still holds in the microcanonical ensemble. In this case, $P_j = 1/\Omega$,

there are Ω accessible states of the system, and (5.23) is reduced to the well-known expression $S = k_B \ln \Omega$. It should be pointed out that there have been several attempts to generalize equation (5.23). Recently, C. Tsallis, under the inspiration of the expressions for the generalized fractal dimensions, proposed the alternative definition

$$S_q = k_B \frac{1}{q-1} \left[1 - \sum_j P_j^q \right],$$ (5.24)

that reduces to the Boltzmann–Gibbs entropy, given by (5.23), in the limit $q \to 1$ [see C. Tsallis, J. Stat. Phys. **52**, 479 (1988)]. The entropy of Tsallis is not additive and thus cannot be applied to problems of thermodynamical interest. However, it is an interesting form, which results in a host of consequences and which may find applications in other domains of science.

Now we may introduce a new postulate of maximization of entropy, given by equation (5.23), for an arbitrary distribution of probabilities, $\{P_j\}$, subjected to certain constraints. For a canonical ensemble, we have the constraints represented by the normalization,

$$\sum_j P_j = 1,$$ (5.25)

and by the value of the thermodynamic average energy,

$$\sum_j E_j P_j = U.$$ (5.26)

Using the technique of the Lagrange multipliers, we maximize the function

$$f(\{P_j\}, \lambda_1, \lambda_2) = -k_B \sum_j P_j \ln P_j - \lambda_1 \left(\sum_j P_j - 1 \right)$$

$$-\lambda_2 \left(\sum_j E_j P_j - U \right),$$ (5.27)

with respect to the set $\{P_j\}$. Hence, we have

$$\frac{\partial f}{\partial P_j} = -k_B \ln P_j - k_B - \lambda_1 - \lambda_2 E_j.$$ (5.28)

Eliminating the multiplier λ_1 through equation (5.25), we have

$$P_j = \frac{\exp(-\lambda_2 E_j / k_B)}{\sum_j \exp(-\lambda_2 E_j / k_B)}.$$ (5.29)

Equation (5.26) leads to the correspondence between the Lagrange multiplier λ_2 and the inverse of the absolute temperature, which reproduces the characteristic probability distribution of the canonical ensemble.

Up to this point, we have adopted a line of purely heuristic arguments, far from any mathematical rigor. We have been able to build the canonical ensemble and to establish the connection with thermodynamics through the Helmholtz free energy. It should be pointed out that, for a large class of intermolecular potentials, it is possible to prove the existence of this free energy in the thermodynamic limit, with the convexity properties required by classical thermodynamics. Also, it is possible to show that the thermodynamic potentials associated with the canonical and the microcanonical ensembles are duly related by a Legendre transformation, giving rise to the same thermodynamic description of a physical system. In the next chapter, we analyze an example of a modern mathematical demonstration. In the remainder of this chapter, however, we just give some examples of applications of the formalism of the canonical ensemble.

5.1 Ideal paramagnet of spin 1/2

Consider again a system of N localized magnetic particles, of spin 1/2, in the presence of an applied magnetic field H, but now in contact with a thermal reservoir at temperature T. The Hamiltonian of the system is given by

$$\mathcal{H} = -\mu_o H \sum_{j=1}^{N} \sigma_j, \qquad (5.30)$$

where $\sigma_j = \pm 1$, for $j = 1, 2, ..., N$. A microscopic state of this system is characterized by the set of values of the spin variables, $\{\sigma_j\}$. Therefore, the canonical partition function may be written as

$$Z = \sum_{\{\sigma_j\}} \exp\left(-\beta\mathcal{H}\right) = \sum_{\sigma_1, \sigma_2, ..., \sigma_N} \exp\left(\beta\mu_o H \sum_{j=1}^{N} \sigma_j\right). \qquad (5.31)$$

Now it is easy to see that this multiple sum fully factorizes, and may be written as

$$Z = \left[\sum_{\sigma_1} \exp\left(\beta\mu_o H\sigma_1\right)\right] \cdots \left[\sum_{\sigma_N} \exp\left(\beta\mu_o H\sigma_N\right)\right] = Z_1^N, \qquad (5.32)$$

where

$$Z_1 = \sum_{\sigma=\pm 1} \exp\left(\beta\mu_o H\sigma\right) = 2\cosh\left(\beta\mu_o H\right). \qquad (5.33)$$

We see that the expression for the partition function is obtained without invoking any combinatorial arguments, which are in general much more difficult to figure out. This same sort of factorization of a sum (or of an integral, for the partition function in classical phase space) normally occurs for classical and semiclassical systems of noninteracting particles. In all of those noninteracting cases, it is enough to calculate a single-particle partition function, Z_1.

From the expression of the partition function, we establish the connection with thermodynamics,

$$g(T, H) = -\frac{1}{\beta} \lim_{N \to \infty} \frac{1}{N} \ln Z = -k_B T \ln \left[2 \cosh \left(\frac{\mu_o H}{k_B T} \right) \right], \quad (5.34)$$

where $g(T, H)$ is a magnetic free energy per particle. Due to the form of Hamiltonian, which reflects our choice of the magnetic energy, including the interaction with the applied magnetic field, this free energy per particle is written as a function of the independent variables T and H. We then use the notation $g = g(T, H)$, in analogy with the Gibbs function of a pure fluid (that depends on the intensive parameters p and T). Now it is easy to obtain the thermodynamic properties of the ideal paramagnet, which should be identical to the results already obtained in the framework of the microcanonical ensemble. The entropy per particle is given by

$$s = -\left(\frac{\partial g}{\partial T} \right)_H = k_B \ln \left[2 \cosh \left(\frac{\mu_o H}{k_B T} \right) \right] - k_B \left(\frac{\mu_o H}{k_B T} \right) \tanh \left(\frac{\mu_o H}{k_B T} \right),$$
$$(5.35)$$

from which we calculate the specific heat at constant field. The magnetization per particle is given by

$$m = -\left(\frac{\partial g}{\partial H} \right)_T = \mu_o \tanh \left(\frac{\mu_o H}{k_B T} \right), \quad (5.36)$$

that is exactly the same expression already obtained in the context of the microcanonical ensemble. As it is expected, we point out that the thermodynamic value of the magnetization per particle does correspond to the probabilistic average in the canonical ensemble. In fact, using the canonical distribution, we have

$$\frac{1}{N} \left\langle \mu_o \sum_{j=1}^{N} \sigma_j \right\rangle = \frac{1}{N\beta} \frac{\partial}{\partial H} \ln Z = \mu_o \tanh (\beta \mu_o H), \quad (5.37)$$

which gives the same expression for the magnetization per particle as equation (5.36). From the expression for $m = m(T, H)$ we obtain the magnetic susceptibility,

$$\chi(T, H) = \left(\frac{\partial m}{\partial H} \right)_T = \frac{\mu_o^2 H}{k_B T} \cosh^{-2} \left(\frac{\mu_o H}{k_B T} \right), \quad (5.38)$$

which leads to the well-known Curie law,

$$\chi(T, H = 0) = \frac{\mu_o^2 H}{k_B T}, \tag{5.39}$$

as we have derived in the context of the microcanonical ensemble. Using equations (5.34) and (5.35), the thermodynamical internal energy per particle can be calculated from the magnetic free energy and the entropy,

$$u = g + Ts = -\mu_o H \tanh\left(\frac{\mu_o H}{k_B T}\right), \tag{5.40}$$

which also coincides with the result of last chapter. We can as well use equation (5.15) to obtain the thermodynamic internal energy from the probabilistic expected value of the energy,

$$\frac{1}{N}\langle E\{\sigma_j\}\rangle = -\frac{1}{N}\frac{\partial}{\partial\beta}\ln Z = -\mu_o H \tanh(\beta\mu_o H) = u. \tag{5.41}$$

At this point, it should be remarked that the partition function can be written as a sum over discrete values of energy. In this case, we use the same combinatorial factor calculated in the context of the microcanonical ensemble. In fact, assuming that there are N_1 spins up and $N_2 = N - N_1$ spins down, the energy of the system can be written as

$$E(N_1) = -\mu_o H N_1 + \mu_o H (N - N_1). \tag{5.42}$$

Therefore, the partition function is given by

$$\begin{aligned}
Z &= \sum_{N_1=0}^{N} \frac{N!}{N_1!(N-N_1)!} \exp\left[\beta\mu_o H N_1 - \beta\mu_o H(N - N_1)\right] \\
&= \left[\exp(\beta\mu_o H) + \exp(-\beta\mu_o H)\right]^N = \left[2\cosh(\beta\mu_o H)\right]^N, \tag{5.43}
\end{aligned}$$

in agreement with the previous result.

5.2 Solid of Einstein

Let us revisit Einstein's model of a solid, as discussed in the last chapter. Consider N localized and noninteracting quantum one-dimensional harmonic oscillators, with the same fundamental frequency ω, but now in contact with a thermal reservoir at temperature T. The microscopic states of this system are characterized by the set of quantum numbers $\{n_1, n_2, ..., n_N\}$, where n_j labels the number of quanta of energy of the jth

oscillator. Given a microscopic state $\{n_j\}$, the energy of this state is written as

$$E\{n_j\} = \sum_{j=1}^{N} \left(n_j + \frac{1}{2}\right) \hbar\omega. \tag{5.44}$$

Therefore, the canonical partition function is given by

$$Z = \sum_{\{n_j\}} \exp\left(-\beta E\{n_j\}\right) = \sum_{n_1, n_2, \dots, n_N} \exp\left[-\sum_{j=1}^{N} \left(n_j + \frac{1}{2}\right)\beta\hbar\omega\right]. \tag{5.45}$$

Again, as there are no interaction terms, the multiple sum fully factorizes, giving rise to a very simple expression,

$$Z = \left\{ \sum_{n=0}^{\infty} \exp\left[-\left(n + \frac{1}{2}\right)\beta\hbar\omega\right] \right\}^{N} = Z_1^N, \tag{5.46}$$

where

$$Z_1 = \sum_{n=0}^{\infty} \exp\left[-\left(n + \frac{1}{2}\right)\beta\hbar\omega\right] = \frac{\exp\left(-\frac{1}{2}\beta\hbar\omega\right)}{1 - \exp\left(-\beta\hbar\omega\right)}. \tag{5.47}$$

Everything works as if we were considering the partition function of a single isolated oscillator.

The Helmholtz free energy per oscillator is given by

$$f = -\frac{1}{\beta} \lim_{N\to\infty} \frac{1}{N} \ln Z = \frac{1}{2}\hbar\omega + k_B T \ln\left[1 - \exp\left(-\frac{\hbar\omega}{k_B T}\right)\right]. \tag{5.48}$$

From this fundamental equation, we can calculate all thermodynamic properties of the model. In particular, the entropy per oscillator as a function of temperature is given by

$$\begin{aligned} s = -\frac{\partial f}{\partial T} &= -k_B \ln\left[1 - \exp\left(-\frac{\hbar\omega}{k_B T}\right)\right] \\ &+ k_B \left(\frac{\hbar\omega}{k_B T}\right) \frac{\exp\left(-\hbar\omega/k_B T\right)}{\left[1 - \exp\left(-\hbar\omega/k_B T\right)\right]}, \end{aligned} \tag{5.49}$$

from which we obtain an expression for the specific heat of the solid of Einstein,

$$c = T\frac{\partial s}{\partial T} = k_B \left(\frac{\hbar\omega}{k_B T}\right)^2 \frac{\exp\left(\hbar\omega/k_B T\right)}{\left[\exp\left(\hbar\omega/k_B T\right) - 1\right]^2}, \tag{5.50}$$

which is the same as the previously calculated result in the context of the microcanonical ensemble.

From equation (5.15), we have the thermodynamic internal energy per oscillator as a function of temperature,

$$u = -\frac{1}{N}\frac{\partial}{\partial\beta}\ln Z = -\frac{\partial}{\partial\beta}\ln Z_1 = \frac{1}{2}\hbar\omega + \frac{\hbar\omega}{\exp\left(\hbar\omega/k_B T\right) - 1}, \quad (5.51)$$

which again corresponds to the same result of last chapter. Later, we shall see that this expression also corresponds to the average energy, as a function of frequency, of the Planck model of oscillators to explain the radiation from black bodies. In the classical limit, $k_B T >> \hbar\omega$, we have $u \to k_B T$, leading to the constant specific heat of Dulong and Petit. In the limit of low temperatures, $k_B T << \hbar\omega$, we have $u \to \hbar\omega/2$, which is the well-known value of energy of point zero (quantum vacuum).

5.3 Particles with two energy levels

Consider now the gas of N semiclassical two-state particles in thermal equilibrium with a reservoir at temperature T. The single-particle states have energies $\epsilon_1 = 0$ and $\epsilon_2 = \epsilon > 0$, respectively, but the total energy of the system is allowed to fluctuate. A microscopic state of this global system is characterized by giving the value of the energy of each one of the particles. We may then introduce a variable t_j, associated with the jth particle, with $j = 1, 2, ..., N$, such that $t_j = 0$, if particle j has zero energy, and $t_j = 1$, if particle j has energy $\epsilon > 0$. A microscopic state of the system, given by the set $\{t_j\}$, has energy

$$E\{t_j\} = \sum_{j=1}^{N} \epsilon t_j. \quad (5.52)$$

Therefore, the canonical partition function may be written as

$$Z = \sum_{\{t_j\}} \exp\left(-\beta E\{t_j\}\right) = \sum_{t_1,...,t_2} \exp\left(-\sum_{j=1}^{N}\beta\epsilon t_j\right), \quad (5.53)$$

which again factorizes to give rise to a much simpler expression,

$$Z = Z_1^N, \text{ where } Z_1 = \sum_t \exp\left(-\beta\epsilon t\right) = 1 + \exp\left(-\beta\epsilon\right). \quad (5.54)$$

It is interesting to remark that the same expression for the partition function can also be obtained from the number of microscopic states in the

microcanonical ensemble. Indeed, we could have written

$$Z = \sum_E \Omega(E, N) \exp(-\beta E) = \sum_{N_1, N_2 = 0,\ N_1 + N_2 = N}^{N} \frac{N!}{N_1! N_2!} \exp(-\beta \epsilon N_2),$$
(5.55)

where N_1 is the number of particles with energy $\epsilon_1 = 0$, and N_2 is the number of particles with energy $\epsilon_2 = \epsilon > 0$. Using Newton's binomial theorem, we recover the results of equation (5.54).

The connection with thermodynamics comes from the Helmholtz free energy per particle,

$$f = f(T) = -\frac{1}{\beta} \lim_{N \to \infty} \frac{1}{N} \ln Z = -k_B T \ln \left[1 + \exp\left(-\frac{\epsilon}{k_B T}\right)\right].$$
(5.56)

From this expression, we have the entropy per particle,

$$s = -\frac{\partial f}{\partial T} = k_B \ln \left[1 + \exp\left(-\frac{\epsilon}{k_B T}\right)\right] + \frac{\epsilon}{T} \frac{\exp(-\epsilon/k_B T)}{1 + \exp(-\epsilon/k_B T)},$$
(5.57)

and the specific heat,

$$c = T \frac{\partial s}{\partial T} = k_B \left(\frac{\epsilon}{k_B T}\right)^2 \frac{\exp(-\epsilon/k_B T)}{[1 + \exp(-\epsilon/k_B T)]^2},$$
(5.58)

which is identical to the previously obtained expression in the framework of the microcanonical ensemble.

Noting that $f = u - Ts$, we use equations (5.56) and (5.57) to write

$$u = \frac{\epsilon \exp(-\epsilon/k_B T)}{1 + \exp(-\epsilon/k_B T)},$$
(5.59)

as obtained in Chapter 4. The factorization of the canonical partition function leads to the probabilistic interpretation of this formula, as discussed in the last chapter. Indeed, if we consider a single particle, the canonical probabilities of finding this particle in states with $\epsilon = 0$ and $\epsilon > 0$ will be respectively given by

$$P_1 = \frac{1}{Z_1} \quad \text{and} \quad P_2 = \frac{\exp(-\beta\epsilon)}{Z_1},$$
(5.60)

where

$$Z_1 = 1 + \exp(-\beta\epsilon),$$
(5.61)

as in equation (5.54). For $\epsilon > 0$ (and positive temperatures!) we have $P_1 > P_2$, that is, the level of lowest energy is always more populated by

particles than the upper level. For $T \to 0$ (that is, $\beta \to \infty$), $P_1 \to 1$ and $P_2 \to 0$, indicating that all particles are in the ground state, with zero energy. For $T \to \infty$ (that is, $\beta \to 0$), $P_1 \to 1/2$ and $P_2 \to 1/2$, indicating that the two energy levels are equally populated (and that the internal energy per particle tends to $\epsilon/2$). We have already referred to the possibility of reversing the sign of ϵ, to mimic a situation at effective negative temperatures, where there is an inversion of the populations of the two levels.

Finally, we note that eliminating the temperature and inserting into equation (5.57), we recover the same fundamental equation already obtained in the context of the microcanonical ensemble,

$$s = s\,(u) = -k_B \left(1 - \frac{u}{\epsilon}\right) \ln \left(1 - \frac{u}{\epsilon}\right) - k_B \frac{u}{\epsilon} \ln \frac{u}{\epsilon}, \qquad (5.62)$$

which gives another example of the equivalence between the ensembles.

5.4 The Boltzmann gas

If the total energy is allowed to fluctuate, and the system is kept in contact with a thermal reservoir, at a well-defined temperature T, it is easy to calculate the canonical partition function associated with the Boltzmann gas of the last chapter. Consider the set of occupation numbers, $\{N_j; j = 1, ..., N\}$, where N_j labels the number of particles with energy ϵ_j. Given the total energy E and the total number of particles N, in the last chapter we had already written the total number of accessible microstates of this system,

$$\Omega\left(\{N_j\}; E, N\right) = \frac{N!}{\prod_j N_j!}, \qquad (5.63)$$

with the restrictions

$$N = \sum_j N_j \qquad (5.64)$$

and

$$E = \sum_j \epsilon_j N_j. \qquad (5.65)$$

Hence, the canonical partition function may be written as

$$Z = \sum_E \Omega\left(\{N_j\}; E, N\right) \exp\left(-\beta E\right) = \sum_{\{N_j\}} \frac{N!}{\prod_j N_j!} \exp\left(-\sum_j \beta \epsilon_j N_j\right),$$

$$(5.66)$$

where the sum over the set $\{N_j\}$ is still restricted by fixing the number N of particles. A direct application of the multinomial theorem (which is a simple generalization of Newton's binomial theorem) leads again to a factorization of the partition function,

$$Z = Z_1^N, \tag{5.67}$$

where

$$Z_1 = \sum_j \exp\left(-\beta\epsilon_j\right), \tag{5.68}$$

which confirms the results obtained in the last chapter within the framework of the microcanonical ensemble.

In the classical context, the Boltzmann gas is just a toy model, since there is no possibility of accounting for the discretization of energy. However, we will see that expressions (5.67) and (5.68), with the introduction of some corrections, do correspond to the classical limit of the quantum statistics for the ideal gas. In this case, index j should be interpreted as a label of a quantum single-particle state (which will be called an orbital, from the language of chemistry); in orbital j, the particle has energy ϵ_j. According to this interpretation, the expression for the canonical partition function should be divided by Boltzmann's correct counting factor, $N!$, to ensure the existence of the thermodynamic limit. These points will become clearer in the following chapters, from the analysis of the classical limit of the thermodynamic expressions for the Bose–Einstein and Fermi–Dirac ideal quantum gases.

Exercises

1. The energy of a system of N localized magnetic ions, at temperature T, in the presence of a field H, may be written as

$$\mathcal{H} = D \sum_{i=1}^{N} S_i^2 - \mu_o H \sum_{i=1}^{N} S_i,$$

where the parameters D, μ_o, and H are positive, and $S_i = +1$, 0, or -1, for all sites i. Obtain expressions for the internal energy, the entropy, and the magnetization per site. In zero field ($H = 0$), sketch graphs of the internal energy, the entropy, and the specific heat versus temperature. Indicate the behavior of these quantities in the limits $T \to 0$ and $T \to \infty$. Calculate the expected value of the "quadrupole moment,"

$$Q = \frac{1}{N} \left\langle \sum_{i=1}^{N} S_i^2 \right\rangle,$$

as a function of field and temperature.

2. Consider a one-dimensional magnetic system of N localized spins, at temperature T, associated with the energy

$$\mathcal{H} = -J \sum_{i=1,3,5,\dots,N-1} \sigma_i \sigma_{i+1} - \mu_o H \sum_{i=1}^{N} \sigma_i,$$

where the parameters J, μ_o, and H are positive, and $\sigma_i = \pm 1$ for all sites i. Assume that N is an even number, and note that the first sum is over odd integers.

(a) Obtain an expression for the canonical partition function and calculate the internal energy per spin, $u = u(T, H)$. Sketch a graph of $u(T, H = 0)$ versus temperature T. Obtain an expression for the entropy per spin, $s = s(T, H)$. Sketch a graph of $s(T, H = 0)$ versus T.

(b) Obtain expressions for the magnetization per particle,

$$m = m(T, H) = \frac{1}{N} \left\langle \mu_o \sum_{i=1}^{N} \sigma_i \right\rangle,$$

and for the magnetic susceptibility,

$$\chi = \chi(T, H) = \left(\frac{\partial m}{\partial H} \right)_T.$$

Sketch a graph of $\chi(T, H = 0)$ versus temperature.

3. Consider a system of N classical and noninteracting particles in contact with a thermal reservoir at temperature T. Each particle may have energies 0, $\epsilon > 0$, or 3ϵ. Obtain an expression for the canonical partition function, and calculate the internal energy per particle, $u = u(T)$. Sketch a graph of u versus T (indicate the values of u in the limits $T \to 0$ and $T \to \infty$). Calculate the entropy per particle, $s = s(T)$, and sketch a graph of s versus T. Sketch a graph of the specific heat versus temperature.

4. A system of N localized and independent quantum oscillators is in contact with a thermal reservoir at temperature T. The energy levels of each oscillator are given by

$$\epsilon_n = \hbar \omega_o \left(n + \frac{1}{2} \right), \text{ with } n = 1, 3, 5, 7, \dots.$$

Note that n is an odd integer.

(a) Obtain an expression for the internal energy u per oscillator as a function of temperature T. What is the form of u in the classical limit ($\hbar\omega_o << k_B T$)?

(b) Obtain an expression for the entropy per oscillator as a function of temperature. Sketch a graph of entropy versus temperature. What is the expression of the entropy in the classical limit?

(c) What is the expression of the specific heat in the classical limit?

5. Consider a system of N noninteracting classical particles. The single-particle states have energies $\epsilon_n = n\epsilon$, and are n times degenerate ($\epsilon > 0$; $n = 1, 2, 3, ...$). Calculate the canonical partition function and the entropy of this system. Obtain expressions for the internal energy and the entropy as a function of temperature. What are the expressions for the entropy and the specific heat in the limit of high temperatures?

6. A set of N classical oscillators in one dimension is given by the Hamiltonian

$$\mathcal{H} = \sum_{i=1}^{N} \left(\frac{1}{2m} p_i^2 + \frac{1}{2} m\omega^2 q_i^2 \right).$$

Using the formalism of the canonical ensemble in classical phase space, obtain expressions for the partition function, the energy per oscillator, the entropy per oscillator, and the specific heat. Compare with the results from the classical limit of the quantum oscillator. Calculate an expression for the quadratic deviation of the energy as a function of temperature.

7. Consider again the preceding problem. The canonical partition function can be written as an integral form,

$$Z(\beta) = \int_0^\infty \Omega(E) \exp(-\beta E)\, dE,$$

where $\Omega(E)$ is the number of accessible microscopic states of the system with energy E. Note that, in the expressions for $Z(\beta)$ and $\Omega(E)$, we are omitting the dependence on the number N of oscillators. Using the expression for $Z(\beta)$ obtained in the last exercise, perform a Laplace transformation to obtain an asymptotic form (in the thermodynamic limit) for $\Omega(E)$. Compare with the expression calculated in the framework of the microcanonical ensemble.

8. A system of N one-dimensional localized oscillators, at a given temperature T, is associated with the Hamiltonian

$$\mathcal{H} = \sum_{i=1}^{N} \left[\frac{1}{2m} p_i^2 + V(q_i) \right],$$

where

$$V(q) = \begin{cases} \frac{1}{2} m \omega^2; & \text{for } q > 0, \\ \frac{1}{2} m \omega^2 + \epsilon; & \text{for } q < 0, \end{cases}$$

with $\epsilon > 0$.

(a) Obtain the canonical partition function of this classical system. Calculate the internal energy per oscillator, $u = u(T)$. What is the form of $u(T)$ in the limits $\epsilon \to 0$ and $\epsilon \to \infty$?

(b) Consider now the quantum analog of this model in the limit $\epsilon \to \infty$. Obtain an expression for the canonical partition function. What is the internal energy per oscillator of this quantum analog?

8. A system of N one-dimensional localized oscillators, at a given temperature T, is associated with the Hamiltonian

$$\mathcal{H} = \sum_{j=1}^{N} \left[\frac{1}{2m} p_j^2 + V(q) \right]$$

where

$$V(q) = \begin{cases} \frac{1}{2} m \omega^2 q^2 & \text{for } q > 0 \\ \frac{1}{2} m \omega^2 q^2 + \epsilon & \text{for } q < 0 \end{cases}$$

with $\epsilon \geq 0$.

(a) Obtain the canonical partition function of this classical system. Calculate the internal energy per oscillator, $u = u(T)$. What is the form of $u(T)$ in the limits $\epsilon \to 0$ and $\epsilon \to \infty$?

(b) Consider now the quantum analog of this model, in the limit $\epsilon \to \infty$. Obtain an expression for the canonical partition function. What is the internal energy per oscillator of this quantum analog?

6

The Classical Gas in the Canonical Formalism

A classical system of N monatomic particles of mass m may be given by the Hamiltonian

$$\mathcal{H} = \sum_{i=1}^{N} \frac{1}{2m}\vec{p}_i^2 + \sum_{i<j} V\left(|\vec{r}_i - \vec{r}_j|\right), \tag{6.1}$$

where $V\left(|\vec{r}|\right)$ is a pair potential, as sketched in figure 6.1. The potential of a realistic system displays a hard core, $V\left(r\right) \to \infty$, for $r = |\vec{r}_i - \vec{r}_j| \to 0$, and a suitably vanishing (for $r \to \infty$) attractive part. The canonical partition function of this classical system, in contact with a heat reservoir at temperature T, in a container of volume V, is given by the integral in phase space,

$$Z_c = \int_V \cdots \int d^3\vec{r}_1 ... d^3\vec{r}_N \int \cdots \int d^3\vec{p}_1 \cdots d^3\vec{p}_N \exp\left(-\beta\mathcal{H}\right), \tag{6.2}$$

where the spatial coordinates are restricted to the region of volume V. The trivial integration over the momentum coordinates may be written as a product of $3N$ Gaussian integrals of the form

$$\int_{-\infty}^{+\infty} dp \exp\left(-\frac{\beta p^2}{2m}\right) = \left(\frac{2\pi m}{\beta}\right)^{1/2}. \tag{6.3}$$

FIGURE 6.1. Interaction potential associated with a pair of classical particles.

Hence, we write

$$Z_c = \left(\frac{2\pi m}{\beta}\right)^{3N/2} Q_N,$$ (6.4)

where the configurational part of this partition function is given by

$$Q_N = \int_V \cdots \int_V d^3\vec{r}_1 \ldots d^3\vec{r}_N \exp\left[-\beta \sum_{i<j} V\left(|\vec{r}_i - \vec{r}_j|\right)\right].$$ (6.5)

In the case of an ideal gas, we discard the potential terms associated with the interactions between pairs of particles. The configuration factor of the partition function is then reduced to the trivial expression

$$Q_N = V^N.$$ (6.6)

Thus, we write the canonical partition function

$$Z_c = \left(\frac{2\pi m}{\beta}\right)^{3N/2} V^N,$$ (6.7)

so that

$$\frac{1}{N} \ln Z_c = \frac{3}{2} \ln\left(\frac{2\pi m}{\beta}\right) + \ln V.$$ (6.8)

As in the context of the microcanonical ensemble, this expression suffers from severe problems in the thermodynamic limit, $N, V \to \infty$, with fixed $V/N = v$. To correct these problems, we use the same trick as before. We just divide the classical partition function by the same **Boltzmann correct counting factor** $N!$. Also, we divide Z_c by the factor h^{3N}, where h is a dimensional (length times momentum) constant, so that the partition

function does not depend on the particular choice of the generalized coordinates of phase space. We shall see that this is fully justified by the classical limit of the quantum monatomic gases. The constant h is identified as the Planck universal constant itself.

Taking into account all of these correction factors, we modify equation (6.4) to write the partition function of the classical gas in the form

$$Z = \frac{1}{N!} \frac{1}{h^{3N}} Z_c = \frac{1}{N!} \left(\frac{2\pi m}{\beta h^2} \right)^{3N/2} Q_N, \qquad (6.9)$$

where $Q_N = V^N$ for the particular case of an ideal gas.

6.1 Ideal classical monatomic gas

To obtain the free energy of an ideal classical monatomic gas of particles, we use equation (6.9), with $Q_N = V^N$. Therefore, we write

$$\frac{1}{N} \ln Z = \frac{3}{2} \ln \left(\frac{2\pi m}{\beta h^2} \right) + \ln \frac{V}{N} + 1 + O \left(\frac{\ln N}{N} \right). \qquad (6.10)$$

In the thermodynamic limit, given by $N, V \to \infty$, with fixed volume per particle, $v = V/N$, we have the Helmholtz free energy,

$$\begin{aligned} f = f(T, v) &= -\frac{1}{\beta} \lim_{\substack{V,N \to \infty \\ (V/N=v)}} \frac{1}{N} \ln Z \\ &= -\frac{3}{2} k_B T \ln T - k_B T \ln v - k_B c T, \qquad (6.11) \end{aligned}$$

where the constant c is given by

$$c = \frac{3}{2} \ln \left(\frac{2\pi m k_B}{h^2} \right) + 1. \qquad (6.12)$$

The equations of state in the Helmholtz representation are given by

$$s = -\left(\frac{\partial f}{\partial T} \right)_v = \frac{3}{2} k_B \ln T + k_B \ln v + k_B c - \frac{3}{2} k_B, \qquad (6.13)$$

which leads to the well-known expression for the specific heat of the classical gas,

$$c_V = T \left(\frac{\partial s}{\partial T} \right)_v = \frac{3}{2} k_B, \qquad (6.14)$$

and

$$p = -\left(\frac{\partial f}{\partial v} \right)_T = \frac{k_B T}{v}, \qquad (6.15)$$

FIGURE 6.2. Entropy versus temperature for an ideal monatomic gas of classical particles.

which corresponds to the Boyle law, $pv = k_BT$. In figure 6.2, we sketch the entropy per particle versus temperature.

At sufficiently low temperatures, the entropy becomes negative, in conflict with the predictions of thermodynamics. This is a defect of the classical calculations. At low temperatures, it is well-known that we have to use quantum statistical mechanics. The correct quantum formulas, at least for realistic models, with a well-defined ground state, do lead to a vanishing entropy at absolute zero.

The internal energy per particle of the classical gas may be obtained from the probabilistic average energy, given by

$$\langle \mathcal{H} \rangle = -\frac{\partial}{\partial \beta} \ln Z = \frac{3}{2} \frac{N}{\beta} = \frac{3}{2} N k_B T, \tag{6.16}$$

from which we have

$$u = \frac{3}{2} k_B T. \tag{6.17}$$

Using the equation of state (6.13), we have the temperature in terms of the entropy, and may write a fundamental equation in the energy representation,

$$u = \frac{3}{2} k_B v^{-2/3} \exp\left[\frac{2}{3}\left(\frac{s}{k_B} - c\right)\right]. \tag{6.18}$$

The inversion of this equation gives the entropy density as a function of u and v,

$$s = s(u, v) = \frac{3}{2} k_B \ln u + k_B \ln v + \text{constant}, \tag{6.19}$$

in full agreement with the fundamental equation that was obtained in Section 4.4, in the framework of the microcanonical ensemble.

6.2 The Maxwell–Boltzmann distribution

From the canonical formalism, we can easily obtain the Maxwell–Boltzmann distribution for the molecular velocities of the ideal monatomic gas of particles. As the partition function factorizes, the canonical probability of finding a particle of the ideal gas with velocity between \vec{v} and $\vec{v} + d\vec{v}$, is given by

$$p(\vec{v})\, d^3\vec{v} = \frac{1}{Z_1} V \exp\left(-\frac{\beta \vec{p}^2}{2m}\right) d^3\vec{p}, \qquad (6.20)$$

where

$$Z_1 = V \int d^3\vec{p} \exp\left(-\frac{\beta \vec{p}^2}{2m}\right) = V \left(\frac{2\pi m}{\beta}\right)^{3/2}. \qquad (6.21)$$

Hence,

$$p(\vec{v}) = \left(\frac{2\pi k_B T}{m}\right)^{-3/2} \exp\left(-\frac{m\vec{v}^2}{2k_B T}\right), \qquad (6.22)$$

in agreement with the result that had already been anticipated in the discussion about the continuum limit of the Boltzmann model for a gas of particles with discrete energy levels. Since the distribution depends on the modulus of the velocity only, it is obvious that $\langle v_x \rangle = \langle v_y \rangle = \langle v_z \rangle = 0$.

Owing to the isotropy of the space of velocities, we may also write

$$p(\vec{v})\, d^3\vec{v} = p(\vec{v})\, 4\pi v^2 dv = p_o(v)\, dv, \qquad (6.23)$$

where the density of probability of the intensity of velocities is given by

$$p_o(v) = 4\pi \left(\frac{2\pi k_B T}{m}\right)^{-3/2} v^2 \exp\left(-\frac{mv^2}{2k_B T}\right). \qquad (6.24)$$

Therefore, the most probable value of the intensity of the molecular velocities,

$$v_m = \left(\frac{2k_B T}{m}\right)^{1/2}, \qquad (6.25)$$

which comes from the maximization of $p_o(v)$, is smaller than the square root of the average value of the square of the intensity,

$$\langle \vec{v}^2 \rangle^{1/2} = \left(\frac{3k_B T}{m}\right)^{1/2}. \qquad (6.26)$$

At room temperature, of about 300 K, and using the molecular mass of nitrogen, $m \approx (28 \times 10^{-3}) / (6 \times 10^{23})$ kg, and Boltzmann's constant,

$k_B = 1,38 \times 10^{-23}$ J/K, the typical value of the intensity of the molecular velocity of a gas (the surrounding atmospheric air, for example) is about 500 m/s, close to the velocity of sound, but definitely away from the relativistic regime. There are clever experiments to confirm the distribution of molecular velocities. The interested reader may check the work of R. C. Miller and P. Kusch, J. Chem. Phys. **25**, 860 (1956).

6.3 The theorem of equipartition of energy

The theorem of equipartition of energy, which has been used since the early investigations of the kinetic theory of gases, roughly states that each quadratic term in the expression for the Hamiltonian of a classical gas produces a contribution of the form $k_B T/2$ to the free energy of the system.

For the classical monatomic gas, given by Hamiltonian (6.1), it is easy to use the canonical formalism to obtain the average value of the kinetic energy,

$$\langle E_{cin} \rangle = \left\langle \sum_{i=1}^{N} \frac{1}{2m} \vec{p}_i^2 \right\rangle = \sum_{i=1}^{N} \left\langle \frac{1}{2m} \left(p_{ix}^2 + p_{iy}^2 + p_{iz}^2 \right) \right\rangle = \frac{3}{2} N k_B T, \quad (6.27)$$

that corresponds to the internal energy for an ideal gas.

For a classical system of N independent harmonic oscillators, given by the Hamiltonian

$$\mathcal{H} = \sum_{i=1}^{N} \left(\frac{1}{2m} p_i^2 + \frac{1}{2} m\omega^2 q_i^2 \right), \quad (6.28)$$

where the position and momentum coordinates may change without any restrictions, we have the expected value

$$\langle \mathcal{H} \rangle = \sum_{i=1}^{N} \left(\left\langle \frac{1}{2m} p_i^2 \right\rangle + \left\langle \frac{1}{2} m\omega^2 q_i^2 \right\rangle \right) = N \left(\frac{1}{2} k_B T + \frac{1}{2} k_B T \right) = N k_B T.$$

$$(6.29)$$

For three-dimensional independent oscillators, we have $U = \langle \mathcal{H} \rangle = 3N k_B T$.

Now we present a more precise statement of the equipartition theorem of energy. Consider a classical system of n degrees of freedom given by the Hamiltonian

$$\mathcal{H}(q_1, ..., q_n, p_1, ..., p_n) = \mathcal{H}_o + \phi p_j^2, \quad (6.30)$$

where (i) \mathcal{H}_o and ϕ are functions that do not depend on the particular coordinate p_j; (ii) the function ϕ is always positive; and (iii) the coordinate

p_j is defined from $-\infty$ to $+\infty$. Hence, we have the expected value

$$\langle \phi p_j^2 \rangle = \frac{1}{2} k_B T. \tag{6.31}$$

To demonstrate this result, let us use the definition of the expected value in the canonical ensemble,

$$\langle \phi p_j^2 \rangle = \frac{\int \cdots \int dq_1 \cdots dp_n \, \phi p_j^2 \exp\left(-\beta \mathcal{H}_o - \beta \phi p_j^2\right)}{\int \cdots \int dq_1 \cdots dp_n \exp\left(-\beta \mathcal{H}_o - \beta \phi p_j^2\right)}. \tag{6.32}$$

Taking into account the restrictions on the functions ϕ and \mathcal{H}_o, we perform the integration over the variable p_j in the numerator,

$$\int_{-\infty}^{+\infty} \phi p_j^2 \exp\left(-\beta \mathcal{H}_o - \beta \phi p_j^2\right) dp_j = \exp\left(-\beta \mathcal{H}_o\right) \left\{ -\frac{\partial}{\partial \beta} \int_{-\infty}^{+\infty} \exp\left(-\beta \phi p_j^2\right) dp_j \right\}$$

$$= \exp\left(-\beta \mathcal{H}_o\right) \left\{ -\frac{\partial}{\partial \beta} \left(\frac{\pi}{\beta \phi}\right)^{1/2} \right\} = \exp\left(-\beta \mathcal{H}_o\right) \frac{1}{2\beta} \left(\frac{\pi}{\beta \phi}\right)^{1/2}$$

$$= \frac{1}{2\beta} \exp\left(-\beta \mathcal{H}_o\right) \int_{-\infty}^{+\infty} \exp\left(-\beta \phi p_j^2\right) dp_j = \frac{1}{2\beta} \int_{-\infty}^{+\infty} \exp\left(-\beta \mathcal{H}_o - \beta \phi p_j^2\right) dp_j. \tag{6.33}$$

Therefore, extracting the factor $1/(2\beta)$, the multiple integrals in both the numerator and the denominator of equation (6.32) are absolutely identical, that is,

$$\langle \phi p_j^2 \rangle = \frac{1}{2\beta} = \frac{1}{2} k_B T, \tag{6.34}$$

which demonstrates the more elaborate formulation of the theorem of equipartition of energy.

6.4 The classical monatomic gas of particles

The difficult step to write the canonical partition function for the classical monatomic gas of particles with pair interactions consists in the calculation of the configurational factor, given by equation (6.5) as a multiple integral, which in general does not factorize. Let us make an attempt to obtain

at least an approximate equation of state that works for sufficiently low densities (that is, for large enough values of $v = V/N$).

It is usual to present the experimental data for real gases as a **virial expansion**, in terms of $1/v$,

$$\frac{p}{k_B T} = \frac{1}{v} + B\frac{1}{v^2} + C\frac{1}{v^3} + \cdots, \qquad (6.35)$$

where the first term gives Boyle's law, and the correction coefficients, B, C,..., are functions of temperature that may be found in tables of thermodynamic data for the more usual gases (see, for example, the *American Institute of Physics Handbook*, published by McGraw-Hill, New York). One of the best known (and more successful) proposals for an equation of state of a real gas is given by the van der Waals equation,

$$\left(p + \frac{a}{v^2}\right)(v - b) = k_B T, \qquad (6.36)$$

where a and b are positive phenomenological parameters. According to the arguments of van der Waals, parameter b is related to the impenetrability of matter, reflecting the hard core of the pair potential, while parameter a is associated with the small attractive part of this potential. Equation (6.36) gives rise to a power series in terms of $1/v$,

$$\frac{p}{k_B T} = \frac{1}{v} + \left(b - \frac{a}{kT}\right)\frac{1}{v^2} + b^2\frac{1}{v^3} + \cdots. \qquad (6.37)$$

Now we present an approximate analysis to find an expression for the second virial coefficient B in terms of the phenomenological parameters of the van der Waals equation.

The configurational part of the canonical partition function of the classical monatomic gas may be written as

$$Q_N = \int d^3\vec{r_1} \cdots \int d^3\vec{r_N} \exp\left[-\beta \sum_{i<j} V\left(|\vec{r_i} - \vec{r_j}|\right)\right]$$

$$= \left(\prod_{i=1}^{N} \int d^3\vec{r_i}\right) \prod_{i<J} \exp\left(-\beta V_{ij}\right) = \left(\prod_{i=1}^{N} \int d^3\vec{r_i}\right) \prod_{i<J}\left(1 + f_{ij}\right), \qquad (6.38)$$

where

$$V_{ij} = V\left(|\vec{r_i} - \vec{r_j}|\right) \quad \text{and} \quad f_{ij} = \exp\left(-\beta V_{ij}\right) - 1. \qquad (6.39)$$

In figure 6.3, we schematically draw the potential $V(r)$ and the function $f(r)$. For a typical potential, we note that $f(r) \to -1$ as $r \to 0$, and

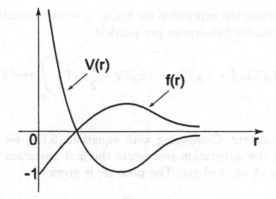

FIGURE 6.3. Schematic representation of the intermolecular potential $V(r)$ and the function $f(r) = \exp(-\beta V) - 1$.

$f(r) \to 0$ as $r \to \infty$. If the attractive part of the potential $V(r)$ is not too strong, $f(r)$ is not so large. Therefore, we can write the expansion

$$\prod_{i<J}(1 + f_{ij}) = 1 + \sum_{i<j} f_{ij} + \text{higher-order terms.} \qquad (6.40)$$

Hence,

$$Q_N = V^N + V^{N-2} \sum_{i<j} \int d^3\vec{r}_i \int d^3\vec{r}_j f_{ij} + ..., \qquad (6.41)$$

from which we have

$$
\begin{aligned}
\ln Q_N - N \ln V &= \ln\left\{1 + \frac{1}{V^2}\sum_{i<j}\int d^3\vec{r}_i \int d^3\vec{r}_j f_{ij} + \cdots\right\} \\
&= \frac{1}{V^2}\sum_{i<j}\int d^3\vec{r}_i \int d^3\vec{r}_j f_{ij} + \cdots \\
&= \frac{1}{2}N(N-1)\frac{1}{V^2}\int d^3\vec{r}_1 \int d^3\vec{r}_2 f_{12} + \ldots . \quad (6.42)
\end{aligned}
$$

To take the thermodynamic limit, we can also write

$$\ln Q_N - N \ln V \to \frac{1}{2}\frac{N^2}{V}\int_0^\infty 4\pi r^2 f(r)\, dr + \ldots . \qquad (6.43)$$

Using equation (6.9), we have

$$\frac{1}{N}\ln Z = \frac{3}{2}\ln\left(\frac{2\pi m}{\beta h^2}\right) + \frac{1}{N}\ln Q_N - \ln N + 1 + O\left(\frac{\ln N}{N}\right). \qquad (6.44)$$

Therefore, inserting the expression for $\ln Q_N$, given by equation (6.41), we obtain the Helmholtz free energy per particle,

$$f(T,v) = -\frac{3}{2}k_B T \ln T - k_B T \ln v - k_B T c - \frac{1}{2}k_B T \frac{1}{v} \int_0^\infty 4\pi r^2 f(r)\, dr + \dots ,$$

(6.45)

where c is a constant. Comparing with equation (6.11), we see that the term involving the integration represents the first deviation with respect to the behavior of an ideal gas. The pressure is given by

$$p = -\left(\frac{\partial f}{\partial v}\right)_T = \frac{k_B T}{v} - \frac{k_B T}{2v^2} \int_0^\infty 4\pi r^2 f(r)\, dr + \dots ,$$

(6.46)

that is,

$$\frac{p}{k_B T} = \frac{1}{v} + B\frac{1}{v^2} + \dots ,$$

(6.47)

with the same form as expansion (6.35). Hence, the virial coefficient B is written as

$$B = -2\pi \int_0^\infty r^2 f(r)\, dr.$$

(6.48)

For an intermolecular potential with a hard core and a small attractive part, as illustrated in figure 6.4, we have

$$B = -2\pi \int_0^\infty r^2 \left[e^{-\beta V(r)} - 1 \right] dr = \frac{2\pi}{3} r_o^3 - \frac{14\pi}{3} r_o^3 \left(e^{\beta V_o} - 1 \right).$$

(6.49)

For very small V_o, we finally have

$$B \to \frac{2\pi r_o^3}{3} - \frac{14\pi r_o^3 V_o}{3k_B} \frac{1}{T},$$

(6.50)

from which we obtain the phenomenological parameters of the van der Waals equation,

$$b = \frac{2\pi r_o^3}{3} \quad \text{and} \quad a = \frac{14\pi r_o^3 V_o}{3}.$$

(6.51)

We should remark that these results confirm the interpretation of the van der Waals parameters in terms of an intermolecular potential, with a hard core and a weak attractive part.

FIGURE 6.4. Intermolecular potential with a hard core and a small attractive part.

6.5 *The thermodynamic limit of a continuum system

We now use the formalism of the canonical ensemble in classical phase space to give an idea of the progress made in setting equilibrium statistical mechanics on a rigorous mathematical basis. To give a more precise definition of thermodynamic limit, consider a sequence of three-dimensional domains Ω_k, with volume V_k, and number of particles N_k, for $k = 1, 2, 3, ...$, with fixed specific volume, $v_k = V_k/N_k$, such that $V_{k+1} > V_k$. The canonical partition function associated with domain Ω_k may be written as

$$Z_k = \frac{1}{\lambda^{3N_k}} Q_k, \tag{6.52}$$

where

$$\lambda = \frac{\sqrt{2\pi m k_B T}}{h} \tag{6.53}$$

and

$$Q_k = \frac{1}{N_k!} \int_{\Omega_k} \cdots \int \exp\left[-\beta \sum_{1 \le i < j \le N_k} V\left(|\vec{r}_i - \vec{r}_j|\right) \right] d^3\vec{r}_1 \cdots d^3\vec{r}_{N_k}. \tag{6.54}$$

To obtain the free energy per particle, let us define the function

$$f_k = \frac{1}{N_k} \ln Z_k, \tag{6.55}$$

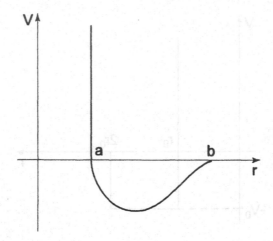

FIGURE 6.5. Intermolecular potential used to prove the existence of the thermodynamic limit in a classical system of particles.

and check for which forms of $V(r)$ and Ω_k that there exists the limit

$$f(T,v) = \lim_{k \to \infty} f_k. \tag{6.56}$$

Almost all forms of the domain Ω_k, such that the surface area does not increase with a power larger than $V_k^{2/3}$, are compatible with the existence of this limit. The central problem is the form of the pair potential.

For simplicity, consider a pair potential of the form

$$V(r) = \begin{cases} \infty, & \text{if } 0 \leq r \leq a, \\ < 0, & \text{if } a < r < b, \\ 0, & \text{if } r \geq b, \end{cases} \tag{6.57}$$

that is bounded from below, that is, such that $-c < V(r) < 0$, with $c > 0$, for $a < r < b$ (see figure 6.5). A potential of this kind has in fact been used by van Hove in his pioneering work to prove the existence of the thermodynamic limit. These attempts were later extended and placed under a more rigorous basis by D. Ruelle, M. E. Fisher, R. B. Griffiths, and others. See, for example, the paper by M. E. Fisher, "The free energy of a macroscopic system," Arch. Rat. Mech. Anal. **17**, 377 (1964).

Now let us assume that each element of the sequence $\{\Omega_k\}$ is formed by a set of cubes of volume V_k, but separated by walls of thickness a. For $k = 1$, domain Ω_1 is formed by a single cube of volume V_1 and walls of thickness $a/2$. Domain Ω_{k+1}, with volume $V_{k+1} = 8V_k$ and number of particles $N_{k+1} = 8N_k$, is formed by eight cubes of volume V_k. In figure 6.6 we illustrate in two dimensions the construction of domain Ω_{k+1} from domain Ω_k.

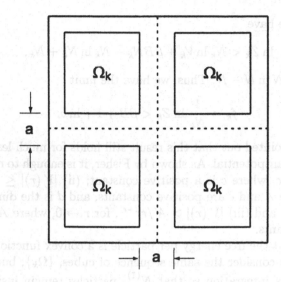

FIGURE 6.6. Formation of the domain Ω_{k+1} from domain Ω_k.

Noting that there are $(8N_{k-1})!/(N_{k-1}!)^8$ ways of arranging N_k particles inside eight identical cubes in domain Ω_{k-1}, it is not difficult to write the inequality

$$Z_k \geq (Z_{k-1})^8. \tag{6.58}$$

Therefore,

$$f_k \geq f_{k-1}, \tag{6.59}$$

which proves that the sequence $\{f_k\}$ is monotonically increasing. To show the existence of the thermodynamic limit, it only remains to carry out an additional step that consists in finding an upper limit for this sequence. The potential of van Hove certainly obeys the stability condition

$$\sum_{1 \leq i < j \leq N} V\left(|\vec{r}_i - \vec{r}_j|\right) > BN, \tag{6.60}$$

for all values of N and all configurations of the system, where the positive constant B is not dependent on N. Since $V(r) = 0$ for $r \geq b$, this may be easily checked. Indeed, given this type of potential, there is only a finite number of particles (of diameter a) that may be packed into a sphere of radius b (that is, particles that interact with a certain particle). Hence, we write

$$Z_k < \frac{1}{N_k!} V_k^{N_k} \exp\left(\beta B N_k\right), \tag{6.61}$$

from which we have

$$\ln Z_k < N_k \ln V_k + \beta B N_k - N_k \ln N_k + N_k, \tag{6.62}$$

since $\ln N! > N \ln N - N$. Thus, we have the limit

$$f_k = \frac{1}{N_k} \ln Z_k < \beta B + 1 + \ln v. \tag{6.63}$$

It should be pointed out that this result still holds for much less restricted forms of the pair potential. As shown by Fisher, it is enough to require that: (i) $V(r) > -c$, where c is a positive constant; (ii) $|V(r)| \leq A/r^{d+\epsilon}$, for $r \to \infty$, where A and ϵ are positive constants, and d is the dimensionality of the system; and (iii) $|V(r)| > A'/r^{d+\epsilon'}$, for $r \to 0$, where A' and ϵ' are positive constants.

To show that the free energy per particle is a convex function of specific volume, let us consider the same sequence of cubes, $\{\Omega_k\}$, but restricting the domains of integration so that $N_{k-1}^{(1)}$ particles remain inside the four cubes belonging to Ω_{k-1}, while the remainder $N_{k-1}^{(2)}$ particles remain inside the other four cubes. Let us also keep fixed the specific volumes $v_1 = V_{k-1}/N_{k-1}^{(1)}$ and $v_2 = V_{k-1}/N_{k-1}^{(2)}$. Thus, it is easy to obtain the inequality

$$Z_k\left(\Omega_k, N_k, T\right) \geq \left[Z_{k-1}\left(\Omega_{k-1}, N_{k-1}^{(1)}, T\right)\right]^4 \left[Z_{k-1}\left(\Omega_{k-1}, N_{k-1}^{(2)}, T\right)\right]^4, \tag{6.64}$$

where

$$N_k = 4N_{k-1}^{(1)} + 4N_{k-1}^{(2)}. \tag{6.65}$$

Therefore, we have

$$\frac{1}{V_k} \ln Z_k \geq \frac{4}{V_k} \ln Z_{k-1}\left(\Omega_{k-1}, N_{k-1}^{(1)}, T\right) + \frac{4}{V_k} \ln Z_{k-1}\left(\Omega_{k-1}, N_{k-1}^{(2)}, T\right). \tag{6.66}$$

Introducing the definition

$$g_k(\rho) = \frac{1}{V_k} \ln Z_k\left(\Omega_k, N_k, T\right), \tag{6.67}$$

with $\rho = N_k/V_k$, and taking into account that $V_k = 8V_{k-1}$, we have

$$g_k(\rho) \geq \frac{1}{2}g_{k-1}(\rho_1) + \frac{1}{2}g_{k-1}(\rho_2), \tag{6.68}$$

where $\rho_1 = 1/v_1$ and $\rho_2 = 1/v_2$. Since

$$\rho = \frac{1}{2}\rho_1 + \frac{1}{2}\rho_2, \tag{6.69}$$

we also have

$$g_k\left(\frac{\rho_1+\rho_2}{2}\right) \geq \frac{1}{2}g_{k-1}\left(\rho_1\right)+\frac{1}{2}g_{k-1}\left(\rho_2\right). \qquad (6.70)$$

In the limit $k \to \infty$, we finally have

$$g\left(\frac{\rho_1+\rho_2}{2}\right) \geq \frac{1}{2}g\left(\rho_1\right)+\frac{1}{2}g\left(\rho_2\right), \qquad (6.71)$$

which shows that $g\left(\rho\right)$ is a concave function of ρ. As

$$f\left(v\right) = vg\left(\frac{1}{v}\right),$$

it is easy to show that $f = f\left(v\right)$ is also a concave function of v. Therefore, the Helmholtz free energy per particle, given by $-f\left(v\right)/\beta$, is a convex function of the specific volume v. A more-detailed discussion of these questions may be found in chapter 3 of the textbook by Colin J. Thompson, *Classical Equilibrium Statistical Mechanics*, Clarendon Press, Oxford, 1988, on which this section was based.

Exercises

1. A system of N classical ultrarelativistic particles, in a container of volume V, at temperature T, is given by the Hamiltonian

$$\mathcal{H} = \sum_{i=1}^{N} c\left|\vec{p_i}\right|,$$

where c is a positive constant. Obtain an expression for the canonical partition function. Calculate the entropy per particle as a function of temperature and specific volume. What is the expression of the specific heat at constant volume?

2. Consider a set of N one-dimensional harmonic oscillators, described by the Hamiltonian

$$\mathcal{H} = \sum_{i=1}^{N} \left[\frac{1}{2m}\vec{p_i}^2 + \frac{1}{2}mw^2 x_i^n\right],$$

where n is a positive and even integer. Use the canonical formalism to obtain an expression for the classical specific heat of this system.

3. Consider a classical system of N very weakly interacting diatomic molecules, in a container of volume V, at a given temperature T. The Hamiltonian of a single molecule is given by

$$\mathcal{H}_m = \frac{1}{2m}\left(\vec{p}_1^2 + \vec{p}_2^2\right) + \frac{1}{2}\kappa\,|\vec{r}_1 - \vec{r}_2|^2,$$

where $\kappa > 0$ is an elastic constant. Obtain an expression for the Helmholtz free energy of this system. Calculate the specific heat at constant volume. Calculate the mean molecular diameter,

$$D = \left\{\left\langle|\vec{r}_1 - \vec{r}_2|^2\right\rangle\right\}^{1/2}.$$

Now consider another Hamiltonian, of the form

$$\mathcal{H}_m = \frac{1}{2m}\left(\vec{p}_1^2 + \vec{p}_2^2\right) + \epsilon\,|r_{12} - r_o|,$$

where ϵ and r_o are positive constants, and $r_{12} = |\vec{r}_1 - \vec{r}_2|$. What are the changes in your previous answers?

4. Neglecting the vibrational motion, a diatomic molecule may be treated as a three-dimensional rigid rotator. The Hamiltonian \mathcal{H}_m of the molecule is written as a sum of a translational, \mathcal{H}_{tr}, plus a rotational, \mathcal{H}_{rot}, term (that is, $\mathcal{H}_m = \mathcal{H}_{tr} + \mathcal{H}_{rot}$). Consider a system of N very weakly interacting molecules of this kind, in a container of volume V, at a given temperature T.

 (a) Obtain an expression for \mathcal{H}_{rot} in spherical coordinates. Show that there is a factorization of the canonical partition function of this system. Obtain an expression for the specific heat at constant volume.

 (b) Now suppose that each molecule has a permanent electric dipole moment $\vec{\mu}$ and that the system is in the presence of an external electric field \vec{E} (with the dipole $\vec{\mu}$ along the axis of the rotor). What is the form of the new rotational part of the Hamiltonian? Obtain an expression for the polarization of the molecule as a function of field and temperature. Calculate the electric susceptibility of this system.

5. Consider a classical gas of N weakly interacting molecules, at temperature T, in an applied electric field \vec{E}. Since there is no permanent electric dipole moment, the polarization of this system comes from the induction by the field. We then suppose that the Hamiltonian of each molecule will be given by the sum of a standard translational term plus an "internal term." This internal term involves an isotropic elastic energy, which tends to preserve the shape of the molecule, and

a term of interaction with the electric field. The configurational part of the internal Hamiltonian is be given by

$$\mathcal{H} = \frac{1}{2}m\omega_o^2 r^2 - q\vec{E} \cdot \vec{r}.$$

Obtain the polarization per molecule as a function of field and temperature. Obtain the electric susceptibility. Compare with the results of the last problem. Make some comments about the main differences between these results. Do you know any physical examples corresponding to these models?

6. The equation of state of gaseous nitrogen at low densities may be written as

$$\frac{pv}{RT} = 1 + \frac{B(T)}{v},$$

where v is a molar volume, R is the universal gas constant, and $B(T)$ is a function of temperature only. In the following table we give some experimental data for the second virial coefficient, $B(T)$, as a function of temperature.

T (K)	B (cm^3/mol)
100	-160.0
273	-10.5
373	6.2
600	21.7

Suppose that the intermolecular potential of gaseous nitrogen is given by

$$V(r) = \begin{cases} \infty, & 0 < r < a, \\ -V_0, & a < r < b, \\ 0, & r > b. \end{cases}$$

Use the experimental data of this table to determine the best values of the parameters a, b, and V_0.

a term of interaction with the electric field. The configurational part of the internal Hamiltonian is be given by

$$\mathcal{H} = \frac{1}{2}m\omega_0^2 z^2 - q\vec{E}\cdot\vec{r}$$

Obtain the polarization per molecule as a function of field and temperature. Obtain the electric susceptibility. Compare with the results of the last problem. Make some comments about the main differences between these results. Do you know any physical examples corresponding to these models?

8. The equation of state of gaseous nitrogen at low densities may be written as

$$\frac{pv}{RT} = 1 + \frac{B(T)}{v}$$

where v is a molar volume, R is the universal gas constant, and $B(T)$ is a function of temperature only. In the following table we give some experimental data for the second virial coefficient, $B(T)$, as a function of temperature.

T (K)	B (cm^3/mol)
100	-160.0
273	-10.5
373	6.2
600	21.7

Suppose that the intermolecular potential of gaseous nitrogen is given by

$$V(r) = \left. \begin{cases} \infty, & 0 < r < a; \\ -V_0, & a < r < b; \\ 0, & r > b \end{cases} \right\}$$

Use the experimental data of this table to determine the best values of the parameters a, b, and V_0.

7
The Grand Canonical and Pressure Ensembles

The canonical ensemble describes a thermodynamic system in contact with a heat reservoir, at a fixed temperature. The canonical partition function of a pure fluid depends on temperature, volume, and number of particles. The connection with thermodynamics is provided by the Helmholtz free energy. In more general cases, we consider a system coupled to a reservoir that fixes additional thermodynamic fields, as pressure or chemical potential.

The **pressure ensemble** is associated with a system in contact with both heat and volume (that is, work) reservoirs. For a pure fluid, the independent variables are temperature, pressure, and number of particles. The connection with thermodynamics is provided by the **Gibbs free energy**. There are some interesting phenomena, as certain kinds of chemical reactions, that occur at fixed temperature and pressure (at atmospheric pressure, for example).

The **grand canonical ensemble**, also known as grand ensemble, is associated with a system in contact with both heat (constant temperature) and particle (constant chemical potential) reservoirs. Therefore, energy and number of particles may fluctuate about their respective average values, but in general the quadratic deviations are very small for sufficiently large systems. For a pure fluid, the independent variables are temperature, volume, and chemical potential. The connection with thermodynamics is provided by the **grand thermodynamic potential**. The grand canonical ensemble is very useful in many circumstances, as in the case of a quantum gas of particles (the grand canonical formalism is particularly suitable to the language of second quantization).

In more-complex cases, we can build more convenient statistical ensembles by allowing the contact between the system under consideration and an appropriate set of reservoirs. In this chapter, we restrict the analysis to the pressure and grand canonical ensembles, with applications to a simple fluid. The same procedures and techniques, however, can be used to deal with more general ensembles.

It should be pointed out that, in the thermodynamic limit, different ensembles for the same system give the same physical results (that is, the corresponding thermodynamic potentials should be related by Legendre transformations). The use of a particular ensemble is a matter of physical or mathematical convenience for the particular problem under consideration.

7.1 The pressure ensemble

Consider a system S in contact with a reservoir R of heat and work (that is, at fixed temperature and pressure). The composite system is isolated, with total energy E_o and total volume V_o. The ideal wall that separates the subsystems is diathermal and free to move, but does not allow the transport of particles. Therefore, the probability of finding system S in a **particular microscopic state** j, with energy E_j and volume V_j, can be written as

$$P_j = c\, \Omega_R\, (E_o - E_j, V_o - V_j),\qquad (7.1)$$

where c is a constant and $\Omega_R(E, V)$ is the number of available microscopic states of reservoir R, with energy E and volume V. We thus write the expansion

$$\ln P_j = \text{constant} + \left(\frac{\partial \ln \Omega_R}{\partial E}\right)_{E_o, V_o} (-E_j) + \left(\frac{\partial \ln \Omega_R}{\partial V}\right)_{E_o, V_o} (-V_j) + \dots .$$
$$(7.2)$$

Using the definition of entropy, given by the second postulate of statistical mechanics, we have

$$\frac{\partial \ln \Omega_R}{\partial E} = \frac{1}{k_B T} \quad \text{and} \quad \frac{\partial \ln \Omega_R}{\partial V} = \frac{p}{k_B T},\qquad (7.3)$$

where T and p are the temperature and the pressure of the reservoir. In the limit of a large enough reservoir, we can neglect higher-order terms in expansion (7.2). Thus, we write

$$\ln P_j = \text{constant} - \frac{E_j}{k_B T} - \frac{p V_j}{k_B T},\qquad (7.4)$$

from which we have

$$P_j = \frac{1}{Y} \exp\left(-\beta E_j - \beta p V_j\right), \tag{7.5}$$

where the partition function Y is given by

$$Y = \sum_j \exp\left(-\beta E_j - \beta p V_j\right). \tag{7.6}$$

The pressure ensemble is defined as the set $\{j, P_j\}$ of microstates and their respective probabilities, given by equation (7.5). For a pure fluid, the partition function Y depends on T, p, and N.

A. Connection with thermodynamics

The sum over all microstates in equation (7.6) may be carried out by first fixing the volume V, adding up all terms corresponding to this value of V, and then summing over V. Hence, we may write

$$Y = \sum_V \exp\left(-\beta p V\right) \sum_j \exp\left[-\beta E_j\left(V\right)\right], \tag{7.7}$$

where the sum over j is restricted to the microstates associated with the volume V. This last sum, however, is just the canonical partition function of a system at temperature T, with volume V. Therefore, using the notation $Z = Z\left(\beta, V\right)$ to emphasize the dependence of the canonical partition function on temperature and volume, we may also write

$$Y = \sum_V \exp\left(-\beta p V\right) Z\left(\beta, V\right). \tag{7.8}$$

To establish the connection with thermodynamics, we substitute the sum in equation (7.8) by its largest term,

$$Y = \sum_V \exp\left(-\beta p V + \ln Z\right) \sim \exp\left[-\beta \min_V\left(-k_B T \ln Z + p V\right)\right]. \tag{7.9}$$

Recalling the identification between $-k_B T \ln Z$ and the Helmholtz free energy F, we have

$$Y \sim \exp\left[-\beta \min_V \left\{F + p V\right\}\right], \tag{7.10}$$

where the minimization is equivalent to a Legendre transformation of the Helmholtz free energy F with respect to volume V, giving rise to the Gibbs free energy function G (which depends on temperature and pressure). These

heuristic arguments suggest that the connection between the pressure ensemble and thermodynamics is given by the correspondence

$$Y \to \exp(-\beta G). \tag{7.11}$$

For a pure fluid, we have

$$G = G(T, p, N) \to -\frac{1}{\beta} \ln Y(T, p, N), \tag{7.12}$$

where we have emphasized the dependence on T, p, and N. In fact, this connection holds in the thermodynamic limit, given in this case by $N \to \infty$, with fixed values of temperature and pressure,

$$g(T, p) = -\frac{1}{\beta} \lim_{N \to \infty} \frac{1}{N} \ln Y(T, p, N). \tag{7.13}$$

Using the Euler relation of thermodynamics, we have $G = U - TS + pV = N\mu$. Therefore, the Gibbs free energy per particle, $g = G/N$, is the chemical potential as a function of temperature and pressure.

B. Fluctuations of energy and volume

The expected values of energy and volume are given by the expressions

$$\langle E_j \rangle = Y^{-1} \sum_j E_j \exp(-\beta E_j - \beta p V_j) = -\frac{\partial}{\partial \beta} \ln Y + \frac{p}{\beta} \frac{\partial}{\partial p} \ln Y \tag{7.14}$$

and

$$\langle V_j \rangle = Y^{-1} \sum_j V_j \exp(-\beta E_j - \beta p V_j) = -\frac{1}{\beta} \frac{\partial}{\partial p} \ln Y. \tag{7.15}$$

From the identification between $-k_B T \ln Y$ and the Gibbs free energy G, it is not difficult to see that these average values do correspond to the thermodynamic values of internal energy and volume of the system.

The quadratic deviation of the volume may be written as

$$\left\langle (\Delta V)^2 \right\rangle = \left\langle (V_j - \langle V_j \rangle)^2 \right\rangle = \langle V_j^2 \rangle - \langle V_j \rangle^2$$

$$= Y^{-1} \sum_j V_j^2 \exp(-\beta E_j - \beta p V_j) - \left[Y^{-1} \sum_j V_j \exp(-\beta E_j - \beta p V_j) \right]^2$$

$$= \frac{1}{\beta^2} Y^{-1} \frac{\partial^2 Y}{\partial p^2} - \frac{1}{\beta^2} \left[Y^{-1} \frac{\partial Y}{\partial p} \right]^2 = \frac{1}{\beta^2} \frac{\partial}{\partial p} \left(Y^{-1} \frac{\partial Y}{\partial p} \right) = \frac{1}{\beta^2} \frac{\partial}{\partial p} \left(\frac{\partial}{\partial p} \ln Y \right).$$

Taking into account the connection with thermodynamics, we have

$$\frac{\partial}{\partial p}\ln Y \rightarrow -\beta\frac{\partial G}{\partial p} = -\beta V.$$

Hence,

$$\left\langle(\Delta V)^2\right\rangle = -\frac{1}{\beta}\left(\frac{\partial V}{\partial p}\right)_{T,N} = k_B T V \kappa_T \geq 0, \tag{7.16}$$

where V is the thermodynamic volume, and κ_T is the isothermal compressibility (which is always a positive quantity). The relative width vanishes as $1/\sqrt{V}$ for a large enough system. In the critical regions, however, there is a dramatic increase of the compressibility, and the volume fluctuations (which are density fluctuations in this case of fixed number of particles) become exaggeratedly large.

C. Example: The classical ideal monatomic gas

In the last chapter, we have already written the canonical partition function of the classical ideal gas of monatomic particles,

$$Z = Z(T, V, N) = \frac{1}{N!}Z_1^N, \tag{7.17}$$

where

$$Z_1 = \left(\frac{2\pi m}{\beta h^2}\right)^{3/2} V, \tag{7.18}$$

with the inclusion of the (quantum) corrections. Therefore, using equation (7.8) with an integral instead of the discrete sum, we write

$$Y(T, p, N) = \int_0^\infty dV \exp(-\beta p V) Z(T, V, N)$$

$$= \frac{1}{N!}\left(\frac{2\pi m}{\beta h^2}\right)^{3N/2} \int_0^\infty V^N \exp(-\beta p V)\, dV. \tag{7.19}$$

Noting that

$$\int_0^\infty x^n e^{-\alpha x} dx = (-1)^n \frac{d^n}{d\alpha^n}\int_0^\infty e^{-\alpha x} dx = (-1)^{2N} N! \alpha^{-N-1}, \tag{7.20}$$

where $\alpha > 0$ and n is an integer, we have

$$Y(T, p, N) = \frac{1}{N!}\left(\frac{2\pi m}{\beta h^2}\right)^{3N/2}\frac{N!}{(\beta p)^{N+1}}. \tag{7.21}$$

Therefore,

$$\lim_{N \to \infty} \frac{1}{N} \ln Y = \frac{3}{2} \ln \left(\frac{2\pi m}{\beta h^2} \right) - \ln (\beta p), \tag{7.22}$$

from which we have the Gibbs free energy per particle,

$$g = g(T, p) = -\frac{3}{2} k_B T \ln \left(\frac{2\pi m k_B T}{h^2} \right) - k_B T \ln \frac{k_B T}{p}. \tag{7.23}$$

It is not difficult to see that this expression indeed corresponds to a Legendre transformation of the Helmholtz free energy per particle, $f = f(T, v)$, as obtained in the last chapter in the framework of the canonical ensemble.

The equations of state in the Gibbs representation are given by

$$s = - \left(\frac{\partial g}{\partial T} \right)_p = \frac{5}{2} k_B \ln T - k_B \ln p + \text{constant}, \tag{7.24}$$

from which we have the specific heat at constant pressure,

$$c_p = T \left(\frac{\partial s}{\partial T} \right)_p = \frac{5}{2} k_B, \tag{7.25}$$

and the specific volume

$$v = \left(\frac{\partial g}{\partial p} \right)_T = \frac{k_B T}{p}, \tag{7.26}$$

which corresponds to Boyle's law.

Before completing this section, it is interesting to note that a change of variables in integral (7.19) yields the expression

$$Y = \frac{1}{N!} \left(\frac{2\pi m}{\beta h^2} \right)^{3N/2} N \exp (N \ln N) \int_0^\infty dv \exp \left[N (\ln v - \beta p v) \right]. \tag{7.27}$$

To obtain Y in the thermodynamic limit ($N \to \infty$), we have to find an asymptotic expression, in the limit of large N, of an integral of the form

$$I(N) = \int_0^\infty dv \exp \left[N f(v) \right]. \tag{7.28}$$

This is another example of application of Laplace's method of asymptotic integration, of enormous relevance in problems of statistical physics. If a function $f(v) = \ln v - \beta p v$ has a maximum at a real positive value, the contribution to the integral $I(N)$ comes essentially from the neighborhood of

this maximum. According to Laplace's method, let us expand the exponent of the integrand about the maximum,

$$f(v) = f(v_o) - \frac{1}{2}\beta^2 p^2 (v - v_o)^2 + \dots, \tag{7.29}$$

where $v_0 = (\beta p)^{-1}$. We then discard higher-order terms and extend the limits of integration to write the asymptotic form

$$I(N) \sim \exp\left[Nf\left(\frac{1}{\beta p}\right)\right] \int_{-\infty}^{+\infty} \exp\left[-\frac{1}{2}N\beta^2 p^2 x^2\right] dx$$

$$= \exp[-N\ln(\beta p) - N]\left(\frac{2\pi}{N\beta^2 p^2}\right)^{1/2}. \tag{7.30}$$

Hence, we have

$$Y \sim \frac{1}{N!}\left(\frac{2\pi m}{\beta h^2}\right)^{3N/2} N \exp[N\ln N - N\ln(\beta p) - N]\left(\frac{2\pi}{N\beta^2 p^2}\right)^{1/2}, \tag{7.31}$$

from which we write

$$\lim_{N\to\infty} \frac{1}{N}\ln Y = \frac{3}{2}\ln\left(\frac{2\pi m}{\beta h^2}\right) - \ln(\beta p), \tag{7.32}$$

in perfect agreement with equation (7.22).

7.2 The grand canonical ensemble

Consider now a system S in contact with a reservoir R of heat and particles (that is, at fixed temperature and chemical potential). The composite system remains isolated, with energy E_o and total number of particles N_o (for simplicity, we consider a pure system, with a single component). The ideal wall that separates the subsystems is diathermal and permeable, but remains fixed, to prevent any changes of volume. Using the fundamental postulate of statistical mechanics, the probability of finding system S in a **particular microscopic state** j, with energy E_j and number of particles N_j, may be written as

$$P_j = c\,\Omega_R(E_o - E_j, N_o - N_j), \tag{7.33}$$

where c is a constant and $\Omega_R(E, N)$ is the number of available microscopic states of reservoir R with energy E and number of particles N (as it remains

fixed, we are omitting the dependence on volume). Now we write the Taylor expansion

$$\ln P_j = \text{constant} + \left(\frac{\partial \ln \Omega_R}{\partial E}\right)_{E_o, N_o} (-E_j) + \left(\frac{\partial \ln \Omega_R}{\partial N}\right)_{E_o, N_o} (-N_j) + \dots . \tag{7.34}$$

Using the definition of entropy given by the second postulate of statistical mechanics, we have

$$\frac{\partial \ln \Omega_R}{\partial E} = \frac{1}{k_B T} \quad \text{and} \quad \frac{\partial \ln \Omega_R}{\partial N} = -\frac{\mu}{k_B T}, \tag{7.35}$$

where T and μ are the temperature and the chemical potential of the reservoir. In the limit of a large enough reservoir, we may neglect higher-order terms in expansion (7.34). Therefore, we write

$$\ln P_j = \text{constant} - \frac{E_j}{k_B T} + \frac{\mu N_j}{k_B T}, \tag{7.36}$$

from which we have

$$P_j = \frac{1}{\Xi} \exp\left(-\beta E_j + \beta \mu N_j\right), \tag{7.37}$$

where the grand partition function Ξ is given by

$$\Xi = \sum_j \exp\left(-\beta E_j + \beta \mu N_j\right). \tag{7.38}$$

The grand canonical ensemble is defined as the set $\{j, P_j\}$ of microstates j, with respective probabilities P_j given by equation (7.37). For a pure fluid, the grand partition function depends on T, V, and μ.

A. Connection with thermodynamics

As in the case of the pressure ensemble, the sum over microstates in equation (7.38) can be rearranged. First, we perform the sum over microstates with a fixed number of particles N. Then, we add over all values of N. Hence, we write

$$\Xi = \sum_N \exp\left(\beta \mu N\right) \sum_j \exp\left[-\beta E_j\left(N\right)\right], \tag{7.39}$$

where the sum over j is restricted to the microstates with a certain value of the number N of particles. This sum over j, however, is a canonical partition function. Then, using the notation $Z = Z\left(\beta, N\right)$ to emphasize the dependence on β and N, we also write

$$\Xi = \sum_N \exp\left(\beta \mu N\right) Z\left(\beta, N\right). \tag{7.40}$$

To establish the connection with thermodynamics, we substitute the sum in equation (7.40) by its largest term,

$$\Xi = \sum_N \exp\left(\beta\mu N + \ln Z\right) \sim \exp\left[-\beta \min_N \left(-k_B T \ln Z - \mu N\right)\right]. \quad (7.41)$$

Recalling the correspondence between $-k_B T \ln Z$ and the Helmholtz free energy F, we can also write

$$\Xi \sim \exp\left[-\beta \min_N \left\{F - \mu N\right\}\right], \quad (7.42)$$

where the minimization is equivalent to performing a Legendre transformation of the Helmholtz free energy F with respect to the number of particles N, which gives rise to the grand thermodynamic potential Φ (which depends on temperature and chemical potential). This heuristic reasoning suggests that the connection with thermodynamics is given by the correspondence

$$\Xi \rightarrow \exp\left(-\beta\Phi\right). \quad (7.43)$$

For a pure fluid,

$$\Phi = \Phi\left(T, V, \mu\right) \rightarrow -\frac{1}{\beta} \ln \Xi\left(T, V, \mu\right), \quad (7.44)$$

where we emphasize the dependence on T, V, and μ. As a matter of fact, the connection with thermodynamics is defined in the thermodynamic limit, which is given in this case by $V \rightarrow \infty$, with fixed values of temperature and chemical potential,

$$\phi\left(T, \mu\right) = -\frac{1}{\beta} \lim_{V \rightarrow \infty} \frac{1}{V} \ln \Xi\left(T, V, \mu\right), \quad (7.45)$$

where $\phi = \phi\left(T, \mu\right)$ is the grand thermodynamic potential per volume. Using the Euler relation of thermodynamics, we have $\Phi = U - TS - N\mu = -pV$. Therefore, the grand thermodynamic potential per volume, $\phi = \Phi/V$, is just minus the pressure (as a function of temperature and chemical potential).

B. Fluctuations of energy and number of particles

In the grand ensemble, the energy and the number of particles can fluctuate about the respective average values,

$$\langle E_j \rangle = \Xi^{-1} \sum_j E_j \exp\left(-\beta E_j + \beta \mu N_j\right) = -\frac{\partial}{\partial \beta} \ln \Xi + \frac{\mu}{\beta} \frac{\partial}{\partial \mu} \ln \Xi \quad (7.46)$$

and

$$\langle N_j \rangle = \Xi^{-1} \sum_j N_j \exp\left(-\beta E_j + \beta \mu N_j\right) = \frac{1}{\beta} \frac{\partial}{\partial \mu} \ln \Xi. \qquad (7.47)$$

The expected value of energy may be written in a simpler fashion, similar to the form adopted in the canonical ensemble. Let us define the fugacity (sometimes called activity),

$$z = \exp\left(\beta \mu\right), \qquad (7.48)$$

to express the grand partition function Ξ in terms of the new independent variables z and β (the volume V remains fixed and will be omitted to simplify the notation). Then, we have

$$\Xi = \Xi\left(z, \beta\right) = \sum_j z^{N_j} \exp\left(-\beta E_j\right) = \sum_{N=0}^{\infty} z^N Z\left(\beta, N\right), \qquad (7.49)$$

where the last sum is a polynomial in the variable z, whose coefficient of the term z^N is a canonical partition function for a system with N particles. In terms of these new variables, we write

$$\langle E_j \rangle = \Xi^{-1} \sum_j E_j z^{N_j} \exp\left(-\beta E_j\right) = -\frac{\partial}{\partial \beta} \ln \Xi\left(z, \beta\right), \qquad (7.50)$$

which is the analog (changing Ξ by Z) of the formula for the expected value of the energy in the canonical ensemble. We also have

$$\langle N_j \rangle = \Xi^{-1} \sum_j N_j z^{N_j} \exp\left(-\beta E_j\right) = z \frac{\partial}{\partial z} \ln \Xi\left(z, \beta\right). \qquad (7.51)$$

From equation (7.44) for the grand thermodynamic potential, it is easy to show that the expected values $\langle E_j \rangle$ and $\langle N_j \rangle$ correspond indeed to the thermodynamic internal energy U and number of particles N of the system under consideration.

The average quadratic deviation of the number of particles is given by

$$\left\langle (\Delta N)^2 \right\rangle = \left\langle (N_j - \langle N_j \rangle)^2 \right\rangle = \frac{1}{\beta^2} \frac{\partial}{\partial \mu} \left[\frac{\partial}{\partial \mu} \ln \Xi \right]. \qquad (7.52)$$

Therefore, from equation (7.44), we have

$$\left\langle (N_j - \langle N_j \rangle)^2 \right\rangle = \frac{1}{\beta} \left(\frac{\partial N}{\partial \mu} \right)_{T,V}, \qquad (7.53)$$

where N is the thermodynamic number of particles (which corresponds to the expected value $\langle N_j \rangle$). As the average quadratic deviation is positive

(and the volume is fixed), the concentration $\rho = N/V$, at a given temperature, is an increasing function of the chemical potential.

Let us now write an expression for the average quadratic deviation of the number of particles in terms of better known quantities. Using the Gibbs–Duhem relation form thermodynamics,

$$d\mu = -\frac{S}{N}dT + \frac{V}{N}dp, \qquad (7.54)$$

we have

$$\left(\frac{\partial\mu}{\partial N}\right)_{T,V} = \frac{V}{N}\left(\frac{\partial p}{\partial N}\right)_{T,V} \quad \text{and} \quad \left(\frac{\partial\mu}{\partial V}\right)_{T,N} = \frac{V}{N}\left(\frac{\partial p}{\partial V}\right)_{T,N}. \qquad (7.55)$$

In the Helmholtz representation, we can write the Maxwell relation

$$\left(\frac{\partial p}{\partial N}\right)_{T,V} = -\left(\frac{\partial\mu}{\partial V}\right)_{T,N}. \qquad (7.56)$$

Hence, we have

$$\left(\frac{\partial\mu}{\partial N}\right)_{T,V} = -\frac{V}{N}\left(\frac{\partial\mu}{\partial V}\right)_{T,N} = -\left(\frac{V}{N}\right)^2\left(\frac{\partial p}{\partial V}\right)_{T,N} = \frac{V}{N^2\kappa_T}, \qquad (7.57)$$

where κ_T is the isothermal compressibility. Therefore, we finally write

$$\left\langle (\Delta N)^2 \right\rangle = \left\langle (N_j - \langle N_j \rangle)^2 \right\rangle = \frac{1}{\beta}\left(\frac{\partial N}{\partial\mu}\right)_{T,V} = N\frac{k_B T\kappa_T}{v} \geq 0, \qquad (7.58)$$

which shows that the positivity of the average quadratic deviation of the number of particles is related to the sign of the isothermal compressibility (and with much deeper requirements of thermodynamic stability). The relative width is given by

$$\frac{\sqrt{\left\langle (\Delta N)^2 \right\rangle}}{\langle N_j \rangle} = \left(\frac{k_B T\kappa_T}{v}\right)^{1/2}\frac{1}{\sqrt{N}}, \qquad (7.59)$$

which vanishes as $1/\sqrt{N}$ for sufficiently large N. The density fluctuations, however, become exaggeratedly large in the neighborhood of the critical points, where $\kappa_T \to \infty$. There is a spectacular phenomenon of critical opalescence which gives an almost direct picture of the density fluctuations in a fluid. In the experiments of opalescence, we send a beam of monochromatic light through a fluid inside a transparent container. Above a certain critical temperature, there is just a homogeneous phase that scatters light isotropically. Below this temperature, the fluid splits into two different phases, with a well-defined meniscus, and the scattered intensity

of light is still isotropic and uniform. The phenomenon of opalescence occurs in the neighborhood of this critical temperature, where the quadratic deviation of density becomes very large. There is then an enormous variety of macroscopic regions of fluid, with different local densities, that scatter light in different ways. All length scales are simultaneously present in this problem. The beam of light, with a well-defined wavelength, will be reflected in different ways by regions of different densities, bringing about an intense glow that is known as critical opalescence.

C. Example: The classical ideal monatomic gas

The canonical partition function of an ideal classical monatomic gas is written as

$$Z = \frac{1}{N!} \left(\frac{2\pi m}{\beta h^2} \right)^{3N/2} V^N. \tag{7.60}$$

Using equation (7.49), we have

$$\Xi = \sum_{N=0}^{\infty} z^N \frac{1}{N!} \left(\frac{2\pi m}{\beta h^2} \right)^{3N/2} V^N = \exp \left[z \left(\frac{2\pi m}{\beta h^2} \right)^{3/2} V \right], \tag{7.61}$$

with the fugacity $z = \exp(\beta \mu)$. Hence,

$$\frac{1}{V} \ln \Xi = z \left(\frac{2\pi m}{\beta h^2} \right)^{3/2}. \tag{7.62}$$

From this expression, we can calculate the expected value of energy,

$$\langle E_j \rangle = -\frac{\partial}{\partial \beta} \ln \Xi \left(\beta, z \right) = \frac{3}{2} z \beta \left(\frac{2\pi m}{\beta h^2} \right)^{3/2} V, \tag{7.63}$$

which is identified as the thermodynamic internal energy U, and the expected value of the number of particles,

$$\langle N_j \rangle = z \frac{\partial}{\partial z} \ln \Xi \left(\beta, z \right) = z \left(\frac{2\pi m}{\beta h^2} \right)^{3/2} V = \ln \Xi, \tag{7.64}$$

which is the thermodynamic number N of particles of the system. From equations (7.63) and (7.64), and assuming that $\langle E_j \rangle = U$ and $\langle N_j \rangle = N$, we recover the equation of state

$$U = \frac{3}{2} N k_B T \tag{7.65}$$

of the classical ideal monatomic gas.

Using equation (7.62), which does not have any dependence on volume, we write the grand thermodynamic potential of the ideal gas,

$$\Phi = -\frac{1}{\beta}\ln\Xi = -V\left(\frac{2\pi m}{h^2}\right)^{3/2}(k_B T)^{5/2}\exp\left(\frac{\mu}{k_B T}\right). \qquad (7.66)$$

The equations of state in the representation of the grand potential are given by

$$S = -\left(\frac{\partial\Phi}{\partial T}\right)_{V,\mu} = V\left(\frac{3}{2} - \frac{\mu}{T}\right)\left(\frac{2\pi m}{h^2}\right)^{3/2}(k_B T)^{3/2}\exp\left(\frac{\mu}{k_B T}\right), \qquad (7.67)$$

$$N = -\left(\frac{\partial\Phi}{\partial\mu}\right)_{T,V} = V\left(\frac{2\pi m}{h^2}\right)^{3/2}(k_B T)^{3/2}\exp\left(\frac{\mu}{k_B T}\right), \qquad (7.68)$$

and

$$p = -\left(\frac{\partial\Phi}{\partial V}\right)_{T,\mu} = \left(\frac{2\pi m}{h^2}\right)^{3/2}(k_B T)^{5/2}\exp\left(\frac{\mu}{k_B T}\right). \qquad (7.69)$$

This last equation, for the pressure in terms of temperature and chemical potential, already displays the status of a fundamental equation (since it could have been obtained from the ratio between the grand potential and the volume). Now it is easy to eliminate the chemical potential to recover the more usual equations of state for the classical ideal gas of monatomic particles.

D. *Complex representation of the grand partition function

We have already seen that the grand partition function can be written as a polynomial in terms of powers of the fugacity z,

$$\Xi = \Xi(\beta, z) = \sum_{N=0}^{\infty} z^N Z(\beta, N), \qquad (7.70)$$

where the coefficient of the generic term z^N is a canonical partition function associated with a system of N particles. Therefore, the grand partition function is a generating function of the canonical partition functions. For a real system, it is important to note that the sum never extends to infinity, since the volume V is fixed and there is always a maximum number of classical particles that we can arrange inside a container of a given volume. Now let us consider this polynomial in the complex z plane. As $Z(\beta, N)$

does not depend on the fugacity z, we may invoke Cauchy's theorem to write

$$Z\left(\beta, N\right) = \frac{1}{2\pi i} \oint_C \frac{\Xi\left(\beta, z\right)}{z^{N+1}} dz, \tag{7.71}$$

where $i = \sqrt{-1}$ and the contour C includes the origin. If we are able to calculate this integral, it is possible to obtain the canonical partition function Z from the expression of the grand canonical partition function Ξ (in very simple situations, as the case of the classical ideal gas of monatomic particles, this integral can be explicitly calculated).

The integral in equation (7.71) may be written in a slightly different form,

$$Z\left(\beta, N\right) = \frac{1}{2\pi i} \oint_C \exp\left\{N\left[\frac{1}{N}\ln\Xi - \frac{N+1}{N}\ln z\right]\right\} dz. \tag{7.72}$$

Using the connection with thermodynamics, and recalling that we are interested in the asymptotic limit of $Z\left(\beta, N\right)$ for very large values of N, we can also write

$$Z\left(\beta, N\right) = \frac{1}{2\pi i} \oint_C \exp\left\{N f\left(z\right)\right\} dz, \tag{7.73}$$

where the function $f\left(z\right)$ has the asymptotic form

$$f\left(z\right) \sim -\beta v\phi\left(\beta, z\right) - \ln z. \tag{7.74}$$

The integral in equation (7.73) is similar to the asymptotic forms that may be treated by the Laplace method [as equation (7.28) for the classical ideal gas in the pressure ensemble]. The generalization of the Laplace method for the complex plane, which is very useful in statistical mechanics, is known as the **saddle point method** (see Appendix A.6; see also Chapter 3 of *Mathematical Methods of Physics*, by J. Mattews and R. L. Walker, W. A. Benjamin, Amsterdam, 1965).

For the particular case of the classical ideal gas of monatomic particles, using the form $\phi\left(\beta, z\right) = \Phi/V$ given by equation (7.66), function $f\left(z\right)$ may be written as

$$f\left(z\right) = v\left(\frac{2\pi m}{\beta h^2}\right)^{3/2} z - \ln z. \tag{7.75}$$

The saddle point z_o is given by

$$f'\left(z_o\right) = v\left(\frac{2\pi m}{\beta h^2}\right)^{3/2} - \frac{1}{z_o} = 0, \tag{7.76}$$

that is,

$$z_o = v \left(\frac{2\pi m}{\beta h^2} \right)^{3/2},$$ (7.77)

which is real and positive. Noting that

$$f''(z_o) = \frac{1}{z_o^2} = \frac{1}{v^2} \left(\frac{\beta h^2}{2\pi m} \right)^3 > 0,$$ (7.78)

the contribution to the saddle point is given by a contour parallel to the imaginary axis, from $z_o - i\infty$ to $z_o + i\infty$ (see Appendix A.6). Hence, we have

$$Z \sim \frac{1}{2\pi i} \int_{z_o - i\infty}^{z_o + i\infty} \exp \left\{ N f(z_o) + N \frac{1}{2} f''(z_o) (z - z_o)^2 \right\} dz$$

$$= [2\pi N |f''(z_o)|]^{-1/2} \exp [N f(z_o)],$$ (7.79)

from which we write

$$\lim_{N \to \infty} \frac{1}{N} \ln Z = f(z_o) = 1 + \ln v - \frac{3}{2} \ln \left(\frac{\beta h^2}{2\pi m} \right),$$ (7.80)

in full agreement with the result that can be directly obtained from the canonical partition function of the classical ideal gas of monatomic particles.

E. *The Yang and Lee theory of phase transitions

Using the formalism of the grand ensemble for a classical gas of interacting particles, C. N. Yang and T. D. Lee [see Phys. Rev. 87, 404 and 419 (1952)] proposed a general theory to explain the occurrence of phase transitions. These two papers are a pioneering work that represents the inauguration of the era of mathematical rigorous results in statistical mechanics.

Consider a classical gas of monatomic particles in a container of volume V, under the action of a hard-sphere potential with an attractive part ($|V_0| < \infty$), as indicated in figures 6.4 or 6.5. According to equation (7.70), the grand partition function may be written as a polynomial in terms of powers of the fugacity,

$$\Xi(\beta, V, z) = \sum_{N=0}^{M(V)} z^N Z(\beta, V, N),$$ (7.81)

FIGURE 7.1. Isotherm (pressure versus specific volume) for a simple fluid below the critical temperature. At a certain pressure, there is coexistence between two distinct phases, with specific volumes v_L and v_G, respectively.

where the coefficients $Z(\beta, V, N)$ are canonical partition functions, and the value $M(V)$ depends on volume (since there is a maximum number of particles that can be piled up in a given volume).

In the pure phase of a fluid system, the isotherms (pressure p versus specific volume v at a given temperature T) are well-behaved analytic functions. In the formalism of the grand ensemble, the equation of state $p = p(T, v)$ is given by the parametric forms

$$\frac{p}{k_B T} = \frac{1}{V} \ln \Xi(\beta, V, z) \qquad (7.82)$$

and

$$\frac{1}{v} = \frac{1}{V} z \frac{\partial}{\partial z} \ln \Xi(\beta, V, z), \qquad (7.83)$$

from which we may eliminate the fugacity z.

Now let us consider the polynomial (7.81) in terms of a variable z that may assume complex values. As the coefficients are all positive, the roots of this polynomial are pairs of complex numbers, and there is no possibility of occurrence of any real root (that is, for a physical value of the fugacity), which might lead to some kind of anomaly in the equation of state. This is the case of a pure phase, in which the $p = p(T, v)$ is a well-behaved function. In the presence of a phase transition, however, the isotherms display a singular behavior. If we plot pressure p versus specific volume v, there may be a flat line (see figure 7.1), at a given value of pressure, which indicates the coexistence of two phases, with distinct values of the specific volume (at the same temperature and pressure). This sort of singularity may occur in the **thermodynamic limit** of equations (7.82) and (7.83).

Yang and Lee showed the existence of the thermodynamic limit,

$$\frac{p}{k_B T} = \lim_{V \to \infty} \frac{1}{V} \ln \Xi(\beta, V, z), \qquad (7.84)$$

FIGURE 7.2. Regions R_1 and R_2, which are free of zeros on the real axis of the z plane, give rise to the two distinct branches of the graph of $1/v$ versus z (from which we explain the isotherm in figure 7.1).

for all $z > 0$, where the pressure is a continuous and monotonically decreasing function of z (hence, with the correct convexity). They have also shown that, given a region R of the complex z plane, including some part of the real axis and free of zeros of the polynomial (7.81), the quantity $\ln \Xi (\beta, V, z) /V$ uniformly converges to the limit value, which leads to the relation

$$\frac{1}{v} = \lim_{V \to \infty} \frac{1}{V} z \frac{\partial}{\partial z} \ln \Xi (\beta, V, z) = z \frac{\partial}{\partial z} \lim_{V \to \infty} \frac{1}{V} \ln \Xi (\beta, V, z), \qquad (7.85)$$

where we can reverse the order of the derivative and the limit. Therefore, on **each region** R of the complex z plane, we can indeed use the thermodynamic limit of the parametric forms (7.84) and (7.85) to obtain the equation of state. If region R includes the whole real axis, the system can exist in a single phase only. If there are two distinct regions R_1 and R_2, separated by a real zero of the polynomial, as indicated in figure 7.2, there may exist distinct parametric forms, giving rise to distinct branches of the equation of state (as it should happen at a phase transition). In exercise 7, we give an example of a situation of this kind, where the zeros of the polynomial cross the real axis in the thermodynamic limit.

Exercises

1. Show that the entropy in the grand canonical ensemble can be written as

$$S = -k_B \sum_j P_j \ln P_j,$$

with probability P_j given by equation (7.37),

$$P_j = \Xi^{-1} \exp \left(-\beta E_j + \beta \mu N_j \right).$$

Show that this same form of entropy still holds in the pressure ensemble (with a suitable probability distribution).

2. Consider a classical ultrarelativistic gas of particles, given by the Hamiltonian

$$\mathcal{H} = \sum_{i=1}^{N} c |\vec{p}_i| ,$$

where c is a positive constant, inside a container of volume V, in contact with a reservoir of heat and particles (at temperature T and chemical potential μ). Obtain the grand partition function and the grand thermodynamic potential. From a Legendre transformation of the grand potential, write an expression for the Helmholtz free energy of this system. To check your result, use the integral in equation (7.71) for obtaining an asymptotic form for the canonical partition function.

3. Obtain the grand partition function of a classical system of particles, inside a container of volume V, given by the Hamiltonian

$$\mathcal{H} = \sum_{i=1}^{N} \left[\frac{1}{2m} \vec{p}_i^2 + u(\vec{r}_i) \right] .$$

Write the equations of state in the representation of the grand potential. For all reasonable forms of the single-particle potential $u(\vec{r})$, show that the energy and the pressure obey typical equations of an ideal gas.

4. Show that the average quadratic deviation of the number of particles in the grand canonical ensemble may be written as

$$\left\langle (\Delta N)^2 \right\rangle = \left\langle N_j^2 \right\rangle - \left\langle N_j \right\rangle^2 = z \frac{\partial}{\partial z} \left[z \frac{\partial}{\partial z} \ln \Xi \left(\beta, z \right) \right] .$$

Obtain an expression for the relative deviation $\sqrt{\left\langle (\Delta N)^2 \right\rangle} / \left\langle N_j \right\rangle$ of an ideal gas of classical monatomic particles.

5. Show that the average quadratic deviation of energy in the grand canonical ensemble may be written as

$$\left\langle (\Delta E)^2 \right\rangle = \left\langle E_j^2 \right\rangle - \left\langle E_j \right\rangle^2 = - \left(\frac{\partial U}{\partial \beta} \right)_{z,V} ,$$

where $U = \left\langle E_j \right\rangle$ is the thermodynamic internal energy in terms of β, z, and V. Hence, show that we may also write

$$\left\langle (\Delta E)^2 \right\rangle = k_B T^2 \left[\left(\frac{\partial U}{\partial T} \right)_{\mu,V} + \frac{\mu}{T} \left(\frac{\partial U}{\partial \mu} \right)_{T,V} \right]$$

(you may use the Jacobian transformations of Appendix A.5). From this last expression, show that

$$\left\langle (\Delta E)^2 \right\rangle = \left\langle (\Delta E)^2 \right\rangle_{can} + k_B T^2 \left[\left(\frac{\partial U}{\partial N} \right)_{T,V} \left(\frac{\partial N}{\partial T} \right)_{V,\mu} \right.$$
$$\left. + \frac{\mu}{T} \left(\frac{\partial U}{\partial N} \right)_{T,V} \left(\frac{\partial N}{\partial \mu} \right)_{T,V} \right],$$

where

$$\left\langle (\Delta E)^2 \right\rangle_{can} = N k_B T^2 c_V$$

is the average quadratic deviation of energy in the canonical ensemble. Finally, show that

$$\left[\left(\frac{\partial U}{\partial N} \right)_{T,V} \left(\frac{\partial N}{\partial T} \right)_{V,\mu} + \frac{\mu}{T} \left(\frac{\partial U}{\partial N} \right)_{T,V} \left(\frac{\partial N}{\partial \mu} \right)_{T,V} \right]$$

$$= \frac{1}{T} \left(\frac{\partial N}{\partial \mu} \right)_{T,V} \left[\left(\frac{\partial U}{\partial N} \right)_{T,V} \right]^2 > 0,$$

since

$$\left\langle (\Delta N)^2 \right\rangle = k_B T \left(\frac{\partial N}{\partial \mu} \right)_{T,V} > 0.$$

6. At a given temperature T, a surface with N_o adsorption centers has $N \le N_o$ adsorbed molecules. Suppose that there are no interactions between molecules. Show that the chemical potential of the adsorbed gas is given by

$$\mu = k_B T \ln \frac{N}{(N_o - N)\, a\,(T)}.$$

What is the meaning of the function $a\,(T)$?

7. The grand partition function for a simplified statistical model is given by the expression

$$\Xi\,(z, V) = (1 + z)^V \left(1 + z^{\alpha V} \right),$$

where α is a positive constant. Write parametric forms of the equation of state. Sketch an isotherm of pressure versus specific volume (draw a graph to eliminate the variable z). Show that this system displays a (first-order) phase transition. Obtain the specific volumes of the coexisting phases at this transition. Find the zeros of the polynomial $\Xi\,(z, V)$ in the complex z plane, and show that there is a zero at $z = 1$ in the limit $V \to \infty$.

(you may use the Jacobian transformations of Appendix A.5). From this last expression, show that

$$\langle(\Delta E)^2\rangle = \langle(\Delta E)^2\rangle_{can} + k_BT^2 \left(\frac{\partial U}{\partial N}\right)_{T,V}^2 \Bigg/ \left[-\frac{1}{T}\left(\frac{\partial U}{\partial N}\right)_{T,V}\left(\frac{\partial N}{\partial N}\right)_{T,V} \right]$$

where

$$\langle(\Delta E)^2\rangle_{can} = Nk_BT^2 c_v$$

is the average quadratic deviation of energy in the canonical ensemble.

Finally show that

$$\left[\left(\frac{\partial U}{\partial N}\right)_{T,V} - \frac{U}{T}\left(\frac{\partial N}{\partial T}\right)_{N,V}\left(\frac{\partial N}{\partial T}\right)_{T,V} + \frac{U}{T}\left(\frac{\partial U}{\partial N}\right)_{T,V}\right]$$

$$= -\frac{1}{T}\left(\frac{\partial^2 U}{\partial T}\right)_{N,V}\left[\left(\frac{\partial U}{\partial N}\right)_{T,V}\right]^2 > 0,$$

since

$$\langle(\Delta N)^2\rangle = k_BT \left(\frac{\partial N}{\partial \mu}\right)_{T,V} > 0$$

6. At a given temperature T, a surface with N_a adsorption centers has $N \leq N_a$ adsorbed molecules. Suppose that there are no interactions between molecules. Show that the chemical potential of the adsorbed gas is given by

$$\mu = k_BT \ln \frac{N}{(N_a - N) a(T)}.$$

What is the meaning of the function $a(T)$?

7. The grand partition function for a simplified statistical model is given by the expression

$$\Xi(z,V) = (1+z)^V (1+z^{\alpha V})$$

where α is a positive constant. Write the parametric forms of the equation of state. Sketch an isotherm of pressure versus specific volume (draw a graph to eliminate the variable z). Show that this system displays a (first-order) phase transition. Obtain the specific volumes of the coexisting phases at this transition. Find the zeros of the polynomial $\Xi(z,V)$ in the complex z plane, and show that there is a zero at $z = 1$ in the limit $V \to \infty$.

8
The Ideal Quantum Gas

A quantum system of N identical particles is described by the wave function

$$\Psi = \Psi\left(q_1, ..., q_N\right), \tag{8.1}$$

where q_j denotes all compatible coordinates of particle j (position and spin, for example). However, not all wave functions of this sort, which satisfy the time-independent Schrödinger equation, are acceptable representations of a quantum system. We also require a symmetry property,

$$\Psi\left(q_1, ..., q_i, ..., q_j, ..., q_N\right) = \pm \Psi\left(q_1, ..., q_j, ..., q_i, ..., q_N\right), \tag{8.2}$$

which indicates that the quantum state of the system does not alter if we change the coordinates of two particles. The **symmetric** wave functions are associated with **integer spin** particles (photons, phonons, magnons, ^4He atoms). These particles are called **bosons**, and obey the **Bose–Einstein statistics**. The **antisymmetric** wave functions are associated with **half-integer spin** particles (electrons, protons, atoms of ^3He). These particles are called **fermions**, and obey the **Fermi–Dirac statistics**.

For example, let us consider a system of two identical and independent particles, described by the Hamiltonian

$$\mathcal{H} = \mathcal{H}_1 + \mathcal{H}_2, \tag{8.3}$$

where

$$\mathcal{H}_j = \frac{1}{2m}\vec{p}_j^{\,2} + V\left(\vec{r}_j\right), \tag{8.4}$$

for $j = 1$ and $j = 2$. The eigenfunctions of Hamiltonian \mathcal{H}, for a given total energy E, may be written as a product, $\psi_{n_1}(\vec{r}_1)\,\psi_{n_2}(\vec{r}_2)$, such that

$$\mathcal{H}_1 \psi_{n_1}(\vec{r}_1) = \epsilon_{n_1} \psi_{n_1}(\vec{r}_1) \tag{8.5}$$

and

$$\mathcal{H}_2 \psi_{n_2}(\vec{r}_2) = \epsilon_{n_2} \psi_{n_2}(\vec{r}_2), \tag{8.6}$$

with

$$E = \epsilon_{n_1} + \epsilon_{n_2}. \tag{8.7}$$

Owing to the symmetry requirements, the accessible quantum states of this system are represented by the symmetric and antisymmetric linear combinations of this product,

$$\Psi_S(\vec{r}_1, \vec{r}_2) = \frac{1}{\sqrt{2}}\left[\psi_{n_1}(\vec{r}_1)\,\psi_{n_2}(\vec{r}_2) + \psi_{n_1}(\vec{r}_2)\,\psi_{n_2}(\vec{r}_1)\right] \tag{8.8}$$

and

$$\Psi_A(\vec{r}_1, \vec{r}_2) = \frac{1}{\sqrt{2}}\left[\psi_{n_1}(\vec{r}_1)\,\psi_{n_2}(\vec{r}_2) - \psi_{n_1}(\vec{r}_2)\,\psi_{n_2}(\vec{r}_1)\right]. \tag{8.9}$$

where $\psi_n(\vec{r})$ is called an **orbital** or **single-particle state**. Note that $\Psi_A = 0$ for $n_1 = n_2$. Therefore, in the case of fermions, there cannot be two particles in the same orbital (that is, with the same set of quantum numbers), in agreement with **Pauli's exclusion principle**. All the experimental evidence support this broad classification of quantum particles as bosons (symmetric wave functions) and fermions (antisymmetric wave functions).

Now we give an explicit example of the set of quantum states of **two** independent particles. Suppose that the quantum numbers of the orbitals (n_1 and n_2) can assume only **three** distinct values (which will be called 1, 2, and 3). Thus, we have the following possibilities:

(i) If the particles are bosons, there are six quantum states, as indicated in the table (where letters A and B denote the particles).

1	2	3	
A, B	–	–	$\psi_1(\vec{r}_1)\,\psi_1(\vec{r}_2)$
–	A, B	–	$\psi_2(\vec{r}_1)\,\psi_2(\vec{r}_2)$
–	–	A, B	$\psi_3(\vec{r}_1)\,\psi_3(\vec{r}_2)$
A	B	–	$\frac{1}{\sqrt{2}}\left[\psi_1(\vec{r}_1)\,\psi_2(\vec{r}_2) + \psi_1(\vec{r}_2)\,\psi_2(\vec{r}_1)\right]$
–	A	B	$\frac{1}{\sqrt{2}}\left[\psi_2(\vec{r}_1)\,\psi_3(\vec{r}_2) + \psi_2(\vec{r}_2)\,\psi_3(\vec{r}_1)\right]$
A	–	B	$\frac{1}{\sqrt{2}}\left[\psi_1(\vec{r}_1)\,\psi_3(\vec{r}_2) + \psi_1(\vec{r}_2)\,\psi_3(\vec{r}_1)\right]$

In this case, although letters A and B have been used to denote the particles, the exchange between A and B does not lead to any new state of the system (since quantum particles are indistinguishable). Therefore, there are only six available quantum states for this system;

(ii) If the particles are fermions, there are only three states, as indicated below.

1	2	3	
A	B	–	$\frac{1}{\sqrt{2}}\left[\psi_1\left(\vec{r}_1\right)\psi_2\left(\vec{r}_2\right)-\psi_1\left(\vec{r}_2\right)\psi_2\left(\vec{r}_1\right)\right]$
–	A	B	$\frac{1}{\sqrt{2}}\left[\psi_2\left(\vec{r}_1\right)\psi_3\left(\vec{r}_2\right)-\psi_2\left(\vec{r}_2\right)\psi_3\left(\vec{r}_1\right)\right]$
A	–	B	$\frac{1}{\sqrt{2}}\left[\psi_1\left(\vec{r}_1\right)\psi_3\left(\vec{r}_2\right)-\psi_1\left(\vec{r}_2\right)\psi_3\left(\vec{r}_1\right)\right]$

As an illustration, we consider a semiclassical case, in which there are no symmetry requirements for the wave functions. In this case the particles are distinguishable, and we have the classical **Maxwell–Boltzmann statistics**. Later, we shall see that the Maxwell–Boltzmann statistics does indeed correspond to the classical limit of either the Bose–Einstein or the Fermi–Dirac statistics. In the table we represent the nine states of this classical gas of two distinguishable particles that may be found in three distinct orbitals.

1	2	3
A, B	–	–
–	A, B	–
–	–	A, B
A	B	–
B	A	–
–	A	B
–	B	A
A	–	B
B	–	A

8.1 Orbitals of a free particle

Consider a particle of mass m, in one dimension, confined to a region of length L (that is, such that the coordinate x may be between 0 and L). Taking into account the position coordinates only, the **orbital** $\psi_n\left(x\right)$ is given by the solution of the time-independent Schrödinger equation,

$$\mathcal{H}\psi_n\left(x\right) = \epsilon_n\psi_n\left(x\right), \qquad (8.10)$$

where

$$\mathcal{H} = \frac{1}{2m}p^2 = -\frac{\hbar^2}{2m}\frac{d^2}{dx^2}. \qquad (8.11)$$

Therefore, we have solutions of the form

$$\psi_n(x) = Ce^{ikx} \text{ with } \epsilon_n = \frac{\hbar^2 k^2}{2m}. \tag{8.12}$$

The quantization of energy comes from imposing boundary conditions. As a matter of convenience, we adopt periodic boundary conditions (as we are interested in the connection with thermodynamics, in the limit of a very large system, the particular boundary conditions should not have any influence on the final physical results). Thus, let us assume that

$$\psi_n(x) = \psi_n(x + L). \tag{8.13}$$

Therefore, $\exp(ikL) = 1$, from which we have $kL = 2\pi n$, that is,

$$k = \frac{2\pi}{L} n, \text{ with } n = 0, \pm 1, \pm 2, \pm 3, \ldots. \tag{8.14}$$

For very large L, the interval between two consecutive values of k is very small (equal to $2\pi/L$). We can then substitute the sum over k by an integral over dk,

$$\sum_k f(k) \rightarrow \int \frac{dk}{\frac{2\pi}{L}} f(k) = \frac{L}{2\pi} \int dk f(k). \tag{8.15}$$

Later, we use this result. Some care should be taken if the function $f(k)$ has some kind of singularity.

These results can be easily generalized for three-dimensional problems. In this case, we have

$$\psi(\vec{r}) = C \exp\left(i\vec{k} \cdot \vec{r}\right), \tag{8.16}$$

with

$$\vec{k} = \frac{2\pi}{L_1} m_1 \vec{e_x} + \frac{2\pi}{L_2} m_2 \vec{e_y} + \frac{2\pi}{L_3} m_3 \vec{e_z}, \tag{8.17}$$

where m_1, m_2, and m_3 are integers, $\vec{e_x}$, $\vec{e_y}$, and $\vec{e_z}$ are unit vectors along the Cartesian directions of space, and the particle is confined to a rectangular box of sides L_1, L_2, and L_3. The energy of the orbital is given by

$$\epsilon_{\vec{k}} = \frac{\hbar^2 k^2}{2m}. \tag{8.18}$$

We can also transform the sum into an integral,

$$\sum_{\vec{k}} f\left(\vec{k}\right) \rightarrow \frac{V}{(2\pi)^3} \int d^3\vec{k} f\left(\vec{k}\right). \tag{8.19}$$

Up to this point, we have dealt with position coordinates only. However, there may be additional degrees of freedom (spin or isospin, for example). To take into account spin degrees of freedom, we characterize the state of the orbital by the symbol $j \equiv \left(\vec{k}, \sigma \right)$, which includes the wave vector \vec{k}, associated with the translational degrees of freedom, and the spin quantum number σ. For free particles, in the absence of spin terms in the Hamiltonian, the energy spectrum is still given by

$$\epsilon_j \equiv \epsilon_{\vec{k},\sigma} \equiv \frac{\hbar^2 k^2}{2m}. \tag{8.20}$$

In the presence of an applied magnetic field H, and taking into account the interaction between the spin and the field, the energy spectrum of a free particle of spin $1/2$ will be given by

$$\epsilon_j \equiv \epsilon_{\vec{k},\sigma} \equiv \frac{\hbar^2 k^2}{2m} - \mu_B H \sigma, \tag{8.21}$$

where the constant μ_B is the Bohr magneton, and the spin variable σ can assume the values ± 1.

The language of orbitals is very convenient to treat situations where the particles have a more complex internal structure, as in the case of a diluted gas of diatomic molecules. The classical models of a gas of diatomic molecules (a rigid rotator in three dimensions, or a rotator with a vibrating axis, as proposed in the exercises) are unable to explain the thermal behavior of some quantities, as the specific heat at constant volume. The Hamiltonian of a diatomic molecule may be written as a sum of several terms, associated with distinct degrees of freedom, and which do not couple in a first approximation,

$$\mathcal{H}_{mol} = \mathcal{H}_{tr} + \mathcal{H}_{el} + \mathcal{H}_{rot} + \mathcal{H}_{vib} + \dots . \tag{8.22}$$

The first term represents the translation of the center of mass, characterized by the wave vector \vec{k}. The second term refers to the electronic states, under the assumption that the nuclei are fixed (in the well-known Born–Oppenheim approximation). In general, the electronic states are well above the ground state, and do not contribute to the thermodynamic free energy. The following term, \mathcal{H}_{rot}, corresponds to the rotations about an axis perpendicular to the normal axis of the molecule, with moment of inertia I (the wave functions are spherical harmonics, characterized by the angular momentum quantum number J and by the projection m_J, with degeneracy $2J+1$). The term \mathcal{H}_{vib} represents the vibrations along the axis of the molecule, with a fundamental frequency ω associated with the shape of the minimum of the molecular potential. Taking into account these terms, we have

$$\epsilon_j = \epsilon_{\vec{k},J,n} = \frac{\hbar^2 k^2}{2m} + \frac{\hbar^2}{2I} J(J+1) + \hbar\omega \left(n + \frac{1}{2} \right), \tag{8.23}$$

where n and J run over zero and all integers. Later, we will examine some consequences of these results.

8.2 Formulation of the statistical problem

Owing to the symmetry properties of the wave function, a quantum state of the ideal gas is fully characterized by the set of occupation numbers

$$\{n_1, n_2, \ldots, n_j, \ldots\} \equiv \{n_j\}, \qquad (8.24)$$

where j labels the quantum state of an orbital and n_j is the number of particles in orbital j. For fermions, it is clear that either $n_j = 0$ or 1, for all j. For bosons, however, n_j may assume all values from 0 to N, where N is the total number of particles. The energy of the system corresponding to the quantum state $\{n_j\}$ is given by

$$E\{n_j\} = \sum_j \epsilon_j n_j, \qquad (8.25)$$

where ϵ_j is the energy of orbital j. The total number of particles is given by

$$N = N\{n_j\} = \sum_j n_j. \qquad (8.26)$$

It is important to remark that in this quantum statistical treatment of the ideal gas we only worry about **how many** particles are in a certain orbital. In the classical case, on the other hand (recall the Boltzmann gas), where the particles are distinguishable, we have to know which particles are in each orbital.

In the **canonical ensemble**, we write the following partition function of the quantum ideal gas,

$$Z = Z(T, V, N) = \sum_{\substack{\{n_j\} \\ (\sum_j n_j = N)}} \exp\left[-\beta \sum_j \epsilon_j n_j\right], \qquad (8.27)$$

where the problem consists in calculating the sum (which does not factorize) over the occupation numbers with the restriction imposed by the fixed total number of particles. This calculation becomes much simpler if we work in the grand canonical ensemble, owing to the removal of this restriction on fixed number of particles.

In the **grand canonical ensemble** we have

$$\Xi = \Xi(T, V, \mu) = \sum_{N=0}^{\infty} \exp(\beta\mu N)\, Z(T, V, N)$$

$$= \sum_{N=0}^{\infty} \exp(\beta\mu N) \sum_{\substack{\{n_j\} \\ (\sum_j n_j = N)}} \exp(-\beta\epsilon_1 n_1 - \beta\epsilon_2 n_2 - \ldots)$$

$$= \sum_{N=0}^{\infty} \sum_{\substack{\{n_j\} \\ (\sum_j n_j = N)}} \exp[-\beta(\epsilon_1 - \mu)\, n_1 - \beta(\epsilon_2 - \mu)\, n_2 - \ldots] \qquad (8.28)$$

As the sum is initially performed over the set of occupation numbers n_1, n_2,..., with the restriction that $N = n_1 + n_2 + \ldots$ is fixed, but there is then another sum over all values of N, we obtain the same final result if we sum over the set of occupation numbers without any restriction (which corresponds to a mere rearrangement of the terms of this multiple sum). Hence, we have

$$\Xi = \sum_{n_1, n_2, \ldots} \exp[-\beta(\epsilon_1 - \mu)\, n_1 - \beta(\epsilon_2 - \mu)\, n_2 - \ldots], \qquad (8.29)$$

which now factorizes, and can be written as

$$\Xi = \left\{ \sum_{n_1} \exp[-\beta(\epsilon_1 - \mu)\, n_1] \right\} \left\{ \sum_{n_2} \exp[-\beta(\epsilon_2 - \mu)\, n_2] \right\} \ldots, \qquad (8.30)$$

that is,

$$\Xi = \Xi(T, V, \mu) = \prod_j \left\{ \sum_n \exp[-\beta(\epsilon_j - \mu)\, n] \right\}. \qquad (8.31)$$

From this expression, and using the formalism of the grand canonical ensemble, we obtain the expected values of energy and number of particles, and write the grand thermodynamic potential. The expected value $\langle n_j \rangle$ of the occupation number of orbital j is given by the relation

$$\langle n_j \rangle = -\frac{1}{\beta}\frac{\partial}{\partial\epsilon_j} \ln \Xi. \qquad (8.32)$$

Now we distinguish between the two physical types of statistics.

A. Bose–Einstein statistics

In this case, the sum over n in equation (8.31) is from 0 to ∞. Hence, we have

$$\sum_{n=0}^{\infty} \exp\left[-\beta\left(\epsilon_j - \mu\right) n\right] = \left\{1 - \exp\left[-\beta\left(\epsilon_j - \mu\right)\right]\right\}^{-1}. \qquad (8.33)$$

Note that this only exists if $\exp\left[-\beta\left(\epsilon_j - \mu\right)\right] < 1$ for all j. Taking into account that the smallest value of ϵ_j is 0, we should have $\exp\left(\beta\mu\right) < 1$, in other words, the **chemical potential μ always has to be negative**. The case where $\mu \to 0-$ is particularly interesting, as it gives rise to the phenomenon of Bose–Einstein condensation, which will be analyzed later in this text. Therefore, we have

$$\ln \Xi\left(T, V, \mu\right) = -\sum_j \ln\left\{1 - \exp\left[-\beta\left(\epsilon_j - \mu\right)\right]\right\}. \qquad (8.34)$$

From equation (8.32), we obtain the average occupation number of orbital j,

$$\langle n_j \rangle = \frac{1}{\exp\left[\beta\left(\epsilon_j - \mu\right)\right] - 1}. \qquad (8.35)$$

The condition $\exp\left[-\beta\left(\epsilon_j - \mu\right)\right] < 1$ is thus related to the physical requirement that $\langle n_j \rangle \geq 0$ for all j.

B. Fermi–Dirac statistics

In this case, n assumes only two values, 0 and 1. Hence, we have

$$\sum_{n=0,1} \exp\left[-\beta\left(\epsilon_j - \mu\right) n\right] = 1 + \exp\left[-\beta\left(\epsilon_j - \mu\right)\right], \qquad (8.36)$$

from which we write

$$\ln \Xi\left(T, V, \mu\right) = \sum_j \ln\left\{1 + \exp\left[-\beta\left(\epsilon_j - \mu\right)\right]\right\}. \qquad (8.37)$$

Therefore,

$$\langle n_j \rangle = \frac{1}{\exp\left[\beta\left(\epsilon_j - \mu\right)\right] + 1},$$

and we see that $0 \leq \langle n_j \rangle \leq 1$, in agreement with the requirements of Pauli's exclusion principle.

For both fermions and bosons, the formulas in the grand canonical ensemble may be cast as follows,

$$\ln \Xi_{FD,BE} = \pm \sum_j \ln \{1 \pm \exp[-\beta(\epsilon_j - \mu)]\} \qquad (8.38)$$

and

$$\langle n_j \rangle_{FD,BE} = \frac{1}{\exp[\beta(\epsilon_j - \mu)] \pm 1}, \qquad (8.39)$$

where FD and BE indicate the statistics of Fermi–Dirac and Bose–Einstein, respectively, with the + sign for fermions and the − sign for bosons. It is important to note that both Ξ and $\langle n_j \rangle$ are expressed in terms of the independent variables T, V, and μ (which may not be the most adequate variables for the physical problem under consideration). Also, it is important to recall the connection between the grand ensemble and thermodynamics,

$$\Xi(T, V, \mu) \rightarrow \exp[-\beta \Phi(T, V, \mu)], \qquad (8.40)$$

in the thermodynamic limit, $V \rightarrow \infty$, with the grand potential given by

$$\Phi(T, V, \mu) = -Vp(T, \mu). \qquad (8.41)$$

8.3 Classical limit

Considering the expression of $\langle n_j \rangle$ at low temperatures (large β), we see that: (i) For fermions, $\langle n_j \rangle \approx 1$ if $\epsilon_j < \mu$, that is, for orbitals associated with the lowest energies, and $\langle n_j \rangle \approx 0$ if $\epsilon_j > \mu$. (ii) For bosons, $\langle n_j \rangle \approx 0$ for the vast majority of orbitals, except at the lowest energies, where $\langle n_j \rangle >> 1$.

In the classical case we do not distinguish between bosons and fermions (that is, we should have $\langle n_j \rangle << 1$ for all j). Therefore, the classical case corresponds to a regime where

$$\exp[\beta(\epsilon_j - \mu)] >> 1, \qquad (8.42)$$

for all j, that is,

$$z = \exp(\beta\mu) << 1. \qquad (8.43)$$

Expanding equations (8.38) and (8.39) for small values of $\exp(\beta\mu)$, we have

$$\ln \Xi_{FD,BE} = \sum_j \exp[-\beta(\epsilon_j - \mu)] \mp \frac{1}{2} \sum_j \exp[-2\beta(\epsilon_j - \mu)] + \dots \quad (8.44)$$

and

$$\langle n_j \rangle_{FD,BE} = \exp[-\beta(\epsilon_j - \mu)] \{1 \mp \exp[-\beta(\epsilon_j - \mu)] + \dots\}. \qquad (8.45)$$

Hence, the classical limit is given by the dominating term of these expansions,

$$\ln \Xi_{cl} = \sum_j \exp\left[-\beta\left(\epsilon_j - \mu\right)\right] \tag{8.46}$$

and

$$\langle n_j \rangle_{cl} = \exp\left[-\beta\left(\epsilon_j - \mu\right)\right]. \tag{8.47}$$

Considering an energy spectrum given by equation (8.20),

$$\epsilon_j \equiv \epsilon_{\vec{k},\sigma} \equiv \frac{\hbar^2 k^2}{2m}, \tag{8.48}$$

we have

$$\ln \Xi_{cl} = \sum_{\vec{k},\sigma} \exp\left[-\beta\left(\frac{\hbar^2 k^2}{2m} - \mu\right)\right]. \tag{8.49}$$

In the thermodynamic limit the sum over the translational degrees of freedom may be substituted by an integral. Hence, we write

$$
\begin{aligned}
\ln \Xi_{cl} &= \gamma \frac{V}{(2\pi)^3} \int d^3 \vec{k} \exp\left[-\beta\left(\frac{\hbar^2 k^2}{2m} - \mu\right)\right] \\
&= \gamma \frac{V}{(2\pi)^3} \exp\left(\beta\mu\right) \left(\frac{2\pi m}{\beta\hbar^2}\right)^{3/2},
\end{aligned} \tag{8.50}
$$

where $\gamma = 2S + 1$ is the multiplicity (factor of degeneracy) of the spin states. The grand thermodynamic potential is given by

$$\Phi_{cl} = -\gamma V \left(\frac{2\pi m}{h^2}\right)^{3/2} (k_B T)^{5/2} \exp\left(\frac{\mu}{k_B T}\right), \tag{8.51}$$

which should be compared with the result obtained in Section 7.2.D of Chapter 7. Thus, we see that the volume normalization in classical phase space by Planck's constant h, and the inclusion of Boltzmann's correct counting factor, are fully justified by the classical limit (the additional factor γ has no classical analog).

If we write equation (8.50) in the form

$$\ln \Xi_{cl} = \gamma V z \left(\frac{2\pi m}{\beta h^2}\right)^{3/2}, \tag{8.52}$$

we have the average number of particles,

$$\left\langle \sum_j n_j \right\rangle = z \frac{\partial}{\partial z} \ln \Xi_{cl} = \gamma V z \left(\frac{2\pi m}{\beta h^2}\right)^{3/2} \to N, \tag{8.53}$$

which corresponds to the thermodynamic number of particles. Therefore, we can write

$$\exp\left(\beta\mu\right) = \frac{N}{V}\,\frac{h^3}{\gamma\left(2\pi mk_BT\right)^{3/2}}. \tag{8.54}$$

Hence, the classical limit corresponds to

$$\frac{N}{V}\,\frac{h^3}{\gamma\left(2\pi mk_BT\right)^{3/2}} \ll 1. \tag{8.55}$$

To understand the physical meaning of this inequality, we note that

$$\left(\frac{V}{N}\right)^{1/3} = a \tag{8.56}$$

represents a typical intermolecular distance, and that

$$\lambda = \frac{h}{\sqrt{2\pi mk_BT}} \tag{8.57}$$

may be associated with the de Broglie thermal wavelength. Indeed, if we consider particles with typical energy $(3/2)\,k_BT$, and typical velocity $v = \sqrt{3k_BT/m}$, we have the de Broglie frequency,

$$\nu = \frac{3k_BT}{2h}, \tag{8.58}$$

from which we write the thermal wavelength

$$\lambda_T = \frac{v}{\nu} = \frac{2h}{\sqrt{3k_BTm}}, \tag{8.59}$$

which is of the same order of magnitude as λ given by equation (8.57). Thus, in the classical limit, we should have

$$a = \left(\frac{V}{N}\right)^{1/3} \gg \lambda = \frac{2h}{\sqrt{3k_BTm}}, \tag{8.60}$$

in agreement with the usual results from quantum mechanics. The classical expressions are used in the diluted regime (small density) and at high enough temperatures.

A. The Maxwell–Boltzmann distribution

Using equation (8.46), we can eliminate the chemical potential in equation (8.47) for the expected value of the occupation number $\langle n_j \rangle$ of orbital j. Let us write equation (8.46) in the form

$$\ln \Xi_{cl} = z \sum_j \exp\left(-\beta\epsilon_j\right). \tag{8.61}$$

Hence, the thermodynamic number of particles is given by

$$N = z\frac{\partial}{\partial z} \ln \Xi_{cl}(\beta, z, V) = z\sum_j \exp(-\beta\epsilon_j). \qquad (8.62)$$

Thus, from equation (8.47), we have

$$\langle n_j \rangle_{cl} = z\exp[-\beta\epsilon_j] = \frac{N\exp[-\beta\epsilon_j]}{\sum_j \exp(-\beta\epsilon_j)}. \qquad (8.63)$$

Therefore,

$$\frac{\langle n_j \rangle_{cl}}{N} = \frac{\exp[-\beta\epsilon_j]}{\sum_j \exp(-\beta\epsilon_j)}, \qquad (8.64)$$

which is the discrete form of the classical Maxwell–Boltzmann distribution. It is important to note that same expression had already been obtained, in the context Boltzmann's model for an ideal classical gas with discrete energy values, without, however, any justification for this discretization of energy. In the thermodynamic limit, we have

$$\frac{\langle n_j \rangle_{cl}}{N} \rightarrow p_o(v)\,dv = \frac{\frac{V}{(2\pi)^3}\exp\left(-\frac{\beta\hbar^2 k^2}{2m}\right)d^3\vec{k}}{\frac{V}{(2\pi)^3}\int \exp\left(-\frac{\beta\hbar^2 k^2}{2m}\right)d^3\vec{k}}. \qquad (8.65)$$

Setting $p = mv = \hbar k$, we can write

$$p_o(v) = 4\pi\left(\frac{2\pi k_B T}{m}\right)^{3/2} v^2 \exp\left(-\frac{mv^2}{2k_B T}\right), \qquad (8.66)$$

as it had already been done in the framework of the classical canonical ensemble.

B. Classical limit in the Helmholtz formalism

The free energy of Helmholtz is given by the Legendre transformation

$$F = \Phi + \mu N, \qquad (8.67)$$

where the chemical potential is eliminated from the equation

$$N = -\left(\frac{\partial\Phi}{\partial\mu}\right)_{T,V}. \qquad (8.68)$$

Using the expression for the grand thermodynamic potential, given by equation (8.51), we have

$$F = -k_B T N \left\{ 1 + \ln \frac{V}{N} + \frac{3}{2} \ln T - \ln \left[\frac{1}{\gamma} \left(\frac{h^2}{2\pi m k_B} \right)^{3/2} \right] \right\}. \quad (8.69)$$

With $\gamma = 1$ (zero spin), this is the expression of the Helmholtz free energy as obtained in the context of the classical canonical ensemble with the introduction of the *ad-hoc* corrections (Boltzmann's correct counting factor $N!$ and the division of the phase space volume element $dqdp$ by Planck's constant h).

The entropy, in terms of temperature, volume, and number of particles, is given by

$$S = - \left(\frac{\partial F}{\partial T} \right)_{V,N} = N k_B \ln \frac{V}{N} + \frac{3}{2} k_B N \ln T + N s_o, \quad (8.70)$$

where

$$s_o = \frac{5}{2} k_B + k_B \ln \left[\frac{1}{\gamma} \left(\frac{h^2}{2\pi m k_B} \right)^{3/2} \right]. \quad (8.71)$$

Using the equation of state $pV = N k_B T$, we may write the entropy as a function of temperature, pressure, and number of particles, to obtain a particularly useful expression to deal with experimental data. In fact, expression (8.70), with the correct constant given by equation (8.71), is known as the Sackur–Tetrode equation, to celebrate the early pioneering works that contributed to the definitive acceptance of the canonical formalism of statistical mechanics. At this time, it was already feasible to use spectroscopic data to calculate the canonical partition function of diluted gases. All the constants in the expression of the entropy are relevant for comparisons between the spectroscopic entropy and the thermodynamic entropy obtained from calorimetric measurements.

C. Classical limit of the canonical partition function

From equation (8.61) we can write

$$\Xi_{cl} = \exp \left[z \sum_j \exp\left(-\beta\epsilon_j\right) \right] = \sum_{N=0}^{\infty} z^N Z_{cl}\left(\beta, V, N\right), \quad (8.72)$$

where $Z_{cl}\left(\beta, V, N\right)$ is the classical canonical partition function in terms of β, the volume V, and the number of particles N. Using the definition

$$Z_1 = \sum_j \exp\left(-\beta\epsilon_j\right), \quad (8.73)$$

we have

$$\exp[zZ_1] = \sum_{N=0}^{\infty} z^N Z_{cl}(\beta, V, N), \tag{8.74}$$

Therefore, the classical limit of the canonical partition function is given by

$$Z_{cl} = \frac{1}{N!} Z_1^N = \frac{1}{N!} \left[\sum_j \exp(-\beta \epsilon_j) \right]^N, \tag{8.75}$$

which is the expression of the canonical partition function associated with Boltzmann's model for a gas of particles, with the inclusion of the correct counting factor.

If the particles have a certain internal structure (as in the case of polyatomic molecules), the energy of orbital j may be written as

$$\epsilon_j = \epsilon_{\vec{k}} + \epsilon_{int} = \frac{\hbar^2 k^2}{2m} + \epsilon_{int}, \tag{8.76}$$

where the second term depends on the internal degrees of freedom. Hence, we have

$$Z_1 = \sum_j \exp(-\beta \epsilon_j) = Z_{int} \sum_{\vec{k}} \exp(-\beta \epsilon_{\vec{k}}). \tag{8.77}$$

Therefore,

$$Z_1 = Z_{int} \frac{V}{(2\pi)^3} \int d^3\vec{k} \exp\left(-\frac{\beta \hbar^2 k^2}{2m}\right) = Z_{int} V \left(\frac{2\pi m k_B T}{h^2}\right)^{3/2}. \tag{8.78}$$

For particles without an internal structure (such that $Z_{int} = 1$), we recover the results for a classical ideal monatomic gas,

$$Z_{cl} = \frac{1}{N!} V^N \left(\frac{2\pi m k_B T}{h^2}\right)^{3N/2}, \tag{8.79}$$

with the inclusion of all quantum corrections associated with the classical limit.

8.4 Diluted gas of diatomic molecules

Neglecting the interactions among the various degrees of freedom, the canonical partition function of a **diluted gas** of diatomic molecules can be written as

$$Z = \frac{1}{N!} V^N \left(\frac{2\pi m k_B T}{h^2}\right)^{3N/2} Z_{int}^N, \tag{8.80}$$

where $m = m_1 + m_2$ is the total mass of the molecule. For most of the heteromolecules (that is, composed of two distinct nuclei), the internal partition function is given by

$$Z_{int} = \gamma_1 \gamma_2 g_e Z_{rot} Z_{vib}, \qquad (8.81)$$

where $\gamma_i = (2S_i + 1)$, S_i is the spin of the nucleus i $(i = 1, 2)$, g_e is the degeneracy of the electronic ground state, and the canonical partition functions of rotation and vibration are given by the expressions

$$Z_{rot} = \sum_{J=0}^{\infty} (2J + 1) \exp\left[-\frac{\beta\hbar^2}{2I} J(J+1)\right], \qquad (8.82)$$

and

$$Z_{vib} = \sum_{n=0}^{\infty} \exp\left[-\beta\hbar\omega\left(n + \frac{1}{2}\right)\right] = \left[2\sinh\left(\frac{\beta\hbar\omega}{2}\right)\right]^{-1}, \qquad (8.83)$$

where the moment of inertia I and the fundamental frequency ω are related to the molecular parameters (and may be obtained, at least in principle, from spectroscopic data). Taking into account these expressions, we define the characteristic temperatures of rotation and vibration,

$$\Theta_r = \frac{\hbar^2}{2k_B I} \text{ and } \Theta_v = \frac{\hbar\omega}{k_B}. \qquad (8.84)$$

Equation (8.83), for the vibrational term, which is identical to the canonical partition function of Einstein's model, gives a contribution to the specific heat at constant volume of the form

$$c_{vib} = k_B \left(\frac{\Theta_v}{T}\right)^2 \frac{\exp(\Theta_v/T)}{[\exp(\Theta_v/T) - 1]^2}. \qquad (8.85)$$

At low temperatures $(T << \Theta_v)$, the specific heat vanishes exponentially. At high temperatures $(T >> \Theta_v)$, we recover the classical result from the equipartition theorem, $c_{vib} = k_B$. For most of the diatomic molecules, the characteristic temperature Θ_v is very high, and we thus have to use quantum results to explain the experimental behavior (see table).

In the limit of low temperatures $(T << \Theta_r)$, the rotational partition function may be represented by an expansion,

$$Z_{rot} = \sum_{J=0}^{\infty} (2J + 1) \exp\left[-\frac{\Theta_r}{T} J(J+1)\right] = 1 + 3\exp\left(-\frac{2\Theta_r}{T}\right) + \cdots, \qquad (8.86)$$

which leads to the free energy

$$f_{rot} = -\frac{1}{\beta} \ln Z_{rot} = -3k_B T \exp\left(-\frac{2\Theta_r}{T}\right) + \cdots. \qquad (8.87)$$

In this limit of low temperatures, the rotational specific heat at constant volume also vanishes exponentially, according to the asymptotic expression

$$c_{rot} = 12k_B \left(\frac{\Theta_r}{T}\right)^2 \exp\left(-\frac{2\Theta_r}{T}\right). \tag{8.88}$$

In the classical limit of high temperatures ($T \gg \Theta_r$), the spacing between rotational energy levels is very small. Therefore, substituting the sum by an integral in equation (8.82), we recover the classical result, $c_{rot} = k_B$, in agreement with the theorem of equipartition of energy. In this regime, however, we are able to make a more careful calculation. Given the well-behaved function $\varphi(x)$, we can write the Euler–MacLaurin expansion (see, for example, Chapter 13 of J. Mathews and R. L. Walker, *Mathematical Methods of Physics*, W. A. Benjamin, New York, 1965),

$$\sum_{n=0}^{\infty} \varphi(x) = \int_0^{\infty} \varphi(x)\,dx + \frac{1}{2}\varphi(0) - \frac{1}{12}\varphi'(0) + \frac{1}{720}\varphi'''(0) + \dots, \tag{8.89}$$

which establishes the asymptotic conditions for the transformation of the sum into an integral. Using this expansion, we can write equation (8.82) in the form

$$Z_{rot} = \frac{T}{\Theta_r} \left\{1 + \frac{1}{3}\left(\frac{\Theta_r}{T}\right) + \frac{1}{15}\left(\frac{\Theta_r}{T}\right)^2 + \dots\right\}. \tag{8.90}$$

Hence, the rotational contribution to the specific heat at constant volume is given by

$$c_{rot} = k_B \left\{1 + \frac{1}{45}\left(\frac{\Theta_r}{T}\right)^2 + \dots\right\}. \tag{8.91}$$

It is interesting to note that the specific heat tends to the classical value k_B from values larger than k_B. Indeed, for most of the diatomic molecules, the experimental graph of the rotational contribution for the specific heat in terms of temperature displays a characteristic broad maximum before reaching the classical value (in figure 8.1, we sketch the rotational specific heat at constant volume versus temperature).

In the following table, we indicate the values of Θ_r and Θ_v for some diatomic molecules.

Gas	Θ_r (K)	Θ_v (K $\times 10^3$)
H_2	85.4	6.10
N_2	2.86	3.34
O_2	2.07	2.23
CO	2.77	3.07
NO	2.42	2.69
HCl	15.2	4.14

FIGURE 8.1. Rotational specific heat versus temperature.

Except in the case of molecules with hydrogen, the characteristic temperature Θ_r is low, which supports the use of classical formulas (at room temperature). In general, Θ_v is high, which indicates the freezing of the vibrational degrees of freedom. For homonuclear molecules (as H_2 and D_2), the analysis is a bit more complicated, since we have to take into account the requirements of symmetry of the quantum states of identical particles (as, for example, in a well-known problem involving the isotopes of hydrogen).

Exercises

1. Obtain the explicit forms for the ground state and of the first excited state of a system of two free bosons, with zero spin, confined to a one-dimensional region of length L. Repeat this problem foe two fermions of spin $1/2$.

2. Show that the entropy of an ideal quantum gas may be written as

$$S = -k_B \sum_j \{f_j \ln f_j \pm (1 \mp f_j) \ln (1 \mp f_j)\},$$

where the upper (lower) sign refers to fermions (bosons), and

$$f_j = \langle n_j \rangle = \frac{1}{\exp\left[\beta\left(\epsilon_j - \mu\right)\right] \pm 1}$$

is the Fermi–Dirac (Bose–Einstein) distribution. Show that we can still use these equations to obtain the usual results for the classical ideal gas.

3. Show that the equation of state

$$pV = \frac{2}{3}U$$

holds for both free bosons and fermions (and also in the classical case). Show that an ideal ultrarelativistic gas, given by the energy spectrum $\epsilon = c\hbar k$, still obeys the same equation of state.

4. Consider a quantum ideal gas inside a cubic vessel of side L, and suppose that the orbitals of the particles are associated with wave functions that vanish at the surfaces of the cube. Find the density of states in \vec{k} space. In the thermodynamic limit, show that we have the same expressions as calculated with periodic boundary conditions.

5. An ideal gas of N atoms of mass m is confined to a vessel of volume V, at a given temperature T. Calculate the classical limit of the chemical potential of this gas.

 Now consider a "two-dimensional" gas of N_A free particles adsorbed on a surface of area A. The energy of an adsorbed particle is given by

 $$\epsilon_A = \frac{1}{2m}\vec{p}^2 - \epsilon_o,$$

 where \vec{p} is the (two-dimensional) momentum, and $\epsilon_o > 0$ is the binding energy that keeps the particle stuck to the surface. In the classical limit, calculate the chemical potential μ_A of the adsorbed gas.

 The condition of equilibrium between the adsorbed particles and the particles of the three-dimensional gas can be expressed in terms of the respective chemical potentials. Use this condition to find the surface density of adsorbed particles as a function of temperature and pressure p of the surrounding gas.

6. Obtain an expression for the entropy per particle, in terms of temperature and density, for a classical ideal monatomic gas of N particles of spin S adsorbed on a surface of area A. Obtain the expected values of \mathcal{H}, \mathcal{H}^2, \mathcal{H}^3, where \mathcal{H} is the Hamiltonian of the system. What are the expressions of the second and third moments of the Hamiltonian with respect to its average value?

7. Consider a homogeneous mixture of two ideal monatomic gases, at temperature T, inside a container of volume V. Suppose that there are N_A particles of gas A and N_B particles of gas B. Write an expression for the grand partition function associated with this system (it should depend on T, V, and the chemical potentials μ_A and μ_B). In the classical limit, obtain expressions for the canonical partition function, the Helmholtz free energy F, and the pressure p of the gas. Show that $p = p_A + p_B$ (Dalton's law), where p_A and p_B are the partial pressures of A and B, respectively.

8. Under some conditions, the amplitudes of vibration of a diatomic molecule may be very large, with a certain degree of anharmonicity. In this case, the vibrational energy levels are given by the approximate expression

$$\epsilon_n = \left(n + \frac{1}{2} \right) \hbar\omega - x \left(n + \frac{1}{2} \right)^2 \hbar\omega,$$

where x is the parameter of anharmonicity. To first order in x, obtain an expression for the vibrational specific heat of this system.

9. The potential energy between atoms of a hydrogen molecule may be described by the Morse potential,

$$V(r) = V_o \left\{ \exp \left[-\frac{2(r - r_o)}{a} \right] - 2 \exp \left[-\frac{r - r_o}{a} \right] \right\},$$

where $V_o = 7 \times 10^{-19}$ J, $r_o = 8 \times 10^{-11}$ m, and $a = 5 \times 10^{-11}$ m. Sketch a graph of $V(r)$ versus r. Calculate the characteristic temperatures of vibration and rotation to compare with experimental data (see table on Section 8.4).

8. Under some conditions, the amplitudes of vibration of a diatomic molecule may be very large, with a certain degree of anharmonicity. In this case, the vibrational energy levels are given by the approximate expression

$$\epsilon_n = \left(n + \frac{1}{2}\right)\hbar\omega - x\left(n + \frac{1}{2}\right)^2 \hbar\omega,$$

where x is the parameter of anharmonicity. To first order in x, obtain an expression for the vibrational specific heat of this system.

9. The potential energy between atoms of a hydrogen molecule may be described by the Morse potential

$$V(r) = V_0 \left\{ \exp\left[-\frac{2(r - r_0)}{a}\right] - 2\exp\left[-\frac{(r - r_0)}{a}\right] \right\}$$

where $V_0 = 7 \times 10^{-19}$ J, $r_0 = 8 \times 10^{-11}$ m, and $a = 5 \times 10^{-11}$ m. Sketch a graph of $V(r)$ versus r. Calculate the characteristic temperatures of vibration and rotation to compare with experimental data (see table on Section 8.4).

9
The Ideal Fermi Gas

Given the volume, the temperature, and the chemical potential, the grand partition function for free fermions may be written in the form

$$\ln \Xi (T, V, \mu) = \sum_j \ln \left\{ 1 + \exp \left[-\beta \left(\epsilon_j - \mu \right) \right] \right\}, \qquad (9.1)$$

where the sum is over all single-particle quantum states. The expected value of the occupation number of an orbital is given by

$$\langle n_j \rangle = \frac{1}{\exp \left[\beta \left(\epsilon_j - \mu \right) \right] + 1}, \qquad (9.2)$$

which is very often called the **Fermi–Dirac distribution**. The connection with thermodynamics, in the limit $V \to \infty$, is provided by the grand thermodynamic potential,

$$\Phi (T, V, \mu) = -\frac{1}{\beta} \ln \Xi (T, V, \mu). \qquad (9.3)$$

Since $\Phi = -pV$, we have

$$p (T, \mu) = k_B T \lim_{V \to \infty} \frac{1}{V} \ln \Xi (T, V, \mu). \qquad (9.4)$$

From the expression of Ξ, written in terms of β, $z = \exp \left(\beta \mu \right)$, and V, it is easy to obtain expressions for the expected value of energy (which corresponds, in the thermodynamic limit, to the internal energy of the system)

and for the expected value of the number of particles (corresponding to the thermodynamic number of particles). Thus, we have the internal energy,

$$U = \sum_j \epsilon_j \langle n_j \rangle = \sum_j \frac{\epsilon_j}{\exp\left[\beta\left(\epsilon_j - \mu\right)\right] + 1},\tag{9.5}$$

and the number of particles,

$$N = \sum_j \langle n_j \rangle = \sum_j \frac{1}{\exp\left[\beta\left(\epsilon_j - \mu\right)\right] + 1}.\tag{9.6}$$

In many cases, it is not convenient to work at fixed chemical potential. So, we may use equation (9.6) to obtain $\mu = \mu\left(T, V, N\right)$ for inserting into the expressions for the internal energy, the pressure, and the expected value of the occupation number of the orbitals. We will later analyze the gas of free electrons in terms of the more convenient variables T, V, and N.

For free fermions, in the absence of electromagnetic fields, the energy spectrum is given by

$$\epsilon_j = \epsilon_{\vec{k},\sigma} = \frac{\hbar^2 k^2}{2m}.\tag{9.7}$$

Thus, in the thermodynamic limit, we can write

$$\ln \Xi = \gamma \frac{V}{(2\pi)^3} \int d^3\vec{k} \ln \left\{ 1 + \exp\left[-\beta \left(\frac{\hbar^2 k^2}{2m} - \mu \right) \right] \right\},\tag{9.8}$$

where $\gamma = 2S + 1$ is the multiplicity of spin. Therefore, we have

$$\langle n_{\vec{k}} \rangle = \left[\exp\left(\frac{\beta\hbar^2 k^2}{2m} - \beta\mu \right) + 1 \right]^{-1},\tag{9.9}$$

$$N = \gamma \frac{V}{(2\pi)^3} \int d^3\vec{k} \left[\exp\left(\frac{\beta\hbar^2 k^2}{2m} - \beta\mu \right) + 1 \right]^{-1},\tag{9.10}$$

and

$$U = \gamma \frac{V}{(2\pi)^3} \int d^3\vec{k} \frac{\hbar^2 k^2}{2m} \left[\exp\left(\frac{\beta\hbar^2 k^2}{2m} - \beta\mu \right) + 1 \right]^{-1}.\tag{9.11}$$

Owing to the spherical symmetry of the energy spectrum of free fermions, we may rewrite these formulas using the energy ϵ as the integration variable. We thus have

$$\ln \Xi = \gamma V \int_0^\infty D\left(\epsilon\right) \ln \left\{ 1 + \exp\left[-\beta\left(\epsilon - \mu\right) \right] \right\} d\epsilon,\tag{9.12}$$

$$N = \gamma V \int_0^\infty D(\epsilon) f(\epsilon) \, d\epsilon, \tag{9.13}$$

and

$$U = \gamma V \int_0^\infty \epsilon D(\epsilon) f(\epsilon) \, d\epsilon, \tag{9.14}$$

where

$$f(\epsilon) = \frac{1}{\exp\left[\beta(\epsilon - \mu)\right] + 1} \tag{9.15}$$

is the Fermi–Dirac distribution function, and

$$D(\epsilon) = \frac{1}{4\pi^2} \left(\frac{2m}{\hbar^2}\right)^{3/2} \epsilon^{1/2} = C\epsilon^{1/2}, \tag{9.16}$$

where the constant C depends on the mass of the fermions. Later, we will show that the structure function $D(\epsilon)$ has a simple physical interpretation: $\gamma D(\epsilon)$ is a density (per energy and per volume) of the available orbitals or single-particle states. It should be pointed out that, with slight modifications, and taking care to treat separately the singular term in the sum (that is, the $\vec{k} = 0$ term), expressions (9.12)–(9.16) can be written as well for free bosons.

Now we obtain a simple result that holds for both free fermions and free bosons. Performing an integration by parts, we can use equation (9.12) to write

$$\ln \Xi = \gamma V C \int_0^\infty \epsilon^{1/2} \ln\left\{1 + e^{-\beta(\epsilon-\mu)}\right\} d\epsilon = \frac{2}{3}\beta\gamma V C \int_0^\infty \epsilon^{3/2} f(\epsilon) \, d\epsilon. \tag{9.17}$$

Therefore,

$$\ln \Xi = \frac{2}{3} \frac{\gamma V}{k_B T} \int_0^\infty \epsilon D(\epsilon) f(\epsilon) \, d\epsilon = \frac{2}{3} \frac{U}{k_B T}. \tag{9.18}$$

Since $\Phi = -kT \ln \Xi = -pV$, we finally have

$$U = \frac{3}{2} pV, \tag{9.19}$$

which holds even for the classical monatomic ideal gas of particles!

FIGURE 9.1. Fermi–Dirac distribution function $f(\epsilon)$ at absolute zero.

9.1 Completely degenerate ideal Fermi gas

We usually refer to a quantum gas in the ground state, at zero temperature, as completely degenerate. For fermions at zero temperature ($\beta \to \infty$), the most probable occupation number of the single-particle states is a step function of energy (see figure 9.1). The chemical potential at zero temperature, ϵ_F, is called **Fermi energy**. In the sketch of figure 9.1, we show that all the orbitals are occupied below the Fermi energy (for $\epsilon < \epsilon_F$), and empty for $\epsilon > \epsilon_F$.

Let us calculate the Fermi energy in terms of the number of particles N and the volume V. In the limit $\beta \to \infty$, equation (9.13) may be written as

$$N = \gamma V \int_0^{\epsilon_F} D(\epsilon) \, d\epsilon. \tag{9.20}$$

Now the meaning of the density $D(\epsilon)$ becomes clear. We just look at the graphs of $D(\epsilon)$ and of $D(\epsilon) f(\epsilon)$ as a function of energy ϵ, at zero temperature. The step function $f(\epsilon)$ cuts the curve $D(\epsilon)$ at the energy ϵ_F. Below the Fermi energy, all of the available single-particle states are occupied. Above the Fermi energy, they are empty. From equation (9.20), we have

$$N = \frac{2}{3} \gamma V \epsilon_F D(\epsilon_F), \tag{9.21}$$

from which we can write

$$\epsilon_F = \frac{\hbar^2}{2m} \left(\frac{6\pi^2}{\gamma} \right)^{2/3} \left(\frac{N}{V} \right)^{2/3}, \tag{9.22}$$

that is, the Fermi energy is inversely proportional to the mass of the particles, and increases with particle density at power 2/3.

From equation (9.14), the internal energy in the ground state is given by

$$U = \frac{2}{5} \gamma V D(\epsilon_F) \epsilon_F^2. \tag{9.23}$$

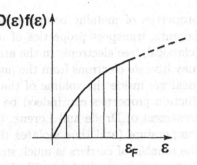

FIGURE 9.2. Graphical representation of the density of states, $D(\epsilon)$, and the function $D(\epsilon) f(\epsilon)$, versus energy ϵ, at absolute zero. The hatched region indicates the occupied orbitals (up to the Fermi energy ϵ_F).

Using equation (9.19), we obtain the pressure in the ground state,

$$p = \frac{2}{5}\frac{N}{V}\epsilon_F = \frac{\hbar^2}{5m}\left(\frac{6\pi^2}{\gamma}\right)^{2/3}\left(\frac{N}{V}\right)^{5/3}, \tag{9.24}$$

which is a particularly interesting result. Even at zero temperature, the Fermi gas is associated with a certain finite pressure, and thus has to be kept inside a container (later, we will see that the pressure vanishes in the ground state of free bosons). This direct consequence of Pauli's exclusion principle is related to the very stability of matter. We cannot explain a stable universe composed of bosons only.

From the Fermi energy, we define the Fermi temperature,

$$T_F = \frac{1}{k_B}\epsilon_F, \tag{9.25}$$

which provides another important characteristic value associated with a particular Fermi gas. Using equation (9.22), we have

$$T_F = \frac{\hbar^2}{2mk_B}\left(\frac{6\pi^2}{\gamma}\right)^{2/3}\left(\frac{N}{V}\right)^{2/3}. \tag{9.26}$$

The condition $T \gg T_F$ corresponds to the classical limit as analyzed in the last chapter (the typical interatomic dimension is much larger than the thermal wavelength). In many cases of interest, however, the Fermi temperature is very high, much larger than room temperature (that is, the system under consideration is very close to the completely degenerate state, in which quantum effects are extremely relevant). In the following section, we will see that in this regime it is possible to write the thermodynamic quantities as an asymptotic power series in terms of the small ratio T/T_F.

The theory of the ideal Fermi gas has many applications of physical interest. One of the best examples is the description of thermal and transport

properties of metallic compounds. According to the ideas of Drude and Lorentz, transport properties of metals may be explained by a model of (classical) free electrons. In the atoms of a metal, the outer electronic shell may liberate electrons from the nuclei to form, at a first approximation, an ideal gas inside the volume of the metallic sample. Although certain conduction properties can indeed be qualitatively described by the classical treatment of Drude and Lorenz, the specific heat of metals changes with temperature (and thus violates the classical equipartition theorem), and the number of carriers is much smaller than the classical predictions. The first quantum results for an ideal gas of spin 1/2 fermions were obtained by Sommerfeld in the 1930s. They gave rise to an explanation for most of the basic properties of metals, including the linear change with temperature of the specific heat at constant volume, and the role of electrons close to the Fermi energy in the conduction processes. Later, there appeared calculations for a gas of electrons in the presence of a crystalline potential, which lead to the concept of bands of energy, to the distinction between metallic conductors and insulators, and to the definition of a semiconductor. In the presence of a crystalline potential, the conduction electrons are not entirely free, since the band structure still reflects a certain relationship with the fixed nuclei.

Besides the metallic systems, there are other physical systems, as nuclear matter or the ionized gas in the interior of some stars, that may be explained by a model of free fermions. The Fermi temperature of an alkaline metal, as lithium or sodium, with just one conduction electron per atom, is typically of the order of 10^4 K (that is, much higher than room temperature). To give an idea of the orders of magnitude in these calculations, it is interesting to recall that $1\,\mathrm{eV}$ of energy corresponds to a temperature of approximately 10^4 K. Therefore, the Fermi energy of a metal is of the same order of magnitude as the ionization energy of an atom of hydrogen. For the electron gas associated with a typical write dwarf star, $T_F \sim 10^9$ K (much bigger than the temperature at the surface of the Sun, of the order of 10^5 K). For nuclear matter, the Fermi temperature is of the order of 10^{11} K.

9.2 The degenerate ideal Fermi gas $(T << T_F)$

In many cases of physical interest, room temperature is much lower than the Fermi temperatures. If we assume the approximate model of an ideal gas to describe conduction electrons in copper, the Fermi temperature is of the order of 8×10^4 K. It is then relevant to make an attempt to obtain analytical results for $T << T_F$.

In the limit $T << T_F$, the Fermi–Dirac distribution $f(\epsilon)$, as sketched in figure 9.3, has the shape of a slightly rounded step. As a function of

FIGURE 9.3. Fermi–Dirac distribution in the limit $T \ll T_F$.

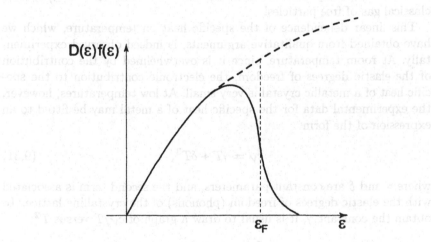

FIGURE 9.4. Function $D(\epsilon)\, f(\epsilon)$ versus energy ϵ in the limit $T \ll T_F$.

T, V, and N, the chemical potential μ should be slightly smaller than its value, ϵ_F, at zero temperature. The function $f(\epsilon) D(\epsilon)$ is sketched in figure 9.4. We can say that some electrons that were occupying states below the Fermi energy have been excited to single-particle states above ϵ_F. This change takes place within a small band of energies, of order $k_B T$, in the neighborhood of ϵ_F. Therefore, the total number of excited electrons will be given by

$$\Delta N \approx \gamma D(\epsilon_F) V k_B T. \tag{9.27}$$

The total change of energy of these electrons may be written as

$$\Delta U \approx k_B T \Delta N \approx \gamma V D(\epsilon_F) (k_B T)^2. \tag{9.28}$$

Thus, we have the specific heat at constant volume,

$$c_V = \frac{1}{N} \left(\frac{\partial U}{\partial T} \right)_{V,N} \approx 2\gamma \frac{V}{N} D(\epsilon_F) k_B^2 T. \tag{9.29}$$

Using equation (9.21), we may also write

$$c_V \approx 3k_B \frac{T}{T_F}, \tag{9.30}$$

which is completely different from the constant value, $c_V = 3k_B/2$, for a classical gas of free particles!

This linear dependence of the specific heat on temperature, which we have obtained from qualitative arguments, is indeed observed experimentally. At room temperature, since it is overwhelmed by the contribution of the elastic degrees of freedom, the electronic contribution to the specific heat of a metallic crystal is very small. At low temperatures, however, the experimental data for the specific heat of a metal may be fitted to an expression of the form

$$c_V = \gamma T + \delta T^3, \tag{9.31}$$

where γ and δ are constant parameters, and the second term is associated with the elastic degrees of freedom (phonons) of the crystalline lattice. To obtain the constant γ, it is usual to draw a graph of c_V/T versus T^2,

$$\frac{1}{T} c_V = \gamma + \delta T^2. \tag{9.32}$$

The linear extrapolation of the experimental data of this graph, in the limit $T^2 \to 0$, gives the value of γ. In the table below, we give theoretical (from calculations for a free electron gas) and experimental results for the constant γ (in units of 10^{-4} cal \times mol^{-1} \times K^{-2}) associated with some

metallic crystals (see Chapter 2 of N. W. Ashcroft and N. D. Mermin, *Solid State Physics*, Holt, Rinehart, and Winston, 1976).

Metal	γ_{theor}	γ_{exp}	$\gamma_{exp}/\gamma_{theor}$
Li	1.8	4.2	2.3
Na	2.6	3.5	1.3
K	4.0	4.7	1.2
Cu	1.2	1.6	1.3
Fe	1.5	12	8.0
Mn	1.5	40	27

For the alkaline metals and copper, the model of free electrons provides a reasonable explanation of the experimental results. Since the constant γ is proportional to the mass of electrons, it is usual in solid-state physics to correct the discrepancies by introducing the concept of an effective mass, associated with the presence of the crystalline potential. Some metals of magnetic character, as iron and manganese, with very large values of the constant γ, are farther away from the ideal behavior (these are the so-called heavy fermions).

We now introduce an expansion proposed by Sommerfeld to obtain an asymptotic form for the specific heat at constant volume of the gas of free fermions. We have to calculate integrals of the form

$$I = \int_0^\infty f(\epsilon)\,\phi(\epsilon)\,d\epsilon, \tag{9.33}$$

where $f(\epsilon)$ is the Fermi–Dirac distribution, $\phi(\epsilon) = A\epsilon^n$, A is a constant, and $n \geq 1/2$. As $f(\epsilon)$ has the approximate shape of a step, the derivative $f'(\epsilon)$ displays a rather pronounced peak at $\epsilon = \mu$. Making an integration by parts, we have

$$I = f(\epsilon)\,\phi(\epsilon)\big|_0^\infty - \int_0^\infty \psi(\epsilon)\,f'(\epsilon)\,d\epsilon, \tag{9.34}$$

with

$$\psi(\epsilon) = \int_0^\infty \phi(\epsilon')\,d\epsilon'. \tag{9.35}$$

Taking into account that $f(\epsilon)$ vanishes exponentially as $\epsilon \to \infty$, we have

$$I = -\int_0^\infty \psi(\epsilon)\,f'(\epsilon)\,d\epsilon. \tag{9.36}$$

Since $f'(\epsilon)$ displays a symmetric peak about $\epsilon = \mu$, we may write the expansion

$$\psi(\epsilon) = \psi(\mu) + \left(\frac{d\psi}{d\epsilon}\right)_{\epsilon=\mu} (\epsilon - \mu) + \ldots = \sum_{k=0}^{\infty} \frac{1}{k!} \left(\frac{d^k \psi}{d\epsilon^k}\right)_{\epsilon=\mu} (\epsilon - \mu)^k.$$

(9.37)

We thus have to calculate integrals of the form

$$I_k = -\int_0^\infty (\epsilon - \mu)^k f'(\epsilon)\, d\epsilon = \frac{1}{\beta^k} \int_{-\beta\mu}^\infty \frac{e^x x^k}{(e^x + 1)^2}\, dx,$$

(9.38)

for $k = 0, 1, 2, \ldots$. At low temperatures ($T \ll T_F$), μ is close to ϵ_F and the lower limit of these integrals may be taken as $-\infty$. Indeed, owing to the form of the integrand, the error in these operations is of exponential order $\exp(-\beta\epsilon_F)$. Therefore, we have

$$I_k = \frac{1}{\beta^k} \int_{-\infty}^\infty \frac{e^x x^k}{(e^x + 1)^2}\, dx + O[\exp(-\beta\epsilon_F)].$$

(9.39)

For odd k, this integral vanishes. For even k, except for exponential corrections, it is easy to show that

$$I_0 = 1 \quad \text{and} \quad I_2 = \frac{\pi^2}{3\beta^2}.$$

(9.40)

Thus, for $T \ll T_F$, we have the asymptotic result

$$I = \int_0^\mu \phi(\epsilon)\, d\epsilon + \frac{\pi^2}{6} (k_B T)^2 \left(\frac{d\phi}{d\epsilon}\right)_{\epsilon=\mu} + \ldots$$

(9.41)

Now we use expansion (9.41) to obtain asymptotic expressions for the internal energy and the thermodynamic number of particles,

$$U = \gamma V \int_0^\infty f(\epsilon)\, C\epsilon^{3/2} d\epsilon = \gamma V C \left\{ \frac{2}{5}\mu^{5/2} + \frac{\pi^2}{4} (k_B T)^2 \mu^{1/2} + \ldots \right\}$$

(9.42)

and

$$N = \gamma V \int_0^\infty f(\epsilon)\, C\epsilon^{1/2} d\epsilon = \gamma V C \left\{ \frac{2}{3}\mu^{3/2} + \frac{\pi^2}{12} (k_B T)^2 \mu^{-1/2} + \ldots \right\}.$$

(9.43)

This last expression may be rewritten in the form

$$\epsilon_F^{3/2} = \mu^{3/2} \left\{ 1 + \frac{\pi^2}{8} \left(\frac{k_B T}{\mu} \right)^2 + \cdots \right\}.$$ (9.44)

From this equation, it is not difficult to write an asymptotic expression for the chemical potential as a function of temperature T and density of particles N/V (which shows up through the expression of the Fermi temperature),

$$\mu = \epsilon_F \left\{ 1 - \frac{\pi^2}{12} \left(\frac{T}{T_F} \right)^2 + \cdots \right\}.$$ (9.45)

At low temperatures, we see that the chemical potential $\mu = \mu(T, V, N)$ is indeed smaller than ϵ_F. Inserting μ into equation (9.42) for the internal energy, we have

$$U = \frac{3}{5} N \epsilon_F \left\{ 1 + \frac{5\pi^2}{12} \left(\frac{T}{T_F} \right)^2 + \cdots \right\}.$$ (9.46)

Therefore, the specific heat at constant volume is given by the asymptotic expression

$$c_V = \frac{\pi^2}{2} k_B \frac{T}{T_F},$$ (9.47)

that differs from the heuristic form of equation (9.30) by just a small numerical prefactor.

9.3 Pauli paramagnetism

In a system of free electrons, the magnetic field couples to the intrinsic (spin) and the orbital angular momenta. Let us consider these contributions separately. We initially analyze the Zeeman effect, referring to the interactions between an external magnetic field and the permanent magnetic moments, which leads to the phenomenon of paramagnetism. The Hamiltonian of the free electron gas in the presence of a magnetic field is then given by

$$\mathcal{H} = \sum_{i=1}^{N} \left[\frac{1}{2m} \vec{p}_i^2 - g\mu_B \vec{H} \cdot \vec{S}_i \right],$$ (9.48)

where \vec{S}_i is an operator of spin $1/2$, $g = 2$ is the gyromagnetic factor, and μ_B is the Bohr magneton. For a single electron, with the z axis along the

direction of the field, we have

$$\mathcal{H}_1 = \frac{1}{2m}\vec{p}^2 - g\mu_B H S_z. \tag{9.49}$$

Therefore, the energy spectrum is given by

$$\epsilon_{\vec{k},\sigma} = \frac{\hbar^2 k^2}{2m} - \mu_B H\sigma, \tag{9.50}$$

with $\sigma = \pm 1$.

The grand partition function is written as

$$\ln \Xi = \sum_{\vec{k}}\sum_{\sigma} \ln\left\{1 + z\exp\left(-\frac{\beta\hbar^2 k^2}{2m} + \beta\mu_B H\sigma\right)\right\}$$

$$= VC\int_0^\infty \epsilon^{1/2}\left\{\sum_\sigma \ln\left[1 + z\exp\left(-\beta\epsilon + \beta\mu_B H\sigma\right)\right]\right\}d\epsilon. \tag{9.51}$$

Thus, we can write

$$\ln \Xi = \ln \Xi_+ + \ln \Xi_-, \tag{9.52}$$

where

$$\ln \Xi_{\pm} = VC\int_0^\infty \epsilon^{1/2}\left\{\ln\left[1 + z\exp\left(-\beta\epsilon \pm \beta\mu_B H\right)\right]\right\}d\epsilon. \tag{9.53}$$

The average number of electron will be given by

$$N = z\frac{\partial}{\partial z}\ln \Xi = \langle N_+ + N_-\rangle, \tag{9.54}$$

where

$$\langle N_{\pm}\rangle = VC\int_0^\infty \epsilon^{1/2}\left\{z^{-1}\exp\left(\beta\epsilon \mp \beta\mu_B H\right) + 1\right\}^{-1}d\epsilon. \tag{9.55}$$

We can also use the same notation to write the expected value of the magnetization,

$$M = \mu_B\langle N_+ - N_-\rangle. \tag{9.56}$$

Ground-state magnetization

From equation (9.55), in the limit $\beta \to \infty$, we have

$$\langle N_+\rangle = VC\int_0^{\epsilon_F + \mu H}\epsilon^{1/2}d\epsilon = VC\int_{-\mu_B H}^{\epsilon_F}(\epsilon - \mu_B H)^{1/2}d\epsilon$$

$$= \frac{2}{3}VC(\epsilon_F + \mu_B H)^{3/2} \tag{9.57}$$

and

$$\langle N_- \rangle = VC \int_0^{\epsilon_F - \mu_B H} \epsilon^{1/2} d\epsilon = VC \int_{\mu_B H}^{\epsilon_F} (\epsilon + \mu_B H)^{1/2} d\epsilon$$

$$= \frac{2}{3} VC (\epsilon_F - \mu_B H)^{3/2} , \qquad (9.58)$$

where ϵ_F is the chemical potential at $T = 0$ (that is, the Fermi energy). To provide a pictorial understanding of this result, we say that the application of a magnetic field displaces the density of states for electrons with spin up ($\sigma = +1$), along the energy axis, by a quantity $-\mu_B H$, while the density of states for electrons with spin down is displaced in the opposite direction, by the same amount, $+\mu_B H$. Therefore, we have the total number of electrons,

$$N = \langle N_+ + N_- \rangle = \frac{2}{3} VC \left[(\epsilon_F + \mu_B H)^{3/2} + (\epsilon_F - \mu_B H)^{3/2} \right] , \qquad (9.59)$$

and the total magnetization,

$$M = \mu_B \langle N_+ - N_- \rangle = \frac{2}{3} VC \mu_B \left[(\epsilon_F + \mu_B H)^{3/2} - (\epsilon_F - \mu_B H)^{3/2} \right] . \qquad (9.60)$$

For weak fields ($\mu_B H \ll \epsilon_F$), we write

$$N = \frac{4}{3} VC \epsilon_F^{3/2} + O \left[\left(\frac{\mu_B H}{\epsilon_F} \right)^2 \right] \qquad (9.61)$$

and

$$M = 2VC \mu_B \epsilon_F^{3/2} \left(\frac{\mu_B H}{\epsilon_F} \right) + O \left[\left(\frac{\mu_B H}{\epsilon_F} \right)^3 \right] . \qquad (9.62)$$

The leading term of the magnetization is given by

$$M = \frac{3}{2} N \mu_B \frac{\mu_B H}{\epsilon_F} , \qquad (9.63)$$

From this expression, we have the ground-state susceptibility in zero field,

$$\chi_o = \left(\frac{\partial M}{\partial H} \right)_{T=0, V, N, H=0} = \frac{3 N \mu_B^2}{2 \epsilon_F} , \qquad (9.64)$$

which is one of the characteristic results of the **Pauli paramagnetism**.

Magnetization in the degenerate limit

At finite temperatures, the magnetization is given by

$$M = \mu_B \langle N_+ - N_- \rangle = -\frac{1}{\beta}\frac{\partial \ln \Xi}{\partial H}$$

$$= \mu_B VC \int_0^\infty \epsilon^{1/2}\left[f\left(\epsilon - \mu_B H\right) - f\left(\epsilon + \mu_B H\right)\right]. \qquad (9.65)$$

Now we make an expansion for weak fields ($\mu_B H << \epsilon_F$). At leading order, we have

$$M = -2VC\mu_B^2 H \int_0^\infty \epsilon^{1/2} f'\left(\epsilon\right) d\epsilon = VC\mu_B^2 H \int_0^\infty \epsilon^{-1/2} f\left(\epsilon\right) d\epsilon. \qquad (9.66)$$

From equation (9.54), at leading order, we also have

$$N = 2VC \int_0^\infty \epsilon^{1/2} f\left(\epsilon\right) d\epsilon. \qquad (9.67)$$

In the grand canonical formalism that we have been using, both the magnetization M and the number of particles N are functions of T, V, H, and μ. In order to calculate the susceptibility, we have to eliminate the chemical potential μ, which can be done from equation (9.67). In the degenerate limit ($T << T_F$), we obtain analytical results using the same expansion as proposed by Sommerfeld to calculate the asymptotic form of the specific heat of free electrons. We thus have

$$M = 2\mu_B^2 VCH\mu^{1/2}\left[1 - \frac{\pi}{24}\left(\frac{k_B T}{\mu}\right)^2 + \dots\right] \qquad (9.68)$$

and

$$N = \frac{4}{3}VC\mu^{3/2}\left[1 + \frac{\pi^2}{8}\left(\frac{k_B T}{\mu}\right)^2 + \dots\right]. \qquad (9.69)$$

From this last expansion, we have

$$\mu = \epsilon_F\left[1 - \frac{\pi^2}{12}\left(\frac{k_B T}{\epsilon_F}\right)^2 + \dots\right]. \qquad (9.70)$$

Inserting into equation (9.68), we write

$$M = \frac{3N\mu_B^2 H}{2\epsilon_F}\left[1 - \frac{\pi^2}{12}\left(\frac{k_B T}{\epsilon_F}\right)^2 + \dots\right], \qquad (9.71)$$

from which we obtain an expansion for the magnetic susceptibility in zero field,

$$\chi_o = \frac{3N\mu_B^2}{2\epsilon_F} \left[1 - \frac{\pi^2}{12} \left(\frac{T}{T_F} \right)^2 + \dots \right], \qquad (9.72)$$

which includes the first correction, due to thermal fluctuations, to the result for the ground state. In his pioneering work, Pauli obtained this type of expression to account for the weak dependence on temperature of the magnetic susceptibility of alkaline metals (whose Fermi temperature T_F is very large).

Classical limit

It is instructive to obtain the classical (high-temperature) limit of the expressions for the magnetization and the number of particles. In this limit, with $z \ll 1$, we have

$$f(\epsilon) = \left[z^{-1} e^{\beta\epsilon} + 1 \right]^{-1} \to z e^{-\beta\epsilon}. \qquad (9.73)$$

Therefore, the limiting expressions for the magnetization and the number of particles are given by

$$M = \mu_B V C z 2 \sinh(\beta\mu_B H) \int_0^\infty \epsilon^{1/2} \exp(-\beta\epsilon) \, d\epsilon \qquad (9.74)$$

and

$$N = 2VCz \cosh(\beta\mu_B H) \int_0^\infty \epsilon^{1/2} \exp(-\beta\epsilon) \, d\epsilon. \qquad (9.75)$$

Eliminating the chemical potential, we have

$$M = N\mu_B \tanh\left(\frac{\mu_B H}{kT} \right), \qquad (9.76)$$

from which we write the susceptibility in zero field,

$$\chi_o = \left(\frac{\partial M}{\partial T} \right)_{H=0} = \frac{N\mu_B^2}{kT}. \qquad (9.77)$$

This is the well-known **Curie law** for paramagnetic crystals. We have to be careful, however, to consider the real meaning of this classical limit. To be more precise, the classical susceptibility vanishes, since Bohr's magneton depends linearly on the Planck's constant h. Expression (9.77) may be regarded as a kind of classical limit where the spin operators have been replaced by permanent magnetic moments of intensity $\mu = \mu_B$.

9.4 Landau diamagnetism

To explain the phenomenon of diamagnetism, let us now take into account the interaction between the external magnetic field and the orbital motion of the electrons. If we disregard the spin term, the Hamiltonian of a particle of mass m and electric charge q, in the presence of a magnetic field \vec{H}, is given by

$$\mathcal{H} = \frac{1}{2m} \left(\vec{p} - \frac{q}{c} \vec{A} \right)^2 , \qquad (9.78)$$

where \vec{A} is the vector potential associated with the field. In the context of classical statistical mechanics, however, there is no possibility of occurrence of diamagnetic effects. This may be easily shown if we write the partition function in classical phase space,

$$Z_1 = \int d^3\vec{r} \int d^3\vec{p} \exp\left[-\frac{\beta}{2m} \left(\vec{p} - \frac{q}{c}\vec{A} \right)^2 \right] . \qquad (9.79)$$

A simple change of variables, $\vec{p} \to \vec{p} - (q/c)\vec{A}$, leads to a trivial form of Z_1 for the classical ideal gas, without any dependence on the magnetic field. Diamagnetism is thus a purely quantum phenomenon. From the Lorentz force, we calculate the frequency of rotation, $\omega = qH/mc$, of a particle in the presence of a uniform magnetic field. The quantization of the orbits of these particles gives rise to the phenomenon of Landau diamagnetism.

Consider a uniform magnetic field along the z direction,

$$\vec{H} = H\vec{k}, \qquad (9.80)$$

and choose the vector potential in the gauge of Landau,

$$\vec{A} = xH\vec{j}. \qquad (9.81)$$

Therefore, the Hamiltonian of a single particle may be written as

$$\mathcal{H} = \frac{1}{2m} \left[p_x^2 + \left(p_y - \frac{qH}{c}x \right)^2 + p_z^2 \right] . \qquad (9.82)$$

If we choose a wave function of the form

$$\psi = \exp\left(ik_y y\right) \exp\left(ik_z z\right) \varphi\left(x\right), \qquad (9.83)$$

the time-independent Schrödinger equation, $\mathcal{H}\psi = \epsilon\psi$, can be written as

$$\frac{1}{2m} \left[p_x^2 + \left(\hbar k_y - \frac{qH}{c}x \right)^2 \right] \varphi\left(x\right) = \left(\epsilon - \frac{\hbar^2 k_z^2}{2m} \right) \varphi\left(x\right). \qquad (9.84)$$

Introducing the change of variables

$$x' = x - \frac{c\hbar}{qH} k_y, \qquad (9.85)$$

equation (9.84) can be rewritten in the form

$$\left[\frac{1}{2m} (p_{x'})^2 + \frac{q^2 H^2}{2mc^2} (x')^2 \right] \varphi(x') = \left(\epsilon - \frac{\hbar^2 k_z^2}{2m} \right) \varphi(x'). \qquad (9.86)$$

The problem is then reduced to the consideration of the Schrödinger equation associated with a one-dimensional harmonic oscillator of fundamental frequency

$$\omega = \frac{|q| H}{mc}, \qquad (9.87)$$

which is identical to the classical Larmor frequency. We thus rewrite equation (9.86) in the form

$$\left[\frac{1}{2m} (p_{x'})^2 + \frac{1}{2} m\omega^2 (x')^2 \right] \varphi_n(x') = \hbar\omega \left(n + \frac{1}{2} \right) \varphi_n(x'), \qquad (9.88)$$

where $n = 0, 1, 2, ...$, and

$$\varphi_n(x') = H_n(x') \exp \left[-\frac{1}{2} \alpha (x')^2 \right], \qquad (9.89)$$

where H_n is a Hermite polynomial, and

$$\alpha = \frac{|q| H}{\hbar c}. \qquad (9.90)$$

From these considerations, we can write the energy spectrum,

$$\epsilon = \epsilon(n, k_z) = \frac{\hbar^2 k_z^2}{2m} + \hbar\omega \left(n + \frac{1}{2} \right), \qquad (9.91)$$

and the functions of Landau,

$$\psi = \psi_n^{k_y, k_z}(x, y, z) = \exp(ik_y y) \exp(ik_z z) \varphi_n \left(x - \frac{1}{\alpha} k_y \right), \qquad (9.92)$$

where it should be noted the enormous degeneracy with respect to k_y.

To calculate the factor of degeneracy, let us compare with the spectrum of free electrons, in the absence of an applied magnetic field,

$$\epsilon_{free} = \frac{\hbar^2}{2m} \left(k_x^2 + k_y^2 + k_z^2 \right). \qquad (9.93)$$

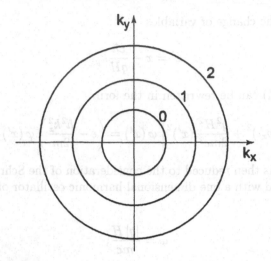

FIGURE 9.5. Available orbits for a system of free electrons in the presence of an applied magnetic field along the z direction.

In the presence of a magnetic field, the states of the free electrons collapse, on the $k_x - k_y$ plane, into a certain number of allowed orbits (see figure 9.5), with the square of the radius given by

$$k_x^2 + k_y^2 = \frac{2m}{\hbar^2}\hbar\omega\left(n + \frac{1}{2}\right) = \frac{2eH}{\hbar c}\left(n + \frac{1}{2}\right), \tag{9.94}$$

where e is the elementary charge of the electron. If we consider the system within a cube of side L, with a surface density of states $2\left(L/2\pi\right)^2$, up to the level $n = 0$, we have

$$2\left(\frac{L}{2\pi}\right)^2 \pi\frac{eH}{\hbar c} = \frac{eHL^2}{hc} \tag{9.95}$$

electronic states. Up to the level $n = 1$, we have

$$2\left(\frac{L}{2\pi}\right)^2 \pi3\frac{eH}{\hbar c} = 3\frac{eHL^2}{hc} \tag{9.96}$$

electronic states. Up to level $n = 2$, we have

$$2\left(\frac{L}{2\pi}\right)^2 \pi5\frac{eH}{\hbar c} = 5\frac{eHL^2}{hc} \tag{9.97}$$

electronic states, and so on. Therefore, each Landau level may be associated with the degeneracy factor

$$g = 2\frac{eHL^2}{hc}. \tag{9.98}$$

Taking into account this degeneracy, we rewrite the energy spectrum in the form

$$\epsilon = \epsilon(n, k_z, \delta) = \frac{\hbar^2 k_z^2}{2m} + \hbar\omega\left(n + \frac{1}{2}\right), \tag{9.99}$$

where $k_z = -\infty, ..., +\infty$, with spacing $L/2\pi$, $n = 0, 1, 2, ...$, and $\delta = 1, 2, ..., g$.

We can finally write the grand canonical partition function,

$$\ln\Xi = 2\frac{eHL^2}{hc}\sum_{n=0}^{\infty}\frac{L}{2\pi}\int_{-\infty}^{+\infty} dk_z \ln\left\{1 + z\exp\left[-\frac{\beta\hbar^2 k_z^2}{2m} - \frac{\beta\hbar eH}{mc}\left(n + \frac{1}{2}\right)\right]\right\}. \tag{9.100}$$

High-temperature limit

In the limit of high temperatures ($z << 1$), the leading term of the grand canonical partition function is given by

$$\ln\Xi = 2\frac{eHL^2}{hc}\frac{L}{2\pi}z\left(\frac{2\pi m}{\beta\hbar^2}\right)^{1/2}\sum_{n=0}^{\infty}\exp\left[-\frac{\beta\hbar eH}{mc}\left(n + \frac{1}{2}\right)\right]. \tag{9.101}$$

Thus,

$$\ln\Xi = \frac{eHL^3}{hc}\lambda z\left[\sinh\left(\beta\mu_B H\right)\right]^{-1}, \tag{9.102}$$

where

$$\mu_B = \frac{e\hbar}{2mc} \tag{9.103}$$

is the Bohr magneton, and

$$\lambda = \frac{h}{(2\pi mk_B T)^{1/2}} \tag{9.104}$$

is the thermal wavelength. The classical (high-temperature) expressions for the number of particles and the magnetization are given by

$$N = z\frac{\partial}{\partial z}\ln\Xi = \frac{eHL^3}{hc\lambda}z\left[\sinh\left(\beta\mu_B H\right)\right]^{-1} \tag{9.105}$$

and

$$M = -\frac{\partial\Phi}{\partial H} = \frac{1}{\beta}\frac{\partial}{\partial H}\ln\Xi = \frac{eL^3 z}{\beta hc\lambda}\left\{\frac{1}{\sinh\left(\beta\mu_B H\right)} - \frac{\beta\mu_B H\cosh\left(\beta\mu_B H\right)}{\sinh^2\left(\beta\mu_B H\right)}\right\}. \tag{9.106}$$

Using equation (9.105) to eliminate the chemical potential, the magnetization can be written as

$$M = -N\mu_B \mathcal{L}(\beta\mu_B H),\tag{9.107}$$

where

$$\mathcal{L}(x) = \coth(x) - \frac{1}{x}\tag{9.108}$$

is the **Langevin function**. It is peculiar that, in the limit of high temperatures, the magnetization is given by the same expression as obtained from the classical Langevin model for paramagnetism, but with the opposite sign. Note that this sign of the magnetization does not depend on the sign of the charge. In the limit of weak fields, $\beta\hbar e H \ll mc$, we have the leading term

$$M = -\frac{N\mu_B^2 H}{k_B T}.\tag{9.109}$$

Therefore, the susceptibility in zero field is given by

$$\chi_o = -\frac{1}{3}\frac{N\mu_B^2}{k_B T}.\tag{9.110}$$

It should be noted that this expression has no classical analog (in fact, χ_o vanishes for $h = 0$). Also, it should be pointed out that χ_o is just one third, with the opposite sign, of the Pauli susceptibility in the regime of high temperatures.

The de Haas–van Alphen effect

We now ignore the k_z direction, and consider the Landau levels on the $k_x - k_y$ plane. Thus we write the energy spectrum

$$\epsilon_n = 2\mu_B H \left(n + \frac{1}{2}\right),\tag{9.111}$$

with degeneracy

$$g = \frac{NH}{H_o},\tag{9.112}$$

where

$$H_o = \frac{hc}{2q}\frac{N}{L^2}.\tag{9.113}$$

In the ground state, with $H > H_o$, the particles may be placed in the lowest Landau level. The total energy will be given by $E_o = N\mu_B H$.

For $H < H_o$, some particles are forced to occupy higher-energy Landau levels. Let us suppose that, given a value of the field H, the first j levels are fully occupied, and the $j + 1$ level is only partially occupied. Then, we have

$$(j+1)\,g < N < (j+2)\,g, \tag{9.114}$$

that is,

$$\frac{H_o}{j+2} < H < \frac{H_o}{j+1}. \tag{9.115}$$

For a magnetic field in this interval, we have the ground-state energy

$$E_o = g \sum_{i=0}^{j} \epsilon_i + [N - (j+1)\,g]\,\epsilon_{j+1} = N\mu_B H \left[2j + 3 - (j+1)(j+2)\frac{H}{H_o} \right]. \tag{9.116}$$

Hence, we have

$$\frac{1}{N} E_o = \mu_B H,$$

for $H > H_o$, and

$$\frac{1}{N} E_o = \mu_B H \left[2j + 3 - (j+1)(j+2)\frac{H}{H_o} \right],$$

for

$$\frac{H_o}{j+2} < H < \frac{H_o}{j+1},$$

with $j = 0, 1, 2,$ Taking derivatives of this expression with respect to magnetic field, we obtain the magnetization and the magnetic susceptibility. In particular, the magnetization per electron versus H/H_o displays the sawtooth shape sketched in the graph of figure 9.6. In terms of H_o/H, there is an oscillatory behavior, with a well-defined period, which is known as the de Haas–van Alphen effect. At very low but finite temperatures, the oscillations are more sinusoidal than sawtooth in form. At higher temperatures, the distribution of electron populations tends to average out these oscillations. A more-detailed study of this oscillatory with the inverse of the magnetic field is, however, beyond the scope of this text. Taking into account the crystalline potential, it can be shown that there appear several periods of oscillation, which are proportional to the extremal sections of the Fermi surface along a direction normal to the magnetic field. In this way the de Haas-van Alphen effect provides a powerful technique to carry out a "tomographic" reconstruction of the Fermi surface of metallic crystals.

FIGURE 9.6. Magnetization per particle versus the inverse of the applied magnetic field in the ground state of the Landau model for diamagnetism.

Exercises

1. What is the compressibility of a gas of free fermions at zero temperature? Obtain the numerical value for electrons with the density of conduction electrons in metallic sodium. Compare your results with experimental data for sodium at room temperature.

2. An ideal gas of fermions, with mass m and Fermi energy ϵ_F, is at rest at zero temperature. Find expressions for the expected values $\langle v_x \rangle$ and $\langle v_x^2 \rangle$, where \vec{v} is the velocity of a fermion.

3. Consider a gas of free electrons, in a d-dimensional space, within a hypercubic container of side L. Sketch graphs of the density of states $D(\epsilon)$ versus energy ϵ for dimensions $d = 1$ and $d = 2$. What is the expression of the Fermi energy in terms of the particle density for $d = 1$ and $d = 2$?

4. Show that the chemical potential of an ideal classical gas of N monatomic particles, in a container of volume V, at temperature T, may be written as

$$\mu = k_B T \ln \left(\frac{\lambda^3}{v} \right),$$

where $v = V/N$ is the volume per particle, and $\lambda = h/\sqrt{2\pi m k_B T}$ is the thermal wavelength. Sketch a graph of $\mu/k_B T$ versus T. Obtain the first quantum correction to this result. That is, show that the chemical potential of the ideal quantum gas may be written as the

expansion

$$\frac{\mu}{k_B T} - \ln\left(\frac{\lambda^3}{v}\right) = A\ln\left(\frac{\lambda^3}{v}\right) + B\ln\left(\frac{\lambda^3}{v}\right)^2 + \dots,$$

and obtain explicit expressions for the prefactor A for fermions and bosons. Sketch a graph of $\mu/k_B T$ versus λ^{-2} (that is, versus the temperature in convenient units) for fermions, bosons, and classical particles.

5. Obtain an asymptotic form, in the limit $T \ll T_F$, for the specific heat of a gas of N free fermions adsorbed on a surface of area A, at a given temperature T.

6. Consider a gas of N free electrons, in a region of volume V, in the ultrarelativistic regime. The energy spectrum is given by

$$\epsilon = \left[p^2 c^2 + m^2 c^4\right]^{1/2} \approx pc,$$

where \vec{p} is the linear momentum.

(a) Calculate the Fermi energy of this system.

(b) What is the total energy in the ground state?

(c) Obtain an asymptotic form for the specific heat at constant volume in the limit $T \ll T_F$.

7. At low temperatures, the internal energy of a system of free electrons may be written as an expansion,

$$U = \frac{3}{5} N\epsilon_F \left\{ 1 + \frac{5\pi^2}{12}\left(\frac{T}{T_F}\right)^2 - A\left(\frac{T}{T_F}\right)^4 + \dots \right\}.$$

Obtain the value of the constant A, and indicate the order of magnitude of the terms that have been discarded.

8. Consider a system of free fermions in d dimensions, with the energy spectrum

$$\epsilon_{\vec{k},\sigma} = c\left|\vec{k}\right|^a,$$

where $c > 0$ and $a > 1$.

(a) Calculate the prefactor A of the relation $pV = AU$.

(b) Calculate the Fermi energy as a function of volume V and number of particles N.

(c) Calculate an asymptotic expression, in the limit $T \ll T_F$, for the specific heat at constant volume.

9. Consider again the gas of N ultrarelativistic free electrons, within a container of volume V, at temperature T, in the presence of a magnetic field \vec{H}. If we neglect the effects of orbital magnetism, the energy spectrum is given by

$$\epsilon_{\vec{p},\sigma} = cp - \mu_B H \sigma,$$

where μ_B is the Bohr magneton and $\sigma = \pm 1$.

(a) Show that the Fermi energy of this system may be written as

$$\epsilon_F = A + BH^2 + O\left(H^4\right).$$

Obtain expressions for the prefactors A and B.

(b) Show that the magnetization in the ground state can be written in the form

$$M = CH + O\left(H^3\right).$$

Obtain an expression for the constant C.

(c) Calculate the susceptibility of the ground state in zero field.

10. In the classical paramagnetic theory of Langevin, proposed before the advent of quantum statistics, we assume a classical Hamiltonian, given by

$$\mathcal{H} = -\sum_{i=1}^{N} \vec{\mu}_i \cdot \vec{H} = -\sum_{i=1}^{N} \mu H \cos \theta_i,$$

where $\vec{\mu}_i$ is the magnetic moment of a localized ion.

(a) Show that the canonical partition function of this system is given by

$$Z = Z_1^N = \left\{ \int d\Omega \exp\left(\beta \mu H \cos \theta\right) \right\}^N,$$

where $d\Omega$ is the elementary solid angle of integration.

(b) Show that the magnetization (along the direction of the field) is given by

$$M = N \langle \mu \cos \theta \rangle = N \mu \mathcal{L}\left(\beta \mu H\right),$$

where

$$\mathcal{L}\left(x\right) = \coth\left(x\right) - \frac{1}{x}$$

is the Langevin function.

(c) Show that the susceptibility in zero field is given by the Curie law,

$$\chi_o = \frac{N\mu^2}{3kT}.$$

11. Obtain an expression for the magnetic susceptibility associated with the orbital motion of free electrons in the presence of a uniform magnetic field H, under conditions of strong degeneracy, $T \ll T_F$, and very weak fields, $\mu_B H \ll k_B T$. To simplify the expression of Ξ, you may use Euler's sum rule,

$$\sum_{n=0}^{\infty} f\left(n + \frac{1}{2}\right) \approx \int_0^{\infty} f(x)\, dx + \frac{1}{24} f'(0).$$

(c) Show that the susceptibility in zero field is given by the Curie law

$$\chi_0 = \frac{N\mu^2}{3kT}$$

11. Obtain an expression for the magnetic susceptibility associated with the orbital motion of free electrons in the presence of a uniform magnetic field B, under conditions of strong degeneracy $T \ll T_F$, and very weak fields, $\mu_B B \ll k_B T$. To simplify the expression of E_n, you may use Euler's sum rule.

$$\sum_{n=0}^{\infty} f\left(n + \frac{1}{2}\right) \approx \int_0^{\infty} f(x)dx + \frac{1}{24}f'(0)$$

10

Free Bosons: Bose–Einstein Condensation; Photon Gas

The thermodynamic properties of an ideal gas of bosons can be obtained from the grand partition function,

$$\ln \Xi (T, V, \mu) = - \sum_j \ln \{1 - \exp[-\beta (\epsilon_j - \mu)]\}, \tag{10.1}$$

where the sum is over all single-particle states. In the thermodynamic limit, the pressure as a function of temperature and chemical potential is given by

$$p(T, \mu) = -k_B T \lim_{V \to \infty} \frac{1}{V} \ln \Xi (T, V, \mu). \tag{10.2}$$

From the grand partition function, we derive the expected value of the occupation number of the orbitals,

$$\langle n_j \rangle = \frac{1}{\exp[\beta (\epsilon_j - \mu) - 1]}, \tag{10.3}$$

the thermodynamic number of particles,

$$N = \sum_j \langle n_j \rangle = \sum_j \frac{1}{\exp[\beta (\epsilon_j - \mu) - 1]}, \tag{10.4}$$

and the internal energy,

$$U = \sum_j \epsilon_j \langle n_j \rangle = \sum_j \frac{\epsilon_j}{\exp[\beta (\epsilon_j - \mu) - 1]}. \tag{10.5}$$

These equations are meaningful for $\epsilon_j - \mu > 0$ only, that is, for a strictly negative chemical potential, $\mu < 0$. At very high temperatures (in the classical limit), it is easy to check that the chemical potential is always negative. In the quantum context, with a fixed number of particles, the chemical potential is still negative, but increases with decreasing temperatures, and may eventually vanish, which gives rise to a peculiar phenomenon known as **Bose–Einstein condensation**. This phase transition resembles the superfluid transition in liquid helium, and the condensation of Cooper pairs in the BCS theory of superconductivity.

In this chapter, we initially study the Bose–Einstein transition in a system of free bosons of mass m. We then proceed to consider some examples of bosonic systems of zero mass (photons, phonons, and magnons), which are characterized by the presence of a quantum vacuum that does not require the conservation of the number of (quasi) particles. In the formalism that we are using, the presence of this quantum vacuum leads to a vanishing chemical potential that simplifies the expression of the grand partition function. A gas of free photons comes from the quantization of the electromagnetic field, and leads to the famous Planck law for the distribution of the black body radiation. The study of the photon gas is an excellent example of the application of the laws of electromagnetism, and of the limitations of the classical canonical formalism. Free phonons and free magnons will be left for the next chapter. Free phonons are quasi-particles associated with the quantization of the crystalline vibrations in the harmonic approximation. The gas of free phonons leads to an explanation of the behavior of the specific heat of solids at low temperatures. Magnons are quasi-particles associated with the quantization of "spin waves" (which are related to the small deviations of the local magnetization of a ferromagnet with respect to the thermodynamic magnetization of the ordered ferromagnetic state).

10.1 Bose–Einstein condensation

From equation (10.4), we have the chemical potential μ in terms of the temperature T and the density of particles $\rho = N/V$. Unfortunately, we cannot write an analytical expression for the function $\mu = \mu(T, \rho)$. However, in the classical limit, that works well at high temperatures, it easy to show that

$$\frac{\mu}{k_B T} = \ln\left[\frac{1}{\gamma}\left(\frac{2\pi\hbar^2}{mk_B}\right)^{3/2}\right] + \ln\left(\frac{N}{V}\right) - \frac{3}{2}\ln T, \qquad (10.6)$$

where $\gamma = 2S + 1$ is the multiplicity of spin. Therefore, at fixed density and high enough temperature, the chemical potential is negative. From a numerical calculation, we can draw the graphs of the chemical potential as a function of temperature, at a given density, for free fermions, bosons,

FIGURE 10.1. Chemical potential versus temperature for free fermions and bosons at fixed density. The solid line corresponds to the classical limit. Temperature T_o defines the onset of Bose–Einstein condensation.

and classical particles, as sketched in figure 10.1. At high temperatures, the three curves are identical. As the temperature decreases, the chemical potential of free fermions may become positive (as well as the chemical potential of classical particles), but the chemical potential of free bosons reaches the limit $\mu \to 0-$, at a given temperature T_o, and "sticks" to the value $\mu = 0$ for all $T \leq T_o$.

To obtain the temperature T_o, let us make $\mu = 0$ in equation (10.4). Using the energy spectrum of free particles,

$$\epsilon_j = \epsilon_{\vec{k},\sigma} = \frac{\hbar^2 k^2}{2m}, \tag{10.7}$$

and recalling that the sum may transformed into a convergent integral (in the thermodynamic limit), we have

$$N = \gamma V C \int_0^\infty \frac{\epsilon^{1/2} d\epsilon}{\exp\left(\beta_o \epsilon\right) - 1}, \tag{10.8}$$

where the constant C is given by

$$C = \frac{1}{4\pi^2} \left(\frac{2m}{\hbar^2}\right)^{3/2}, \tag{10.9}$$

as in the case of fermions. Introducing the change of variables $x = \beta_o \epsilon$, and using the mathematical result

$$\int\limits_0^\infty \frac{x^{1/2}dx}{e^x - 1} = \Gamma\left(\frac{3}{2}\right)\zeta\left(\frac{3}{2}\right),$$

(10.10)

we have

$$T_o = \frac{\hbar^2}{2mk_B}\left[\frac{4\pi^2}{\gamma\Gamma\left(\frac{3}{2}\right)\zeta\left(\frac{3}{2}\right)}\right]^{3/2}\left(\frac{N}{V}\right)^{2/3},$$

(10.11)

known as the **Bose–Einstein temperature**. Below T_o, the gas of free bosons displays some very peculiar features, which have been associated with the superfluid behavior of liquid helium, although it may not be so reasonable to neglect the interactions between particles in such a dense and strongly interacting system as liquid helium. Recently, it has been detected the existence of a Bose–Einstein condensate in a diluted gas of magnetically confined atoms of alkaline elements (that are cooled by evaporation below temperatures of the order of 200 nanokelvins!). A diluted system of this kind is closer to a free gas of bosons than either superfluid helium or the Cooper pairs in metallic superconductors [see Eric A. Cornell and Carl E. Wieman, "The Bose–Einstein condensate", Scientific American **278**, 26 (1998)].

We can calculate the order of magnitude of the Bose–Einstein temperature from a purely heuristic argument. Let us suppose that $\epsilon_o \approx k_B T_o$ is the zero-point energy to localize a particle in a region of volume V/N, with typical dimension

$$\Delta x \approx \left(\frac{V}{N}\right)^{1/3}.$$

As $\Delta x \Delta p \approx h$, we have

$$\Delta p \approx h\left(\frac{N}{V}\right)^{1/3}.$$

From the zero-point energy,

$$\epsilon_o \approx k_B T_o \approx \frac{1}{2m}\left(\Delta p\right)^2,$$

we have

$$T_o \approx \frac{h^2}{2mk_B}\left(\frac{N}{V}\right)^{2/3},$$

which is of the same order of magnitude as the Bose–Einstein temperature, given by equation (10.11).

To investigate the behavior in the limit $\mu \to 0-$, with $T \leq T_o$, let us look again at the expression for the number of particles, given by the sum in equation (10.4). As a matter of fact, we should have written this equation in the form

$$\frac{N}{V} = \left[\frac{1}{V}\frac{z}{1-z}\right] + \frac{1}{V}\sum_{j \neq 0}\frac{1}{z^{-1}\exp(\beta\epsilon_j) - 1}. \tag{10.12}$$

In this expression, the limit $z = \exp(\beta\mu) \to 1$ should be taken together with the thermodynamic limit, $V \to \infty$. We have already seen that there is no difficulty to transform this sum into an integral (the lower limit of integration, for small energies, does not pose any problem). At fixed ratio N/V, for $T \leq T_o$, in the limit $\mu \to 0$ and $V \to \infty$, we then write

$$\left[\frac{1}{V}\frac{z}{1-z}\right] \to \frac{N_o}{V}, \tag{10.13}$$

where N_o/V is the density of particles in a state of zero energy (that is, in the **condensate of Bose–Einstein**), and

$$\frac{1}{V}\sum_{j \neq 0}\frac{1}{z^{-1}\exp(\beta\epsilon_j) - 1} \to \gamma C \int_0^\infty \frac{\epsilon^{1/2}d\epsilon}{\exp(\beta\epsilon) - 1} = \frac{N_e}{V}, \tag{10.14}$$

where N_e/V is the density of particles in the excited states. Therefore, we have

$$N = N_o + N_e. \tag{10.15}$$

As equation (10.8) fixes the total number of particles, we can also write

$$\frac{N}{V} = \gamma C \int_0^\infty \frac{\epsilon^{1/2}d\epsilon}{\exp(\beta_o\epsilon) - 1}, \tag{10.16}$$

where $1/\beta_o = k_B T_o$. Thus, we also have

$$N_o = N\left[1 - \left(\frac{T}{T_o}\right)^{3/2}\right]. \tag{10.17}$$

In figure 10.2 we draw N_o/N versus temperature T, along the coexistence region between a macroscopic condensate and the excited states (that is, for $\mu = 0$, with $T \leq T_o$, in a phase diagram in terms of μ versus T). We

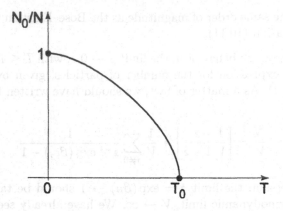

FIGURE 10.2. Fraction of particles belonging to the Bose–Einstein condensate as a function of temperature along the coexistence region ($\mu = 0$, with $T < T_o$).

should note that $N_o \to N$ for $T \to 0$, and $N_o \to 0$ for $T \to T_o$. In the neighborhood of T_o, we still can write

$$\frac{N_o}{N} \sim \frac{3}{2} \frac{T_o - T}{T_o}. \tag{10.18}$$

A more-sophisticated approach to the gas of bosons indicates that $(N_o/N)^{1/2}$ is proportional to the "order parameter" of the transition. This quantity vanishes for $T \geq T_o$, and behaves according to the asymptotic law $(T_o - T)^\beta$, with the critical exponent $\beta = 1/2$, as $T \to T_o-$. As it will be shown later (see Chapter 12), this asymptotic behavior is associated with second-order phase transitions in the mean-field approximation. The temperature T_o, however, depends on the density of particles, and does not define a standard critical point (at least in the usual way, as in the case of the liquid–gas phase transition). If we draw a phase diagram in terms of the thermodynamic fields μ and T, there is a line of coexistence (that is, of first-order phase transitions) between the Bose–Einstein condensate and the excited states, along the axis $\mu = 0$, up to the temperature T_o (see figure 10.3). This line of first-order transitions, however, cannot be crossed, since the gas of free bosons has no physical meaning for $\mu < 0$ (although we may expect that the introduction of interactions between particles will correct these non-physical aspects of the ideal model). It is important to point out that, in contrast to the case of the standard condensation of a fluid in real space, the Bose–Einstein transition represents a condensation of particles in momentum space (there is a macroscopic occupation of single-particle states with zero momentum).

(a) (b)

FIGURE 10.3. In figure (a), we sketch the phase diagram of a simple fluid, in the neighborhood of the critical point, in terms of chemical potential and temperature (the solid line represents the coexistence between two distinct phases, which become identical at the critical point). In figure (b), we sketch the same type of diagram for a gas of free bosons (it makes no sense to cross the coexistence line, which is now identified with the $\mu = 0$ axis).

Phase diagram of liquid helium

Helium is the only substance that does not solidify, at atmospheric pressure, even at the lowest temperatures. Liquid helium displays a phase transition between a normal liquid phase and a superfluid phase, which has been historically associated with the Bose–Einstein condensation (see, for example, the classical text of Fritz London, *Superfluids*, volume I, Dover, New York, 1961). We are always referring to the more common isotope of helium, ^4He, which is a boson. The less common isotope, ^3He, which is a fermion, does not condense, but displays a number of very peculiar properties (at much lower temperatures).

 In figure 10.4, we sketch the phase diagram of helium in terms of the intensive variables pressure and temperature. The solid lines are transitions of first order, that is, lines of coexistence of two distinct thermodynamic phases, which are characterized by different densities, but by the same intensive thermodynamic parameters (and the same Gibbs free energy per particle). The dashed line represents a continuous, second-order, transition between the normal liquid phase (I) and the superfluid phase (II). In this diagram we indicate the standard liquid–gas critical point (C), with critical temperature $T_c \approx 5.2\,\mathrm{K}$ and critical pressure $p_c \approx 2\,\mathrm{atm}$, and the so-called λ-point, given by $T_\lambda \approx 2.17\,\mathrm{K}$ and $p_\lambda \approx 1\,\mathrm{atm}$, with the specific volume $v_\lambda \approx 46.2\,\text{Å}^3$. Unlike the standard first-order liquid-gas transitions, which are associated with a latent heat of transition, and refer to phases with distinct densities, the superfluid transition is of second order, since there is a continuous transformation of phase II into phase I. Extremely careful measurements of the specific heat at constant volume in the neighborhood of the λ-point were carried out by Buckingham and Fairbanks in the 1950s

FIGURE 10.4. Schematic representation of the phase diagram of helium in terms of pressure and temperature. The dashed line corresponds to a second-order (continuous) transition between a normal (He I) and a superfluid (He II) phase, and C is a standard critical point.

(see figure 10.5). The shape of these experimental curves is associated with the name of the λ-point. In the immediate vicinity of T_λ, even more precise measurements of the specific heat have been fitted to an asymptotic expression of the form

$$c_V = A + B \ln |T - T_\lambda|,$$

where the coefficients may be different above and below the critical temperature. It should be pointed out that this fitting of the specific heat data to a diverging law is quite impressive. It does work from $t = |T - T_\lambda|/T_\lambda$ of the order 10^{-1} down to values of the order $10^{-6} - 10^{-5}$ (see, for example, M. J. Buckingham and W. M. Fairbanks, in *Progress in Low Temperature Physics*, edited by C. J. Gorter, North Holland, Amsterdam 1961).

As a curiosity, let us calculate the Bose–Einstein temperature T_o, given by equation (10.11), and using the available data for helium at the λ-point ($m = 4M$, where M is the mass of a proton, and $V/N = v_\lambda \approx 46.2$ Å3). With these values, we have $T_o \approx 3.14\,\mathrm{K}$, which is quite close to the temperature of the λ-point. There is no doubt that this is not a mere coincidence. In fact, there are indeed similarities between the superfluid phase of helium and the condensed phase of the ideal Bose gas (that is, the Bose gas at zero chemical potential and temperature below T_o). Also, there are large differences, as we have already pointed out, since it is not realistic to neglect the interactions between particles in a quantum liquid. More realistic models for the superfluid transition in liquid helium, however, are far beyond the scope of this text.

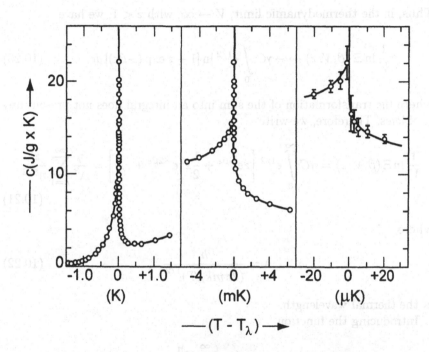

FIGURE 10.5. Measurements of the specific heat at constant volume of helium in the neighborhood of the λ-point (the curves display indeed a λ shape). Note that the critical temperature is sharply defined. This figure is based on the work of M. J. Buckingham and W. M. Fairbank [see *Progress in low temperature physics*, volume 3, edited by C. J. Gorter, North Holland, Amsterdam, 1961].

Free bosons in the normal region ($\mu < 0$)

For a strictly negative chemical potential, we can write the logarithm of the grand partition function of the ideal Bose gas as a power series of the fugacity $z = \exp(\beta\mu)$. From equation (10.1), we have

$$\frac{1}{V}\ln\Xi\left(\beta, V, z\right) = -\frac{1}{V}\ln\left(1 - z\right) - \frac{1}{V}\sum_{j\neq0}\ln\left[1 - z\exp\left(-\beta\epsilon_j\right)\right]. \quad (10.19)$$

Thus, in the thermodynamic limit, $V \to \infty$, with $z < 1$, we have

$$\frac{1}{V}\ln\Xi\left(\beta, V, z\right) \to -\gamma C\int_0^\infty \epsilon^{1/2}\ln\left[1 - z\exp\left(-\beta\epsilon\right)\right]d\epsilon, \quad (10.20)$$

where the transformation of the sum into an integral does not present any problems. Therefore, we write

$$\frac{1}{V}\ln\Xi\left(\beta, V, z\right) = \gamma C\int_0^\infty \epsilon^{1/2}\left[ze^{-\beta\epsilon} + \frac{1}{2}z^2 e^{-2\beta\epsilon} + \ldots\right] = \frac{\gamma}{\lambda^3}\sum_{n=1}^\infty \frac{z^n}{n^{5/2}}, \quad (10.21)$$

where

$$\lambda = \frac{h}{\left(2\pi m k_B T\right)^{1/2}} \quad (10.22)$$

is the thermal wavelength.

 Introducing the function

$$g_\alpha\left(z\right) = \sum_{n=1}^\infty \frac{z^n}{n^\alpha}, \quad (10.23)$$

we can write the more compact form

$$\frac{1}{V}\ln\Xi\left(\beta, V, z\right) = \frac{\gamma}{\lambda^3}g_{5/2}\left(z\right). \quad (10.24)$$

Therefore,

$$N = z\frac{\partial}{\partial z}\ln\Xi\left(\beta, V, z\right) = \frac{\gamma V}{\lambda^3}g_{3/2}\left(z\right) \quad (10.25)$$

and

$$U = -\frac{\partial}{\partial\beta}\ln\Xi\left(\beta, V, z\right) = \frac{3\gamma V}{2\beta\lambda^3}g_{5/2}\left(z\right). \quad (10.26)$$

To obtain the internal energy U as a function of the more-convenient variables T, V, and N, we use equation (10.25) to eliminate z (which can be done numerically).

As an example, let us calculate the specific heat at constant volume, given by

$$c_V = c_V(T, v) = \frac{1}{N}\left(\frac{\partial U}{\partial T}\right)_{V,N} = -\frac{k_B \beta^2}{N}\left(\frac{\partial U}{\partial \beta}\right)_{V,N}. \qquad (10.27)$$

To carry out this calculation, we need the function U in terms of β, V, and N. However, we just have U as a function of β, V, and the fugacity z (and of N as a function of β, V, and z). Now we may use the technique of Jacobians (see Appendix A.4). If we omit the dependence on volume, which is always fixed, we may write

$$\left(\frac{\partial U}{\partial \beta}\right)_N = \frac{\partial(U,N)}{\partial(\beta,N)} = \frac{\partial(U,N)}{\partial(\beta,z)}\frac{\partial(\beta,z)}{\partial(\beta,N)} = \left(\frac{\partial U}{\partial \beta}\right)_z - \left(\frac{\partial U}{\partial z}\right)_\beta \frac{\left(\frac{\partial N}{\partial \beta}\right)_z}{\left(\frac{\partial N}{\partial z}\right)_\beta}. \qquad (10.28)$$

Using equations (10.25) and (10.26), we are prepared to calculate all of these derivatives. Thus, we have

$$\left(\frac{\partial U}{\partial \beta}\right)_z = -\frac{15\gamma V}{4\beta^2 \lambda^3}g_{5/2}(z); \quad \left(\frac{\partial U}{\partial z}\right)_\beta = \frac{3\gamma V}{2\beta\lambda^3 z}g_{3/2}(z); \qquad (10.29)$$

$$\left(\frac{\partial N}{\partial \beta}\right)_z = -\frac{3\gamma V}{2\beta\lambda^3}g_{3/2}(z); \quad \left(\frac{\partial N}{\partial z}\right)_\beta = \frac{\gamma V}{\lambda^3 z}g_{1/2}(z). \qquad (10.30)$$

Hence, the specific heat at constant volume is given by

$$c_V = \frac{3}{2}k_B\left\{\frac{5}{2}\frac{g_{5/2}(z)}{g_{3/2}(z)} - \frac{3}{2}\frac{g_{3/2}(z)}{g_{1/2}(z)}\right\}, \qquad (10.31)$$

where equation (10.25) should be used to eliminate the fugacity. In the classical limit, $g(z) \approx z$. Therefore, $c_V \approx (3/2)k_B$, in agreement with the theorem of equipartition of energy. At the Bose–Einstein transition ($z = 1$, $T = T_o$), c_V is finite, since $g_{1/2}(1) \to \infty$, $g_{3/2}(1) = \xi(3/2) = 2.612...$, and $g_{5/2}(1) = \xi(5/2) = 1.342...$.

As an additional example, let us obtain an expression for the entropy as a function T, V, and the fugacity z. In the formalism of the grand thermodynamic potential, we have

$$S = -\left(\frac{\partial \Phi}{\partial T}\right)_{V,\mu} = V\left(\frac{\partial p}{\partial T}\right)_\mu. \qquad (10.32)$$

From equation (10.24), we have the pressure in terms of temperature and chemical potential,

$$p = p(T, \mu) = \frac{\gamma}{\lambda^3 \beta} g_{5/2}(z).$$ (10.33)

Therefore,

$$S = \frac{k_B \gamma V}{\lambda^3} \left\{ \frac{5}{2} g_{5/2}(z) - g_{3/2}(z) \ln z \right\}.$$ (10.34)

Using equation (10.25), we can also write

$$S = k_B N \left\{ \frac{5}{2} \frac{g_{5/2}(z)}{g_{3/2}(z)} - \ln z \right\}.$$ (10.35)

Free bosons in the coexistence region ($\mu = 0$, $T < T_o$)

As the condensate has zero energy, the calculations along the coexistence line are rather simple. From equations (10.25) and (10.26), the internal energy of the system and the number of particles in the excited states are given by

$$U = \frac{3\gamma V}{2\beta\lambda^3} g_{5/2}(1)$$ (10.36)

and

$$N_e = \frac{\gamma V}{\lambda^3} g_{3/2}(1).$$ (10.37)

The specific heat at constant volume is written as

$$c_V = \frac{1}{N} \left(\frac{\partial U}{\partial T} \right)_{V,N} = \frac{15\gamma V k_B}{4\lambda^3 N} g_{5/2}(1) = c \left(\frac{V}{N} \right) T^{3/2},$$ (10.38)

where c is a constant prefactor. For $T \to 0$, c_V behaves as $T^{3/2}$. The specific heat of helium, on the other hand, is characterized by an asymptotic law of the form T^3, as in the Debye model for crystalline solids (see Chapter 11). From equation (10.38), we also see that there is no discontinuity in the curve of c_V versus T as we go through the temperature of the Bose–Einstein transition.

The pressure may be obtained from the logarithm of the grand partition function, given by equation (10.19). This calculation, however, requires some extra attention. For $\mu \to 0$ and $V \to \infty$, according to equation (10.13), which gives the density of particles in the condensate, we have

$$\frac{1}{V} \frac{z}{1-z} \to \frac{1}{V} \frac{1}{1-z} \to \frac{N_o}{V},$$ (10.39)

where N_o/V assumes a finite value. Therefore, we also have

$$\frac{1}{V} \ln (1 - z) \to 0. \tag{10.40}$$

Thus, if we consider equation (10.19), the pressure along the coexistence is given by

$$p = \frac{1}{\beta V} \ln \Xi (\beta, V, z) \to \frac{\gamma C}{\beta} \int_0^\infty \epsilon^{1/2} \ln \left[1 - e^{-\beta \epsilon} \right] d\epsilon = \frac{\gamma}{\beta \lambda^3} g_{5/2} (1) .$$

$$\tag{10.41}$$

In contrast to the case of a Fermi gas, the pressure of bosons vanishes at absolute zero.

To calculate the entropy, we write

$$S = - \left(\frac{\partial \Phi}{\partial T} \right)_{V, \mu} = V \left(\frac{\partial p}{\partial T} \right)_\mu . \tag{10.42}$$

Thus,

$$S (T, V, \mu = 0) = V \left(\frac{\partial p}{\partial T} \right)_{\mu = 0} = \frac{5 k_B \gamma V}{2 \lambda^3} g_{5/2} (1) , \tag{10.43}$$

which also comes from a direct integration of equation (10.38) for the specific heat. The entropy vanishes as $T^{3/2}$ at absolute zero, and the condensate has no entropy.

10.2 Photon gas. Planck statistics

The origins of quantum theory can be traced to the **problem of the black body radiation**, consisting in the attempt to explain the distribution of the intensity of electromagnetic radiation emitted from a small cavity. Let us define $u(\nu)$ such that $u(\nu) d\nu$ is the quantity of energy per volume emitted by the cavity within the interval of frequencies between ν and $\nu + d\nu$. It is reasonable to assume that, as far as there is a situation of equilibrium, $u(\nu)$ does not depend on the way the walls of the cavity emit and absorb electromagnetic energy. Indeed, by the end of the nineteenth century, it was experimentally known that $u(\nu)$ depends on the equilibrium temperature T, but does depend on the geometrical form of the cavity. The radiation is detected if we make a small hole in the cavity, and measure the flux of energy, $I(\nu) = c u(\nu)$, where c is the velocity of light. We then have the typical experimental curve sketched in figure 10.6, where we also indicate the theoretical predictions from the Rayleigh-Jeans law.

FIGURE 10.6. Density of energy as a function of frequency for the black body radiation. The dashed line corresponds to the Rayleigh–Jeans law.

Within the theoretical framework of electromagnetism and classical statistical mechanics, it is possible to derive the Rayleigh-Jeans law, according to which $u(\nu)$ is proportional to ν^2, in agreement with the experimental results at low frequencies, but which leads to deep trouble. As the integral of $u(\nu)$ over the frequency ν diverges, there would be an infinite amount of energy within the cavity! This discrepancy between the Rayleigh–Jeans law and the experimental results has been known as the **ultraviolet catastrophe**.

From thermodynamic arguments, of a phenomenological character, it is possible to show that

$$u(\nu) = \nu^3 f\left(\frac{\nu}{T}\right),\tag{10.44}$$

but it is not possible to establish the form of the function $f(\nu/T)$. Using the equations of state,

$$pV = \frac{2}{3}U \text{ and } U = Vu(T),\tag{10.45}$$

where U is the total internal energy of the system, it is also possible to derive the Stefan–Boltzmann law,

$$u(T) = \int_0^\infty u(\nu)\, d\nu = \sigma T^4,\tag{10.46}$$

where σ is a constant. Besides these thermodynamic results, by the end of the nineteenth century, it was also known the Wien formula,

$$u(\nu) = \nu^3 \exp\left(-\frac{h\nu}{k_B T}\right),\tag{10.47}$$

where the constants h and k_B were later associated with Planck and Boltzmann, respectively. Wien's empirical formula provided a good account of $u(\nu)$ in the region of higher frequencies.

Spectral decomposition of the electromagnetic field

Now we use Maxwellian electromagnetism to obtain the spectral decomposition of the energy associated with the electromagnetic field. The energy within an empty box of volume V is given by

$$\mathcal{H} = \frac{1}{8\pi} \int_V \left(\vec{E}^2 + \vec{H}^2 \right) d^3\vec{r}, \tag{10.48}$$

where the fields \vec{E} and \vec{H} are functions of position \vec{r} and time t, and obey the Maxwell equations,

$$\nabla \times \vec{E} = -\frac{1}{c}\frac{\partial \vec{H}}{\partial t}; \quad \nabla \cdot \vec{E} = 0, \tag{10.49}$$

and

$$\nabla \times \vec{H} = \frac{1}{c}\frac{\partial \vec{E}}{\partial t}; \quad \nabla \cdot \vec{H} = 0. \tag{10.50}$$

As a first step, let us determine how the energy is distributed among the frequencies within the cavity at a given temperature T.

Introducing the Hertz potentials, we write

$$\vec{H} = \nabla \times \vec{A} \tag{10.51}$$

and

$$\vec{E} = -\frac{1}{c}\frac{\partial \vec{A}}{\partial t} - \nabla\phi. \tag{10.52}$$

Maxwell's equations for the divergent of the magnetic field and the rotational of the electric field are automatically fulfilled. The remaining equations are then used to define the potentials \vec{A} and ϕ. We should note, however, that these potentials are defined except for a change of gauge. Indeed, the new potentials,

$$\vec{A} \to \vec{A} - \nabla\psi \quad \text{and} \quad \phi \to \phi + \frac{1}{c}\frac{\partial \psi}{\partial t}, \tag{10.53}$$

where ψ is an arbitrary function, lead to exactly the same fields \vec{E} and \vec{H}. As we have an enormous freedom to choose any function ψ, it is always possible to require that

$$\nabla \cdot \vec{A} = 0, \tag{10.54}$$

which is known as the **gauge of Coulomb**. With this kind of choice, we have

$$\nabla^2 \vec{A} - \frac{1}{c^2}\frac{\partial^2 \vec{A}}{\partial t^2} = \frac{1}{c}\frac{\partial}{\partial t}\nabla\phi \tag{10.55}$$

and

$$\nabla^2 \phi = 0. \tag{10.56}$$

As there are no charges within the cavity, we take the solution $\phi = 0$, so that

$$\nabla^2 \vec{A} - \frac{1}{c^2} \frac{\partial^2 \vec{A}}{\partial t^2} = 0, \tag{10.57}$$

and the fields are given by

$$\vec{H} = \nabla \times \vec{A} \quad \text{and} \quad \vec{E} = -\frac{1}{c} \frac{\partial \vec{A}}{\partial t}. \tag{10.58}$$

As we have already mentioned, experimental data indicate that neither the geometry nor the particular materials of the cavity have any influence on the equilibrium results for $u(\nu)$. Therefore, let us consider a rectangular cavity of sides L_1, L_2, and L_3. Also, let us assume periodic boundary conditions. It all works as if the space has been filled out by a collection of similar boxes, such that the fields have the same values in the corresponding points of different boxes. From the thermodynamic point of view, periodic boundary conditions are justified because the walls are just meant to restrict the loss (and gain) of energy. The periodicity of the fields has precisely the same effect, since this way the rectangular cavities cannot gain energy from other cavities. With these periodic boundary conditions, we can perform a Fourier analysis of the problem. We thus write the Fourier expansion

$$\vec{A} = \sum_{\vec{k}} \vec{A}_{\vec{k}} \exp \left(i \vec{k} \cdot \vec{r} \right), \tag{10.59}$$

where the wave vector \vec{k} is given by

$$\vec{k} = (k_x, k_y, k_z) = \left(\frac{2\pi m}{L_1}, \frac{2\pi n}{L_2}, \frac{2\pi l}{L_3} \right), \tag{10.60}$$

with m, n, $l = 0, \pm 1, \pm 2, \ldots$. As the vector potential is real, we also write

$$\vec{A}_{\vec{k}} = \vec{A}^*_{-\vec{k}}. \tag{10.61}$$

From the condition of transversality, $\nabla \cdot \vec{A} = 0$, we have

$$\vec{k} \cdot \vec{A}_{\vec{k}} = 0, \tag{10.62}$$

that is, the complex vector $\vec{A}_{\vec{k}}$ is normal to the wave vector \vec{k}. We then have the wave equation,

$$\frac{d^2 \vec{A}_{\vec{k}}}{dt^2} + c^2 k^2 \vec{A}_{\vec{k}} = 0, \tag{10.63}$$

from which we recognize the harmonic character of the vibrations of the electromagnetic field.

The solutions of equation (10.63) may be written as

$$\vec{A}_{\vec{k}} = \vec{a}_{\vec{k}} \exp\left(i\omega_{\vec{k}}t\right) + \vec{b}_{\vec{k}} \exp\left(-i\omega_{\vec{k}}t\right), \tag{10.64}$$

where the frequency spectrum is given by

$$\omega_{\vec{k}} = ck. \tag{10.65}$$

Taking into account the restrictions imposed by equation (10.61), we can write equation (10.64) as

$$\vec{A}_{\vec{k}} = \vec{a}_{\vec{k}} \exp\left(i\omega_{\vec{k}}t\right) + \vec{a}^{*}_{-\vec{k}} \exp\left(-i\omega_{\vec{k}}t\right). \tag{10.66}$$

Therefore, inserting into equation (10.59), the vector potential is given by the real expression

$$\vec{A} = \sum_{\vec{k}} \left[\vec{a}_{\vec{k}} \exp\left(i\omega_{\vec{k}}t + i\vec{k}\cdot\vec{r}\right) + c.c.\right], \tag{10.67}$$

where c.c. means the complex conjugate of the previous term. The fields \vec{E} and \vec{H} are finally given by

$$\vec{E} = -\frac{1}{c}\frac{\partial\vec{A}}{\partial t} = \sum_{\vec{k}} \left[-ik\vec{a}_{\vec{k}} \exp\left(i\omega_{\vec{k}}t + i\vec{k}\cdot\vec{r}\right) + c.c.\right] \tag{10.68}$$

and

$$\vec{H} = \nabla\times\vec{A} = \sum_{\vec{k}} \left[i\left(\vec{k}\times\vec{a}_{\vec{k}}\right) \exp\left(i\omega_{\vec{k}}t + i\vec{k}\cdot\vec{r}\right) + c.c.\right]. \tag{10.69}$$

Now we have to obtain \vec{E}^{2} and \vec{H}^{2}, and to integrate over the volume. Using the normalization property

$$\int_{V} \exp\left(i\vec{k}\cdot\vec{r} + i\vec{k}'\cdot\vec{r}\right) d^{3}\vec{r} = V\delta_{\vec{k},-\vec{k}'}, \tag{10.70}$$

it is not difficult to show that

$$\int_{V} \vec{E}^{2}d^{3}\vec{r} = \sum_{\vec{k}} \left[-Vk^{2}\vec{a}_{\vec{k}}\cdot\vec{a}_{-\vec{k}} \exp\left(2i\omega_{\vec{k}}t\right) + Vk^{2}\vec{a}_{\vec{k}}\cdot\vec{a}^{*}_{\vec{k}}\right.$$
$$\left. + Vk^{2}\vec{a}^{*}_{\vec{k}}\cdot\vec{a}_{\vec{k}} - Vk^{2}\vec{a}^{*}_{\vec{k}}\cdot\vec{a}^{*}_{-\vec{k}} \exp\left(-2i\omega_{\vec{k}}t\right)\right]. \tag{10.71}$$

Also, using properties of the mixed vector product, and the transversality of the electromagnetic fields, we can show that

$$
\int_V \vec{H}^2 d^3\vec{r} = \sum_{\vec{k}} \left[V k^2 \vec{a}_{\vec{k}} \cdot \vec{a}_{-\vec{k}} \exp\left(2i\omega_{\vec{k}} t\right) + V k^2 \vec{a}_{\vec{k}} \cdot \vec{a}_{\vec{k}}^* \right.
$$
$$
\left. + V k^2 \vec{a}_{\vec{k}}^* \cdot \vec{a}_{\vec{k}} + V k^2 \vec{a}_{\vec{k}}^* \cdot \vec{a}_{-\vec{k}}^* \exp\left(-2i\omega_{\vec{k}} t\right) \right]. \quad (10.72)
$$

Thus, we finally write

$$
\mathcal{H} = \frac{1}{8\pi} \int_V \left(\vec{E}^2 + \vec{H}^2 \right) d^3\vec{r} = \frac{V}{2\pi} \sum_{\vec{k}} k^2 \vec{a}_{\vec{k}} \cdot \vec{a}_{\vec{k}}^*, \quad (10.73)
$$

which gives the spectral decomposition of the electromagnetic field within the cavity of volume V.

The classical solution

To use the classical canonical formalism, let us define the generalized coordinates of position,

$$
\vec{Q}_{\vec{k}}(t) = \alpha \left[\vec{a}_{\vec{k}} \exp\left(i\omega_{\vec{k}} t\right) + \vec{a}_{\vec{k}}^* \exp\left(-i\omega_{\vec{k}} t\right) \right], \quad (10.74)
$$

where α is a real constant (to guarantee that the generalized coordinates are also real), and the generalized coordinates of momentum,

$$
\vec{P}_{\vec{k}}(t) = \frac{d}{dt} \vec{Q}_{\vec{k}}(t) = \alpha \left[i\omega_{\vec{k}} \vec{a}_{\vec{k}} \exp\left(i\omega_{\vec{k}} t\right) - i\omega_{\vec{k}} \vec{a}_{\vec{k}}^* \exp\left(-i\omega_{\vec{k}} t\right) \right]. \quad (10.75)
$$

If we write $\vec{a}_{\vec{k}}$ and $\vec{a}_{\vec{k}}^*$ in terms of these generalized coordinates of position and momentum, and insert into equation (10.73), we have

$$
\mathcal{H} = \frac{V}{8\pi\alpha^2 c^2} \sum_{\vec{k}} \left[\vec{P}_{\vec{k}}^2 + k^2 c^2 \vec{Q}_{\vec{k}}^2 \right]. \quad (10.76)
$$

Hamilton's canonical equations,

$$
\frac{d}{dt} \vec{Q}_{\vec{k}} = \frac{\partial \mathcal{H}}{\partial \vec{P}_{\vec{k}}} \quad \text{and} \quad \frac{d}{dt} \vec{P}_{\vec{k}} = -\frac{\partial \mathcal{H}}{\partial \vec{Q}_{\vec{k}}}, \quad (10.77)
$$

will be immediately fulfilled with the choice

$$
\alpha = \left(\frac{V}{4\pi c^2} \right)^{1/2}. \quad (10.78)
$$

Given these canonical forms, we can use the standard formalism of classical statistical mechanics in phase space. Owing to the transversality of the electromagnetic fields, it is important to remark that

$$\vec{Q}_{\vec{k}} \cdot \vec{k} = \vec{P}_{\vec{k}} \cdot \vec{k} = 0. \tag{10.79}$$

Thus, both $\vec{Q}_{\vec{k}}$ and $\vec{P}_{\vec{k}}$ have just two components, and the Hamiltonian of this system may be written as

$$\mathcal{H} = \frac{1}{2} \sum_{\vec{k},j} \left[P_{\vec{k},j}^2 + \omega_{\vec{k}}^2 Q_{\vec{k},j}^2 \right] = \sum_{\vec{k},j} \mathcal{H}_{\vec{k},j}, \tag{10.80}$$

where $j = 1, 2$, and $\mathcal{H}_{\vec{k},j}$ is the Hamiltonian of a single one-dimensional harmonic oscillator of characteristic angular frequency $\omega_{\vec{k},j} = kc$.

The canonical partition function is given by

$$Z = \prod_{\vec{k},j} \int\!\!\int_{-\infty}^{+\infty} dQ_{\vec{k},j} dP_{\vec{k},j} \exp\left\{ -\frac{\beta}{2} \sum_{\vec{k},j} \left[P_{\vec{k},j}^2 + \omega_{\vec{k}}^2 Q_{\vec{k},j}^2 \right] \right\}. \tag{10.81}$$

Hence,

$$\ln Z = \sum_{\vec{k},j} \ln \left(\frac{2\pi}{\beta k c} \right). \tag{10.82}$$

Thus, we write the internal energy

$$U = -\frac{\partial}{\partial \beta} \ln Z = \sum_{\vec{k},j} \frac{1}{\beta} = 2 \frac{V}{(2\pi)^3} \int \frac{1}{\beta} d^3 k \to \infty, \tag{10.83}$$

whose divergence is the well-known ultraviolet catastrophe. At least from a formal point of view, it is interesting to write

$$U = \frac{V}{4\pi^3} k_B T \int_0^\infty 4\pi k^2 dk = \frac{8\pi}{c^3} V k_B T \int_0^\infty \nu^2 d\nu, \tag{10.84}$$

where

$$\nu = \frac{1}{2\pi} \omega_{\vec{k}} = \frac{kc}{2\pi}. \tag{10.85}$$

Thus, we have the well-known **Rayleigh–Jeans law**,

$$u(\nu) = 8\pi \frac{k_B T}{c^3} \nu^2, \tag{10.86}$$

which gives a good account of the experimental results in the region of low frequencies.

Planck's law

At the end of 1900, Planck proposed that the energy of an oscillator of frequency ν should be restricted to integer multiples of a basic quantity $h\nu$. Therefore, the canonical partition function would be given by

$$Z = \prod_{\vec{k},j} Z_{\vec{k},j}, \tag{10.87}$$

where

$$Z_{\vec{k},j} = \sum_{n=0}^{\infty} \exp\left(-\beta \frac{h}{2\pi}\omega_{\vec{k}} n\right) = \left[1 - \exp\left(-\beta\hbar\omega_{\vec{k}}\right)\right]^{-1}. \tag{10.88}$$

Thus, we have

$$\ln Z = \sum_{\vec{k},j} \ln Z_{\vec{k},j} = -2\frac{V}{(2\pi)^3} \int d^3\vec{k} \ln\left[1 - \exp\left(-\beta\hbar\omega_{\vec{k}}\right)\right]. \tag{10.89}$$

The resulting internal energy is finite and given by

$$U = -\frac{\partial}{\partial\beta} \ln Z = 2\frac{V}{(2\pi)^3} \int d^3\vec{k} \frac{\hbar k c}{\exp\left(\beta\hbar kc\right) - 1}, \tag{10.90}$$

from which we obtain the spectral density of energy,

$$u(\nu) = \frac{8\pi h}{c^3} \frac{\nu^3}{\exp\left(\beta h\nu\right) - 1}, \tag{10.91}$$

which is the famous **Planck law**.

In the limit of low frequencies, $\nu \to 0$, equation (10.91) reproduces the formula of Rayleigh and Jeans,

$$u(\nu) \to 8\pi\frac{k_B T}{c^3}\nu^2, \tag{10.92}$$

in full agreement with equation (10.86). From equation (10.91), however, the total energy does not present any divergences. In fact, we have

$$\frac{U}{V} = \int_0^{\infty} u(\nu)\, d\nu = \frac{8\pi h}{c^3} \int_0^{\infty} \frac{\nu^3 d\nu}{\exp\left(\beta h\nu\right) - 1}. \tag{10.93}$$

Introducing the change of variables $\beta h\nu = x$, we obtain the Stefan–Boltzmann law,

$$\frac{U}{V} = \frac{8\pi}{(hc)^3} \left[\int_0^{\infty} \frac{x^3 dx}{e^x - 1}\right] (k_B T)^4 = \sigma T^4, \tag{10.94}$$

where σ is a constant.

Quantization of the electromagnetic field

In 1928, Dirac suggested that the classical canonical variables $Q_{\vec{k},j}$ and $P_{\vec{k},j}$ should be treated as operators, with the commutation relations

$$\left[Q_{\vec{k},j}, P_{\vec{k}',j'}\right] = i\hbar\delta_{\vec{k},\vec{k}'}\delta_{j,j'} \tag{10.95}$$

and

$$\left[Q_{\vec{k},j}, Q_{\vec{k}',j'}\right] = \left[P_{\vec{k},j}, P_{\vec{k}',j'}\right] = 0. \tag{10.96}$$

Dirac has also introduced the operators

$$a_m = \alpha_1 Q_m + i\alpha_2 P_m \tag{10.97}$$

and

$$a_m^\dagger = \alpha_1 Q - i\alpha_2 P_m, \tag{10.98}$$

such that

$$\left[a_m, a_n^\dagger\right] = 2\hbar\alpha_1\alpha_2\delta_{m,n} \tag{10.99}$$

and

$$\left[a_m^\dagger, a_n^\dagger\right] = [a_m, a_n] = 0. \tag{10.100}$$

With a choice of the constants α_1 and α_2 such that

$$2\hbar\alpha_1\alpha_2 = 1 \tag{10.101}$$

and

$$\omega_{\vec{k}}^2 = 4\hbar^2\alpha_1^4, \tag{10.102}$$

the operator Hamiltonian can be written in the characteristic language of second quantization,

$$H = \sum_{\vec{k},j} \hbar\omega_{\vec{k},j}\left(a_{\vec{k},j}^\dagger a_{\vec{k},j} + \frac{1}{2}\right), \tag{10.103}$$

which includes the term of zero-point energy (that is, the quantum vacuum).

Within this formalism of Dirac, it is natural to associate with the electromagnetic field some particles, called photons, that behave according to the Bose–Einstein statistics. The spin of a photon is 1, although, owing to the transversality of the fields, j assumes two values only. Of course, we recover all the results of Planck. In the introductory texts of quantum mechanics,

there is always a chapter on the quantum treatment of the harmonic oscillator, in which one considers the properties of operators of creation and annihilation, a and a^\dagger, respectively, and of the operator number of particles, $a^\dagger a$, in the representation of occupation numbers (associated with the so-called Fock space). For this particular gas of photons, the thermal equilibrium is guaranteed by the absorption and emission of photons by matter. As the number of photons is not fixed, the chemical potential should be zero. We will come back to this quantum problem in the next chapter (in the context of ideal systems of phonons and magnons).

Exercises

1. Consider a system of ideal bosons of zero spin ($\gamma = 1$), within a container of volume V.

 (a) Show that the entropy above the condensation temperature T_o is given by

 $$S = k_B \frac{V}{\lambda^3} \left[\frac{5}{2} g_{5/2}(z) - \frac{\mu}{k_B T} g_{3/2}(z) \right],$$

 where

 $$\lambda = \frac{h}{\sqrt{2\pi m k_B T}}$$

 and

 $$g_\alpha(z) = \sum_{n=1}^{\infty} \frac{z^n}{n^\alpha}.$$

 (b) Given N and V, what is the expression of the entropy below T_o? What is the entropy associated with the particles of the condensate?

 (c) From the expression for the entropy, show that the specific heat at constant volume above T_o is given by

 $$c_V = \frac{3}{4} k_B \left[5 \frac{g_{5/2}(z)}{g_{3/2}(z)} - 3 \frac{g_{3/2}(z)}{g_{1/2}(z)} \right].$$

 (d) Below T_o, show that the specific heat at constant volume is given by

 $$c_V = \frac{15}{4} k_B \frac{v}{\lambda^3} g_{5/2}(1).$$

(e) Given the specific volume v, sketch a graph of c_V/k_B versus λ^{-2} (that is, in terms of the temperature in convenient units). Obtain the asymptotic expressions of the specific heat for $T \to 0$ and $T \to \infty$. Obtain the value of the specific heat at $T = T_o$.

2. Consider again the same problem for an ideal gas of two-dimensional bosons confined to a surface of area A. What are the changes in the expressions of item (a)? Show that there is no Bose–Einstein condensation in two dimensions (that is, show that in this case the Bose–Einstein temperature vanishes).

3. Consider an ideal gas of bosons with internal degrees of freedom. Suppose that, besides the ground state with zero energy ($\epsilon_o = 0$), we have to take into account the first excited state, with internal energy $\epsilon_1 > 0$. In other words, assume that the energy spectrum is given by

$$\epsilon_j = \epsilon_{\vec{k},\sigma} = \frac{\hbar^2 k^2}{2m} + \epsilon_1 \sigma,$$

where $\sigma = 0, 1$. Obtain an expression for the Bose–Einstein condensation temperature as a function of the parameter ϵ_1.

4. Consider a gas of non-interacting bosons associated with the energy spectrum

$$\epsilon = \hbar c \left| \vec{k} \right|,$$

where \hbar and c are constants, and \vec{k} is a wave vector. Calculate the pressure of this gas at zero chemical potential. What is the pressure of radiation of a gas of photons?

5. *The Hamiltonian of a gas of photons within an empty cavity of volume V is given by the expression

$$\mathcal{H} = \sum_{\vec{k},j} \hbar \omega_{\vec{k},j} \left(a^\dagger_{\vec{k},j} a_{\vec{k},j} + \frac{1}{2} \right),$$

where $j = 1, 2$ indicates the polarization, \vec{k} is the wave vector,

$$\omega_{\vec{k},j} = ck,$$

c is the velocity of light, and $k = \left| \vec{k} \right|$.

(a) Use the formalism of the occupation numbers (second quantization) to obtain the canonical partition function associated with this system.

(b) Show that the internal energy is given by

$$U = \sigma V T^n.$$

Obtain the value of the constants n and σ.

(c) Consider the Sun as a black body at temperature $T \approx 5800\,\mathrm{K}$. The solar diameter and the distance between the Sun and the Earth are of the order of $10^9\,\mathrm{m}$ and $10^{11}\,\mathrm{m}$, respectively. Obtain the intensity of the total radiation that reaches the surface of Earth. What is the value of the pressure of this radiation?

11

Phonons and Magnons

We now discuss some additional examples of Bose systems of interest in condensed matter physics. Phonons are quasi-particles produced by the quantization of the elastic vibrations of a crystalline lattice. Magnons are quasi-particles associated with the elementary magnetic excitations above a ferromagnetic ground state (although there are magnons associated with antiferromagnetic and even more complex magnetic orderings). Models of noninteracting quasi-particles are useful to explain transport and thermodynamic properties at low temperatures (as the behavior of the specific heat of crystalline solids). Noninteracting phonons are related to harmonic elastic potentials. Free magnons correspond to an approximate treatment, presumably correct at low temperatures, which takes into account only the lowest excited energy states above a magnetically ordered ground state. In the last section of this chapter, we examine the role of interactions in a Bose system. Just as a curiosity, we sketch a theory to account for superfluidity in liquid helium.

11.1 Crystalline phonons

The thermal properties of a crystalline solid may be described by a model of vibrating ionic particles about fixed equilibrium positions (which define the so-called Bravais lattice). As the electrons move too quickly, we adopt an adiabatic approximation. It consists in the consideration of the ionic motions only, without taking into account the interactions with additional

degrees of freedom. The choice of a quadratic pair potential between ions, which may be regarded as a first-order approximation for real crystalline potentials, already leads to the description of several physical properties at low temperatures. Free phonons are quasi-particles associated with the quantization of the elastic vibrations of this model of harmonic potentials. At higher temperatures, for larger excursions about the equilibrium positions, it is necessary to consider a gas of interacting phonons, whose study is certainly beyond the limits of this book.

Instead of considering a more realistic model, on a three-dimensional Bravais lattice, as it should be properly done in courses on solid-state physics, we initially discuss a very simple harmonic model, in one dimension, but that already displays all the ingredients of the original physical problem.

Elastic vibrations in one dimension

Let us consider a chain of N ions of mass m, connected by ideal springs of elastic constant k, with kinetic energy

$$T = \sum_{j=1}^{N} \frac{1}{2} m \dot{x}_j^2, \tag{11.1}$$

and potential energy

$$V = \sum_{j=1}^{N} \frac{1}{2} k \left(x_{j+1} - x_j - a_o\right)^2, \tag{11.2}$$

where a_o is a positive constant, and x_j is the position of the jth crystalline ion. In one dimension, we can write

$$x_j = ja + u_j, \tag{11.3}$$

where a is the equilibrium interionic distance, $j = 1, 2, ..., N$, and u_j labels a (small) excursion about the equilibrium position. If we assume periodic boundary conditions,

$$x_{N+1} = x_1, \tag{11.4}$$

it is possible to use the new set of variables $\{u_j\}$ to write

$$T = \sum_{j=1}^{N} \frac{1}{2} m \dot{u}_j^2 \tag{11.5}$$

and

$$V = \frac{1}{2} kN \left(a - a_o\right)^2 + \sum_{j=1}^{N} \frac{1}{2} k \left(u_{j+1} - u_j\right)^2. \tag{11.6}$$

Now it is convenient to write the displacements in terms of the normal modes of vibration (that is, Fourier components) of the system,

$$u_j = \frac{1}{\sqrt{N}} \sum_q \hat{u}_q \exp(iqja). \tag{11.7}$$

With periodic boundary conditions, and taking N as an even integer, the wave number q can assume N discrete values,

$$q = 0, \pm\frac{2\pi}{Na}, \pm\frac{2\pi}{Na}2, \dots, \pm\frac{2\pi}{Na}\left(\frac{N}{2} - 1\right), \frac{\pi}{a}, \tag{11.8}$$

which form the first **Brillouin zone** of the one-dimensional crystal. Furthermore, as the displacements are real numbers, we have the relation

$$\hat{u}_q = \hat{u}^*_{-q}. \tag{11.9}$$

From the orthogonalization properties,

$$\sum_{j=1}^{N} \exp[i(q_1 + q_2)ja] = N\delta_{q_1,-q_2}, \tag{11.10}$$

we can write the total energy,

$$E = T + V = \sum_q \left\{ \frac{1}{2}m\dot{\hat{u}}_q\dot{\hat{u}}_{-q} + \frac{1}{2}m\omega_q^2\hat{u}_q\hat{u}_{-q} \right\} + \frac{1}{2}kN(a - a_o)^2, \tag{11.11}$$

where

$$\omega_q^2 = \frac{2k}{m}(1 - \cos qa), \tag{11.12}$$

and the last term is just a trivial elastic contribution.

In analogy with the previous treatment of photons, we write the Fourier components of the displacements in the form

$$\hat{u}_q = a_q \exp(i\omega_q t) + a^*_{-q} \exp(-i\omega_q t), \tag{11.13}$$

where it should be emphasized that a_q and a_{-q} do not depend on time (and that we are preserving the relation $\hat{u}_q = \hat{u}^*_{-q}$). It is thus easy to see that the harmonic energy associated with small vibrations is given by the spectral decomposition

$$\mathcal{H} = E - \frac{1}{2}kN(a - a_o)^2 = \sum_q m\omega_q^2 a_q a^*_q, \tag{11.14}$$

where the wave number q belongs to the first Brillouin zone. Now we introduce the real generalized coordinates,

$$Q_q = \alpha \left[a_q \exp\left(i\omega_q t\right) + a_q^* \exp\left(-i\omega_q t\right) \right] \tag{11.15}$$

and

$$P_q = \dot{Q}_q = \alpha i \omega_q \left[a_q \exp\left(i\omega_q t\right) - a_q^* \exp\left(-i\omega_q t\right) \right], \tag{11.16}$$

where α is a positive constant. In terms of these new coordinates, we have the energy

$$\mathcal{H} = \sum_q \left\{ \frac{m}{4\alpha^2} P_q^2 + \frac{m\omega_q^2}{4\alpha^2} Q_q^2 \right\}, \tag{11.17}$$

where the constant α should be chosen so that the definitions (11.15) and (11.16) become compatible with Hamilton's canonical equations,

$$\frac{\partial \mathcal{H}}{\partial Q_q} = \frac{m\omega_q^2}{2\alpha^2} Q_q = -\dot{P}_q \tag{11.18}$$

and

$$\frac{\partial \mathcal{H}}{\partial P_q} = \frac{m}{2\alpha^2} P_q = \dot{Q}_q. \tag{11.19}$$

It can be seen immediately that the choice

$$\alpha^2 = \frac{m}{2} \tag{11.20}$$

gives the Hamiltonian of the system,

$$\mathcal{H} = \sum_q \left\{ \frac{1}{2} P_q^2 + \frac{1}{2} \omega_q^2 Q_q^2 \right\}. \tag{11.21}$$

In the classical canonical formalism, the partition function factorizes. If we discard the uniform term of elastic vibrations, which gives a trivial contribution to the free energy, we have

$$Z = \prod_q Z_q, \tag{11.22}$$

where

$$Z_q = \int\limits_{-\infty}^{+\infty} dQ_q \int\limits_{-\infty}^{+\infty} dP_q \exp\left\{ -\frac{\beta}{2} P_q^2 - \frac{\beta}{2} \omega_q^2 Q_q^2 \right\} = \frac{2\pi}{\beta\omega_q}, \tag{11.23}$$

in which we have excluded the $q = 0$ mode (that corresponds to a uniform translation of the system). In the thermodynamic limit, we have

$$\frac{1}{N} \ln Z = \frac{1}{N} \sum_{q \neq 0} \ln \left(\frac{2\pi}{\beta \omega_q} \right) = \frac{a}{4\pi} \int\limits_{-\pi/a}^{+\pi/a} \ln \left[\frac{2\pi^2 m}{k\beta^2 (1 - \cos qa)} \right] dq. \quad (11.24)$$

Calculating the integral, we finally have

$$\frac{1}{N} \ln Z = \frac{1}{2} \ln \left(\frac{2\pi^2 m}{k\beta^2} \right) - \frac{1}{4\pi} \int\limits_{-\pi}^{+\pi} \ln (1 - \cos \theta) \, d\theta = \frac{1}{2} \ln \left(\frac{4\pi^2 m}{k\beta^2} \right),$$

$$(11.25)$$

from which we obtain the classical results for the internal energy per ion, $u = k_B T$, and for the specific heat at constant volume, $c_V = (\partial u / \partial T)_a = k_B$, in agreement with the results from the classical equipartition theorem for a one-dimensional system. Therefore, the inclusion of harmonic interactions between ions, within a classical context, is not enough to explain the change with temperature of the specific heat of solids!

Quantization of the one-dimensional model

Now we proceed along the usual steps to quantize the classical harmonic oscillator. If we introduce the momentum operator,

$$P_q = \frac{\hbar}{i} \frac{\partial}{\partial Q_q}, \quad (11.26)$$

equation (11.21) can be written as

$$\mathcal{H} = \sum_q \frac{1}{2} \hbar \omega_q \left[-\frac{\hbar}{\omega_q} \frac{\partial^2}{\partial Q_q^2} + \frac{\omega_q}{\hbar} Q_q^2 \right]. \quad (11.27)$$

Redefining the position coordinates,

$$Q_q = \left(\frac{\hbar}{\omega_q} \right)^{1/2} \xi_q, \quad (11.28)$$

we have a more-convenient form,

$$\mathcal{H} = \sum_q \frac{1}{2} \hbar \omega_q \left[-\frac{\partial^2}{\partial \xi_q^2} + \xi_q^2 \right]. \quad (11.29)$$

Introducing the well-known annihilation and creation operators,

$$a_q = \frac{1}{\sqrt{2}} \left(\xi_q + \frac{\partial}{\partial \xi_q} \right) \quad (11.30)$$

and

$$a_q^\dagger = \frac{1}{\sqrt{2}} \left(\xi_q - \frac{\partial}{\partial \xi_q} \right),$$ (11.31)

with the commuting properties of bosons,

$$[a_{q_1}, a_{q_2}] = [a_{q_1}^\dagger, a_{q_2}^\dagger] = 0 \quad \text{and} \quad [a_{q_1}, a_{q_2}^\dagger] = \delta_{q_1, q_2},$$ (11.32)

we finally have

$$\mathcal{H} = \sum_q \left[\hbar \omega_q a_q^\dagger a_q + \frac{1}{2} \hbar \omega_q \right],$$ (11.33)

where the frequency spectrum ω_q is given by equation (11.12).

It is now interesting to recall some results from the standard quantization of the harmonic oscillator. Consider a system of a single particle. The state of the system (the wave function is a Hermite polynomial) is fully characterized by a quantum number n. The single-particle wave function ψ_n represents a situation with n quanta (phonons, in the present case) and energy $\hbar n \omega$. It is then convenient to introduce the notation $\psi_n \equiv |n\rangle$. The creation and annihilation operators act on these states according to the following rules,

$$a^\dagger |n\rangle = \sqrt{n+1} |n+1\rangle,$$

$$a |n\rangle = \sqrt{n} |n-1\rangle.$$

Thus, we have

$$a^\dagger a |n\rangle = n |n\rangle,$$

which means that the operator $a^\dagger a$ counts the number of quanta (phonons) in state $|n\rangle$. Therefore, the Hamiltonian is diagonal in this representation of occupation numbers (known as second quantization). A single-particle state with n quanta comes from the successive application of the creation operator on the quantum vacuum,

$$|n\rangle = \frac{1}{\sqrt{n}} \left(a^\dagger \right)^n |0\rangle.$$

The Hamiltonian (11.33) represents a set of N localized quantum oscillators. The corresponding wave function represents a state with n_q phonons associated with the qth oscillator, for all values of q, and will be given by the direct product

$$|n_0, n_1, ...\rangle = \prod_q |n_q\rangle,$$

FIGURE 11.1. Frequency spectrum for a one-dimensional gas of phonons. Note the linear dependence, $\omega = cq$, for small q.

which guarantees the diagonal form of the Hamiltonian in this space of occupation numbers (also known as Fock space).

The (canonical) partition function is given by the sum

$$Z = \text{Tr} \exp\left(-\beta \mathcal{H}\right) = \sum_{\{n_q\}} \exp\left\{-\beta \sum_q \left[\hbar\omega_q n_q + \frac{1}{2}\hbar\omega_q\right]\right\}, \quad (11.34)$$

where there are no restrictions on the values of the occupation numbers n_q. In the language of second quantization, the trace is performed over the states of the Fock space (as the Hamiltonian \mathcal{H} is diagonal, the result is quite simple). Thus, we have

$$Z = \prod_q \left\{\sum_{n=0}^{\infty} \exp\left[-\beta\hbar\omega_q n - \frac{1}{2}\beta\hbar\omega_q\right]\right\} = \prod_q \frac{\exp\left(-\frac{1}{2}\beta\hbar\omega_q\right)}{1 - \exp\left(-\beta\hbar\omega_q\right)}. \quad (11.35)$$

Therefore, the internal energy per particle can be written as

$$u = -\frac{1}{N}\frac{\partial}{\partial\beta}\ln Z = \frac{1}{N}\sum_q \frac{1}{2}\hbar\omega_q + \frac{1}{N}\sum_q \frac{\hbar\omega_q}{\exp\left(\beta\hbar\omega_q\right) - 1}, \quad (11.36)$$

where the first term represents the zero-point energy and the second term is the analog of Planck's energy of the photon gas. In the thermodynamic limit, the average energy of the elementary excitations above the quantum vacuum is given by

$$u = \frac{a}{2\pi} \int_{-\pi/a}^{+\pi/a} \frac{\hbar\omega_q}{\exp\left(\beta\hbar\omega_q\right) - 1} dq. \quad (11.37)$$

From equation (11.37), we have the specific heat at constant volume,

$$c_V = \left(\frac{\partial u}{\partial T}\right)_a = \frac{ak_B}{\pi} \int_0^{\pi/a} \frac{\left(\beta\hbar\omega_q\right)^2 \exp\left(\beta\hbar\omega_q\right)}{\left[\exp\left(\beta\hbar\omega_q\right) + 1\right]^2} dq. \quad (11.38)$$

At high temperatures ($\beta\hbar\omega_q \ll 1$), we can write the expansion

$$\frac{(\beta\hbar\omega_q)^2 \exp(\beta\hbar\omega_q)}{[\exp(\beta\hbar\omega_q) + 1]^2} = 1 + O(\beta\hbar\omega_q). \tag{11.39}$$

Therefore, $c_V \to k_B$, according to the classical equipartition theorem. At low temperatures ($\beta\hbar\omega_q \gg 1$), and taking into account the shape of the frequency spectrum (see figure 11.1), we obtain a good approximation by keeping just the leading terms, for small values of the wave number q, in the expression of the integrand. In fact, from equation (11.12), we have

$$\omega_q^2 = \frac{2k}{m}(1 - \cos qa) = c^2 q^2 + O(q^4), \tag{11.40}$$

where the constant $c = (ka/m)$ may be interpreted as the sound velocity in the crystal. Inserting this expansion into the formula for the specific heat, we have the limit

$$c_V \to \frac{ak_B}{\pi} \int_0^{\pi/a} \frac{(\beta\hbar cq)^2 \exp(\beta\hbar cq)}{[\exp(\beta\hbar cq) + 1]^2} dq = \frac{ak_B}{\pi}\left(\frac{k_B T}{\hbar c}\right) \int_0^{\beta\hbar c\pi/a} \frac{x^2 e^x}{(e^x + 1)^2} dx. \tag{11.41}$$

At low temperatures, the upper limit of integration may be replaced by $+\infty$, up to an exponentially vanishing error. Therefore, we have the asymptotic form

$$c_V = \frac{ak_B}{\pi}\left(\frac{k_B T}{\hbar c}\right) \int_0^{\infty} \frac{x^2 e^x}{(e^x + 1)^2} dx = \frac{ak_B}{\pi}\zeta(2)\left(\frac{k_B T}{\hbar c}\right). \tag{11.42}$$

In contrast with the Einstein model, in which c_V vanishes exponentially, in this model of elastically interacting particles in one dimension, the specific heat at constant volume vanishes with a (much slower) first power of the temperature T.

Now we introduce the definition of **Debye temperature**. In the first Brillouin zone, there are N normal modes of vibration of the crystal. The largest momentum, called the Debye momentum, is given by $q_D = \pi/a$. The Debye energy is defined as

$$\epsilon_D = \hbar\omega_D = \hbar c q_D = \frac{\hbar c\pi}{a}. \tag{11.43}$$

We then have the Debye temperature,

$$T_D = \frac{\hbar\omega_D}{k_B} = \frac{\hbar c\pi}{k_B a}, \tag{11.44}$$

and the asymptotic form of the specific heat,

$$c_V = \zeta(2) k_B \left(\frac{T}{T_D}\right), \tag{11.45}$$

in the quantum limit, $T \ll T_D$.

Specific heat in three dimensions

Although the notation is somewhat more complicated, the specific heat of a more-realistic model, for a three-dimensional crystal with one atom per primitive cell, can be obtained according to the same procedures already used in the one-dimensional case. The quantum Hamiltonian is written as

$$\mathcal{H} = \sum_{\vec{q},j} \left[\hbar \omega_{\vec{q},j} a^{\dagger}_{\vec{q},j} a_{\vec{q},j} + \frac{1}{2} \hbar \omega_{\vec{q},j} \right],$$ (11.46)

where $j = 1, 2, 3$ labels the three branches of the energy spectrum (a longitudinal branch, and two transverse branches), and $\omega_{\vec{q},j}$ certainly displays a more complicated form than given by the simple equation (11.12). However, we still can write the internal energy per atom (with respect to the zero-point energy),

$$u = \frac{1}{N} \sum_{\vec{q},j} \frac{\hbar \omega_{\vec{q},j}}{\exp\left(\beta \hbar \omega_{\vec{q},j}\right) - 1}.$$ (11.47)

The **Debye model**, which works at low temperatures, consists in the introduction of a linear approximation for the frequency spectrum,

$$\omega_{\vec{q},j} = cq,$$ (11.48)

where the constant c is the velocity of sound. Also, the first Brillouin zone is approximated by a sphere of radius q_D. At low temperatures, the specific heat at constant volume is given by the asymptotic form

$$c_V = \frac{3}{N} \left(\frac{V}{8\pi^3} \right) \frac{\partial}{\partial T} \int_0^{q_D} \frac{(\hbar cq) \, 4\pi q^2 dq}{\exp\left(\beta \hbar cq\right) - 1} = \frac{3 k_B v}{2\pi^2 \left(\beta \hbar c\right)^3} \int_0^{\beta \hbar c q_D} \frac{x^4 e^x dx}{\left(e^x - 1\right)^2},$$ (11.49)

which already indicates the well-known dependence with the cubic power of temperature. In three dimensions, the Debye momentum is defined by the relation

$$N = \frac{V}{8\pi^3} \int_0^{q_D} 4\pi q^2 dq,$$ (11.50)

that is,

$$q_D = \left(\frac{6\pi^2}{v} \right)^{1/3}.$$ (11.51)

Therefore, the Debye temperature is given by

$$T_D = \frac{\hbar c q_D}{k_B} = \frac{\hbar c}{k_B} \left(\frac{6\pi^2}{v} \right)^{1/3}.$$ (11.52)

At low temperatures, $\beta \hbar c q_D \gg 1$, except for an exponentially small error, we can calculate the integral in equation (11.49),

$$\int_0^\infty \frac{x^4 e^x dx}{(e^x - 1)^2} = 24\zeta(4) = \frac{4\pi^4}{15}. \tag{11.53}$$

We then have a compact asymptotic form, the **Debye law**, for the specific heat at constant volume at low temperatures,

$$c_V = \frac{12\pi^4}{5} k_B \left(\frac{T}{T_D}\right)^3. \tag{11.54}$$

In the following table, we give some typical values of the Debye temperature.

Crystal	$T_D(K)$
Li	400
Na	150
K	100
Si	625
Pb	88
C	1860

Except for carbon (that is, diamond), the typical Debye temperature is of the order of $100\,K$.

11.2 Ferromagnetic magnons

As discussed in previous chapters, the paramagnetic behavior of crystalline compounds may be explained by a model of noninteracting and localized spins in the presence of an external magnetic field. As we have already pointed out, this model of noninteracting spins is unable to account for a ferromagnetically ordered state (in which there is a spontaneous magnetization in the absence of an external applied field). It then seems reasonable to include dipole–dipole interactions as the new ingredient to describe the existence of ferromagnetic ordering. However, magnetic interactions are too weak, and it has been very difficult to invoke these dipole–dipole interactions to explain the persistence of magnetic ordering in many compounds up to quite high temperatures. In the 1930s, Dirac, Heisenberg, and some others, proposed another mechanism, on the basis of Coulomb interactions and quantum arguments, which can bring about strong enough interactions to account for a ferromagnetic state even at the highest observed temperatures.

The exchange interaction and the Heisenberg model

To present a justification of the **exchange interaction** introduced by Dirac and Heisenberg, let us discuss a perturbative calculation of the Coulomb interaction energy between two electrons, $-e^2/r_{12}$, where e is the elementary charge and r_{12} is the distance between electrons. There are two globally antisymmetric wave functions associated with the system of two free electrons,

$$\psi_1 = \phi_S(\vec{r}_1, \vec{r}_2)\, \chi_A(\sigma_1, \sigma_2), \tag{11.55}$$

with a symmetric positional factor and an antisymmetric spin factor, and

$$\psi_2 = \phi_A(\vec{r}_1, \vec{r}_2)\, \chi_S(\sigma_1, \sigma_2), \tag{11.56}$$

with an antisymmetric positional factor and a symmetric spin factor. The positional factors may written as

$$\phi_{S,A}(\vec{r}_1, \vec{r}_2) = \varphi_1(\vec{r}_1)\, \varphi_2(\vec{r}_2) \pm \varphi_2(\vec{r}_1)\, \varphi_1(\vec{r}_2), \tag{11.57}$$

where $\varphi_{1,2}$ are the standard single-particle wave functions. As the interaction energy does not depend on spin, we have the (first-order) perturbative contributions

$$\left\langle \psi_1 \left| -\frac{e^2}{r_{12}} \right| \psi_1 \right\rangle = A + B \tag{11.58}$$

and

$$\left\langle \psi_2 \left| -\frac{e^2}{r_{12}} \right| \psi_2 \right\rangle = A - B, \tag{11.59}$$

where

$$A = 2 \left\langle \varphi_1(\vec{r}_1)\, \varphi_2(\vec{r}_2) \left| -\frac{e^2}{r_{12}} \right| \varphi_1(\vec{r}_1)\, \varphi_2(\vec{r}_2) \right\rangle \tag{11.60}$$

is the **direct term**, and

$$B = 2 \left\langle \varphi_1(\vec{r}_1)\, \varphi_2(\vec{r}_2) \left| -\frac{e^2}{r_{12}} \right| \varphi_2(\vec{r}_1)\, \varphi_1(\vec{r}_2) \right\rangle \tag{11.61}$$

is the **exchange term**. We thus have

$$\left\langle \psi_1 \left| -\frac{e^2}{r_{12}} \right| \psi_1 \right\rangle - \left\langle \psi_2 \left| -\frac{e^2}{r_{12}} \right| \psi_2 \right\rangle = 2B, \tag{11.62}$$

which shows that the energy difference between these two perturbative levels depends on the exchange term. It should be emphasized that the splitting of these levels comes from the symmetry requirements of quantum

mechanics. The splitting is strong enough, since it involves a Coulomb term, but the interactions are short ranged, since the coefficients A and B depend on spatial overlap integrals of positional wave functions.

Now let us show that an **effective spin Hamiltonian**, of the form

$$\mathcal{H} = C\vec{S}_1 \cdot \vec{S}_2 + D, \qquad (11.63)$$

where C and D are constant parameters, simulates the existence of two distinct energy levels. The spin antisymmetric wave function,

$$\chi_A = \frac{1}{\sqrt{2}} \left(\alpha\beta - \beta\alpha \right), \qquad (11.64)$$

is a singlet with zero (total) spin. Note that we use the standard notation of quantum mechanics; α and β refer to single-particle spin states; the direct product $\alpha\beta$ means that spin 1 is in state α, and spin 2 is in state β. The symmetric wave function is given by the triplet

$$\chi_S = \begin{cases} \alpha\alpha, \\ \beta\beta, \\ \frac{1}{\sqrt{2}} \left(\alpha\beta + \beta\alpha \right), \end{cases} \qquad (11.65)$$

with total spin 1. If we write equation (11.63) as

$$\mathcal{H} = C\left\{ \frac{1}{2} \left(\vec{S}_1 + \vec{S}_2 \right)^2 - \frac{1}{2}\vec{S}_1^2 - \frac{1}{2}\vec{S}_2^2 \right\} + D, \qquad (11.66)$$

and note that

$$\left(\vec{S}_1 + \vec{S}_2 \right)^2 \chi_A = 0, \quad \left(\vec{S}_1 + \vec{S}_2 \right)^2 \chi_S = 2\chi_S, \qquad (11.67)$$

$$\vec{S}_{1,2}^2 \chi_A = \frac{3}{4}\chi_A, \quad \vec{S}_{1,2}^2 \chi_S = \frac{3}{4}\chi_S, \qquad (11.68)$$

we have

$$\langle \chi_S | \mathcal{H} | \chi_S \rangle = \frac{1}{4}C + D \qquad (11.69)$$

and

$$\langle \chi_A | \mathcal{H} | \chi_A \rangle = -\frac{3}{4}C + D. \qquad (11.70)$$

Therefore,

$$\langle \chi_S | \mathcal{H} | \chi_S \rangle - \langle \chi_A | \mathcal{H} | \chi_A \rangle = C, \qquad (11.71)$$

which leads to the identification between the prefactor C (of the scalar product of the vector spin operators) and the exchange term of Coulomb nature.

On the basis of these arguments, we now introduce the **Heisenberg spin Hamiltonian**,

$$\mathcal{H} = -J\sum_{(jk)} \vec{S}_j \cdot \vec{S}_k - g\mu_B \sum_{j=1}^{N} \vec{H} \cdot \vec{S}_j, \tag{11.72}$$

where J is the exchange parameter, the first sum is over pairs of nearest-neighbor sites on a crystalline lattice (recall that the exchange term involves an overlap integral between wave functions, which precludes long-range interactions), and the second term represents the interaction of the spins with an external applied magnetic field. In this chapter, we assume $J > 0$, which leads to a ferromagnetic ground state.

Spin waves in the Heisenberg ferromagnet

For $J > 0$, $H > 0$, and with the magnetic field along the z direction, the Heisenberg Hamiltonian can be written as

$$\mathcal{H} = -J\sum_{(jk)} \vec{S}_j \cdot \vec{S}_k - g\mu_B H \sum_{j=1}^{N} S_j^z. \tag{11.73}$$

Taking into account the definitions

$$S^+ = S^x + iS^y \text{ and } S^- = S^x - iS^y, \tag{11.74}$$

and the commutation rules for angular momentum operators,

$$[S^x, S^y] = S^z, \tag{11.75}$$

with the corresponding cyclic permutations, equation (11.73) can be written as

$$\mathcal{H} = -J\sum_{(jk)} \left[S_j^z S_k^z + \frac{1}{2} \left(S_j^+ S_k^- + S_j^- S_k^+ \right) \right] - g\mu_B H \sum_{j=1}^{N} S_j^z. \tag{11.76}$$

A simple inspection of this last equation shows that the ground state is associated with the wave function

$$\Psi_o = \prod_j |S\rangle_j, \tag{11.77}$$

where each ion belongs to a state in which $S^z = +S$. It is easy to see that

$$\mathcal{H}\Psi_o = \left[-JdNS^2 - g\mu_B HNS \right] \Psi_o, \tag{11.78}$$

where d is the dimension of a hypercubic lattice. Thus, the energy of the ferromagnetic ground state is given by

$$E_o = -JdNS^2 - g\mu_B HNS. \tag{11.79}$$

It should be pointed out that both the total angular momentum and its component along the z direction are constants of the motion (since they both commute with the Hamiltonian \mathcal{H}).

To simplify the analysis of the excited sates above this ground state, let us consider the problem in the absence of an applied magnetic field. If the z component of the spin associated with the nth ion decreases by just one unit, we have a reasonable candidate to be an excited state,

$$\Psi_n = S_n^- \prod_j |S\rangle_j. \tag{11.80}$$

To further simplify the problem, let us consider a Heisenberg Hamiltonian in one dimension,

$$\mathcal{H}_1 = -J \sum_{j=1}^{N} \left[S_j^z S_{j+1}^z + \frac{1}{2} \left(S_j^+ S_{j+1}^- + S_j^- S_{j+1}^+ \right) \right], \tag{11.81}$$

and assume periodic boundary conditions. Thus, it is easy to show that

$$\mathcal{H}_1 \Psi_n = -J \left(NS^2 - 2S \right) \Psi_n - JS\Psi_{n+1} - JS\Psi_{n-1}, \tag{11.82}$$

which means that Ψ_n is **not** an eigenfunction of the Hamiltonian operator!

Let us then make another attempt to find an acceptable wave function associated with the excited states. We may try a superposition of spin deviations,

$$\Psi = \sum_n \Psi_n \exp\left(iqan\right), \tag{11.83}$$

where n runs over all lattice sites and a is the lattice spacing between nearest neighbors. For the one-dimensional Heisenberg Hamiltonian, it is not difficult to see that

$$\mathcal{H}_1 \Psi = \left[-J \left(NS^2 - 2S \right) - JS\exp\left(-iqa\right) - JS\exp\left(iqa\right) \right] \Psi, \tag{11.84}$$

which shows that the new wave function Ψ is **an eigenstate** of the Hamiltonian, with the energy eigenvalue

$$E_q = -JNS^2 + 2JS\left(1 - \cos qa\right). \tag{11.85}$$

The elementary excitations represented by this eigenstate, which are called **spin waves**, are characterized by the energy spectrum

$$\epsilon_q = E_q + JNS^2 = 2JS\left(1 - \cos qa\right) \tag{11.86}$$

above the ferromagnetic ground state. For small values of the wave number q, the energy spectrum of this one-dimensional model is given by the quadratic asymptotic form

$$\epsilon_q = JSa^2q^2 = \frac{\hbar^2 q^2}{2m^*},\tag{11.87}$$

with the **effective mass**

$$m^* = \frac{\hbar^2}{2JSa^2}.\tag{11.88}$$

What is the meaning of these excitations? Are they similar to quasi-particles of Bose character? To consider these questions, let us introduce the Holstein–Primakoff transformation.

The Holstein–Primakoff transformation

In order to write the Heisenberg spin Hamiltonian (11.76) in terms of Bose operators, we introduce the transformations

$$S_j^+ = (2S)^{1/2} \left[1 - \frac{a_j^\dagger a_j}{2S}\right]^{1/2} a_j \tag{11.89}$$

and

$$S_j^- = (2S)^{1/2} a_j^\dagger \left[1 - \frac{a_j^\dagger a_j}{2S}\right]^{1/2},\tag{11.90}$$

where the operators a_j and a_j^\dagger obey Bose commutation rules,

$$\left[a_j, a_k^\dagger\right] = \delta_{j,k}\tag{11.91}$$

and

$$[a_j, a_k] = \left[a_j^\dagger, a_k^\dagger\right] = 0.\tag{11.92}$$

It is a straightforward calculation to show that

$$S_j^+ S_j^- = 2S \left\{1 - \frac{a_j^\dagger a_j}{2S} + a_j^\dagger a_j \left[1 - \frac{a_j^\dagger a_j}{2S}\right]\right\},\tag{11.93}$$

and that

$$S_j^- S_j^+ = 2S \left\{a_j^\dagger a_j - \frac{1}{2S} a_j^\dagger a_j^\dagger a_j a_j\right\}.\tag{11.94}$$

We also have the result

$$S_j^z = \frac{1}{2}\left[S_j^+, S_j^-\right] = S - a_j^\dagger a_j, \tag{11.95}$$

which indicates that the number operator of quasi-particles, $a_j^\dagger a_j$, is associated with the decrease of the z component of spin on the jth lattice site.

Now we consider the Heisenberg Hamiltonian on a cubic lattice in d dimensions,

$$\mathcal{H} = -\frac{1}{2}J\sum_{R,\delta}\left[S_R^z S_{R+\delta}^z + \frac{1}{2}\left(S_R^+ S_{R+\delta}^- + S_R^- S_{R+\delta}^+\right)\right], \tag{11.96}$$

where the vector R labels a lattice site and the vector δ refers to one of its first neighbors. Instead of dealing with the algebraic problem of the Holstein–Primakoff transformation in its full generality, let us introduce an expansion for large values of the spin S. The square root in equations (11.89) and (11.90) is then expanded in powers of $1/S$. If we keep the leading terms, we have

$$S_j^+ = (2S)^{1/2}\,a_j \text{ and } S_j^- = (2S)^{1/2}\,a_j^\dagger. \tag{11.97}$$

In this limit, the Heisenberg Hamiltonian (11.96) is reduced to two-body terms,

$$\mathcal{H} = -\frac{1}{2}J\sum_{R,\delta}\left[S^2 - S\left(a_R^\dagger a_R + a_{R+\delta}^\dagger a_{R+\delta}\right) + \frac{S}{2}a_R a_{R+\delta}^\dagger + \frac{S}{2}a_R^\dagger a_{R+\delta}\right]. \tag{11.98}$$

In order to diagonalize this Hamiltonian, we assume periodic boundary conditions, and perform a Fourier transformation,

$$b_q = \frac{1}{\sqrt{N}}\sum_R a_R \exp\left(iq\cdot R\right) \tag{11.99}$$

and

$$b_q^\dagger = \frac{1}{\sqrt{N}}\sum_R a_R^\dagger \exp\left(-iq\cdot R\right), \tag{11.100}$$

where the wave vector q belongs to the first Brillouin zone. Although both R and q are vectors, there are no ambiguities if we omit any vector notation. As the Fourier transformation preserves the canonical commutation relations of the involved operators, we can use equations (11.99) and

(11.100), and the expression for the Heisenberg Hamiltonian given by equation (11.98), to write

$$\mathcal{H} = -\frac{1}{2}J\left\{2dNS^2 - 4dS\sum_q b_q^\dagger b_q + S\sum_{q,\delta}e^{iq\delta}b_q^\dagger b_q + S\sum_{q,\delta}e^{-iq\delta}b_q^\dagger b_q\right\}$$

$$= -JdNS^2 + \sum_q\left\{2JSd - 2JS\left(\cos aq_1 + \ldots + \cos aq_d\right)\right\}b_q^\dagger b_q,$$

$$(11.101)$$

which gives the energy spectrum of the quasi-particles,

$$\epsilon_q = 2JSd - 2JS\left(\cos aq_1 + \ldots + \cos aq_d\right). \qquad (11.102)$$

For $d = 1$, we recover equation (11.86). For long wavelengths, we have again an asymptotic quadratic form,

$$\epsilon_q \approx JSa^2q^2. \qquad (11.103)$$

To summarize, as in the analysis of phonons in the context of an elastic medium, we have obtained an approximate Hamiltonian (which is good for large S; which is presumable reasonable at low temperatures) for a system of noninteracting spin waves,

$$\mathcal{H} = -JdNS^2 + \sum_q \epsilon_q b_q^\dagger b_q, \qquad (11.104)$$

where the sum is over all wave vectors belonging to the first Brillouin zone. The detailed study of the Holstein-Primakoff transformation, which has been pioneered by a celebrated work of F. J. Dyson, is beyond the scope of this book (the interested reader is referred to the book by D. C. Mattis, *Theory of Magnetism I: Statics and Dynamics*, Springer-Verlag, New York, 1988).

Magnetization of the Heisenberg model

In the Heisenberg model, the component of the magnetization along the direction of the field, in dimensionless units, is given by

$$M_z = \sum_R \langle S_R^z\rangle = NS - \sum_R \left\langle a_R^\dagger a_R\right\rangle = NS - \sum_q \langle b_q^\dagger b_q\rangle, \qquad (11.105)$$

where we omit the vector notation.

To calculate thermal averages in the spin-wave approximation, we write the partition function associated with Hamiltonian (11.104),

$$Z = \text{Tr}\exp\left(-\sum_q \beta\epsilon_q b_q^\dagger b_q\right). \qquad (11.106)$$

Since there is a factorization of the trace over states belonging to Fock space, we have

$$Z = \prod_q \left[\sum_n \exp\left(-\beta\epsilon_q n\right) \right] = \prod_q \left[1 - \exp\left(-\beta\epsilon_q\right)\right]^{-1}. \tag{11.107}$$

Therefore,

$$\ln Z = -\sum_q \ln\left[1 - \exp\left(-\beta\epsilon_q\right)\right], \tag{11.108}$$

with the energy spectrum given by equation (11.102). From this expression, we obtain

$$\langle b_q^\dagger b_q \rangle = -\frac{1}{\beta}\frac{\partial}{\partial\epsilon_q}\ln Z = \frac{1}{\exp\left(\beta\epsilon_q\right) - 1}, \tag{11.109}$$

which leads to the magnetization,

$$M_z = NS - \sum_{\vec{q}} \frac{1}{\exp\left(\beta\epsilon_{\vec{q}}\right) - 1}, \tag{11.110}$$

where the sum is over the first Brillouin zone.

This expression may have serious problems of convergence in the thermodynamic limit. In fact, if we consider a d-dimensional hypercubic lattice, the contribution of the small momentum terms ($|\vec{q}| \to 0$) may be written as

$$\sum_{\vec{q}} \frac{1}{\exp\left(\beta\epsilon_{\vec{q}}\right) - 1} \to \frac{V}{(2\pi)^d}\int d^d\vec{q}\,\frac{1}{\exp\left(\beta JSq^2 + \ldots\right) - 1} \to \int_0 \frac{q^{d-1}dq}{q^2}, \tag{11.111}$$

which exists for $d > 2$ only. Therefore, we have clear indications that the Heisenberg model in two (and one) dimensions does not display a ferromagnetically ordered state at finite temperatures. This result, which is known as the Mermin–Wagner theorem, has been rigorously shown in the 1970s. On the other hand, in three dimensions, at low enough temperatures, there are no divergencies, and we can write

$$M_z \sim NS - \frac{V}{(2\pi)^3}\int_0^\infty \frac{4\pi q^2 dq}{\exp\left(\beta JSq^2\right) - 1}. \tag{11.112}$$

Introducing the change of variables $q = \sqrt{\beta}x$, we finally have

$$\frac{1}{V}M_z - \frac{NS}{V} \sim AT^{3/2}, \tag{11.113}$$

where A is a constant prefactor. This asymptotic law at low temperatures has indeed been obtained either from experiments or from (more-refined) theoretical calculations.

11.3 Sketch of a theory of superfluidity

In a dilute Bose gas, at low temperatures, most of the particles are in the zero-momentum ground state. Only binary collisions, associated with small momentum transfers, should have some relevance. Let us then consider a gas of bosons of mass m, given by the following Hamiltonian in the occupation number representation,

$$\mathcal{H} = \sum_k \epsilon_k^o a_k^\dagger a_k + \frac{g}{2V} \sum_{k_1,k_2,k_3,k_4} a_{k_1}^\dagger a_{k_2}^\dagger a_{k_3} a_{k_4} \delta_{k_1+k_2,k_3+k_4}, \qquad (11.114)$$

where the constant g plays the role of a pseudopotential. According to the interaction term, two particles, of momenta k_3 and k_4, are destroyed, and two other particles, of momenta k_1 and k_2, are created, such that the total momentum of the system remains unchanged.

At low temperatures, in the thermodynamic limit, as almost all particles belong to the condensate $(N - N_o << N)$, it is reasonable to substitute the operators a_o^\dagger and a_o by $\sqrt{N_o}$. Let us then order the interaction terms of this Hamiltonian according to the number of times that we have the operators a_o^\dagger and a_o. Keeping the dominant terms only, we write

$$\mathcal{H}_{int} = \frac{g}{2V} \sum_{k_1,k_2,k_3,k_4} a_{k_1}^\dagger a_{k_2}^\dagger a_{k_3} a_{k_4} \delta_{k_1+k_2,k_3+k_4} \approx \frac{g}{2V} a_o^\dagger a_o^\dagger a_o a_o$$

$$+ \frac{g}{2V} \sum_{k \neq 0} \left\{ 4 a_k^\dagger a_k a_o^\dagger a_o + a_o^\dagger a_o^\dagger a_k a_{-k} + a_k^\dagger a_{-k}^\dagger a_o a_o \right\}, (11.115)$$

where we have discarded all interactions that do not involve particles in the condensate. Thus, we have

$$\mathcal{H}_{int} = \frac{g}{2V} \left\{ N_o^2 + 2N_o \sum_{k \neq 0} \left(a_k^\dagger a_k + a_{-k}^\dagger a_{-k} \right) + N_o \sum_{k \neq 0} \left(a_k a_{-k} + a_k^\dagger a_{-k}^\dagger \right) \right\}.$$

$$(11.116)$$

Within this approximation, operator N_{op} associated with the number of particles may be written as

$$N_{op} = \sum_k a_k^\dagger a_k \approx N_o + \frac{1}{2} \sum_{k \neq 0} \left(a_k^\dagger a_k + a_{-k}^\dagger a_{-k} \right). \qquad (11.117)$$

Given the number N of particles, we can eliminate N_o from this Hamiltonian. Discarding terms of higher order than N^2, we have

$$\mathcal{H} = \frac{1}{2} V g n^2 + \frac{1}{2} \sum_{k \neq 0} \left\{ (\epsilon_k^o + ng) \left(a_k^\dagger a_k + a_{-k}^\dagger a_{-k} \right) + ng \left(a_k a_{-k} + a_k^\dagger a_{-k}^\dagger \right) \right\},$$

$$(11.118)$$

where $n = N/V$ and ϵ_k^o is the energy spectrum of free particles. This quadratic form, known as the **Bogoliubov Hamiltonian**, may be diagonalized by the (Bogoliubov) transformation,

$$a_k = u_k \alpha_k - v_k \alpha_{-k}^\dagger \qquad (11.119)$$

and

$$a_k^\dagger = u_k \alpha_k^\dagger - v_k \alpha_{-k}, \qquad (11.120)$$

where u_k and v_k are real and spherically symmetric functions of the vector k. Furthermore, if we impose the condition

$$u_k^2 - v_k^2 = 1, \qquad (11.121)$$

the transformation is duly canonical, and the set of operators $\{\alpha_k\}$ obey the usual Bose commutation relations,

$$\left[\alpha_{k_1}, \alpha_{k_2}^\dagger\right] = \delta_{k_1, k_2} \qquad (11.122)$$

and

$$\left[\alpha_{k_1}^\dagger, \alpha_{k_2}^\dagger\right] = [\alpha_{k_1}, \alpha_{k_2}] = 0. \qquad (11.123)$$

In terms of these new operators, we can write the Hamiltonian

$$\mathcal{H} = \frac{1}{2} V g n^2 - \frac{1}{2} \sum_{k \neq 0} (\epsilon_k^o + ng - E_k) + \frac{1}{2} \sum_{k \neq 0} E_k \left(\alpha_k^\dagger \alpha_k + \alpha_{-k}^\dagger \alpha_{-k}\right), \qquad (11.124)$$

where

$$E_k = \left[(\epsilon_k^o + ng)^2 - (ng)^2\right]^{1/2}. \qquad (11.125)$$

We now analyze the energy spectrum given by equation (11.125). For small values of the momentum k, we have the asymptotic form

$$E_k \rightarrow (2ng\epsilon_k^o)^{1/2} = \left(\frac{ng}{m}\right)^{1/2} \hbar k = c\hbar k. \qquad (11.126)$$

In this limit, E_k is linear with k, which indicates the presence of excitations similar to phonons in an elastic crystal. The constant c may be regarded as the velocity of propagation of sound in superfluid helium. In the limit $k \rightarrow \infty$, we have

$$E_k \rightarrow \epsilon_k^o + ng, \qquad (11.127)$$

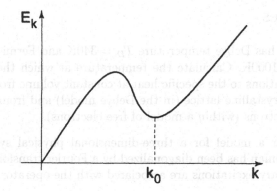

FIGURE 11.2. Energy spectrum of the quasi-particles in superfluid helium. For small k, the spectrum is linear; for large k, it is quadratic.

from which we recover the quadratic spectrum of a free particle. A graph of E_k versus the momentum k should display the shape sketched in figure 11.2, with a minimum at a certain value k_o. This simple theory, however, is too crude to display the minimum at k_o. The elementary excitations in the neighborhood of k_o define a new quasi-particle, called **roton**, with several characteristic properties (but whose analysis is beyond the scope of this book).

There is a well-known criterion proposed by Landau to decide whether a given system may undergo a superfluid transition. According to Landau, there may be superfluidity if

$$\min_k \left\{ \frac{E_k}{\hbar k} \right\} \equiv v_{cr} > 0. \tag{11.128}$$

This Landau criterion is clearly satisfied by the spectrum (11.125), but it cannot be satisfied by the spectrum of a free particle. The argument of Landau (see L. D. Landau and E. M. Lifshitz, *Statistical Physics*, Pergamon Press, London, 1958) is based on an elementary calculation for the flux of a fluid system, with velocity v, through a small capillary tube, at absolute zero. Owing to the viscosity, the kinetic energy of this fluid is gradually dissipated, and the motion is finally stopped. However, friction also works to excite the "internal degrees of freedom" of this fluid system (which should be associated with the appearance of quasi-particles). Landau's criterion may be easily obtained if we write the transformation equations of energy and velocity between reference frames associated with the moving fluid and the capillary, respectively.

Exercises

1. Copper has Debye temperature $T_D = 340K$ and Fermi temperature $T_F = 81000K$. Calculate the temperature at which there are equal contributions to the specific heat at constant volume from vibrations of the crystalline lattice (in the Debye model) and from the conduction electrons (within a model of free electrons).

2. Consider a model for a three-dimensional physical system, whose Hamiltonian has been diagonalized by a Fourier transformation. The elementary excitations are associated with the operator

$$\mathcal{H} = \sum_{\vec{k}} \epsilon_{\vec{k}} a_{\vec{k}}^{\dagger} a_{\vec{k}},$$

where the sum is over the first Brillouin zone, and the creation and annihilation operators obey Bose commutation rules. The energy spectrum is given by

$$\epsilon_{\vec{k}} = A \left| \vec{k} \right|^n,$$

where A is a positive constant and n is an integer (for example, $n = 1$, for phonons in the Debye model, and $n = 2$, for ferromagnetic magnons in the limit of long wave lengths). Obtain an asymptotic expression for the constant volume specific heat at low temperatures. What is the new result for this kind of system on a hypercubic lattice in $d \geq 3$ dimensions?

3. For $T \ll T_D$, where T_D is the Debye temperature, show that the difference between the specific heats at constant volume and at constant pressure behaves as T^7 (note that the ratio c_p/c_V tends to unit).

4. Consider a one-dimensional Heisenberg antiferromagnet consisting of a chain of N localized ions of spin $1/2$. The lowest-lying energy states are represented by spin waves that behave as phonons belonging to a crystalline lattice. The spin wave with momentum $q = 2\pi m/N$, where m is an integer between $-N/2$ and $+N/2$, has energy

$$\epsilon_q = A \left| \sin q \right|,$$

where A is a positive constant. Sketch a graph of ϵ_q versus the momentum q. Owing to the peculiarities of this problem, and in contrast with phonons, there might be only one spin wave associated with each value of momentum. The energy levels of the global system (that is, the chain), given by

$$E\{n_q\} = \sum_q n_q \epsilon_q + E_o,$$

where $E_o > 0$ is the energy of the ground state, are characterized by the set of occupation numbers $\{n_q; n_q = 0, 1\}$.

(a) Write the partition function in order to obtain expressions for the internal energy, the entropy, and the specific heat of this system. In the thermodynamic limit, the answers are given as integrals whose limits of integration should be clearly indicated.

(b) Show that the asymptotic behavior of the entropy at low temperatures is given by the expression

$$S \to S_o + BT^a.$$

Obtain the constants S_o, B, and a.

5. Consider the energy spectrum of the Bogoliubov theory of superfluidity,

$$E_k = \left[\left(\frac{\hbar^2 k^2}{2m} + ng \right)^2 - (ng)^2 \right]^{1/2}.$$

Suppose that g is a function of k such that $g \to g_o$ as $k \to 0$ and $k \to \infty$, and $g \sim a/k$, with $a > 0$, for $k \sim k_o$, where k_o is the roton minimum. Write an expansion for E_k in the neighborhood of k_o. Keeping terms up to second order in this expansion, calculate the contribution of the rotons for the entropy and the specific heat.

where $B_o > 0$ is the energy of the ground state, are characterized by the set of occupation numbers $\{n_o, n_q = 0,1\}$.

(a) Write the partition function in order to obtain expressions for the internal energy, the entropy, and the specific heat of this system. In the thermodynamic limit, the answers are given as integrals whose limits of integration should be clearly indicated.

(b) Show that the asymptotic behavior of the entropy at low temperatures is given by the expression

$$S = S_o + BT^\alpha$$

Obtain the constants S_o, B, and α.

5. Consider the energy spectrum of the Bogoliubov theory of superfluidity,

$$E_k = \left[\left(\frac{\hbar^2 k^2}{2m} + ng \right)^2 - (ng)^2 \right]^{1/2}$$

Suppose that g is a function of k such that $g \to g_o$ as $k \to 0$ and $k \to \infty$, and $g = a/k$, with $k > 0$, for $k \to k_o$, where k_o is the roton minimum. Write an expansion for E_k in the neighborhood of k_o. Keeping terms up to second order in this expansion, calculate the contribution of the rotons for the entropy and the specific heat.

12
Phase Transitions and Critical Phenomena: Classical Theories

Phase transitions and critical phenomena are usual events associated with an enormous variety of physical systems (simple fluids and mixtures of fluids, magnetic materials, metallic alloys, ferroelectric materials, superfluids and superconductors, liquid crystals, etc.). The doctoral dissertation of van der Waals, published in 1873, contains the first successful theory to account for the "continuity of the liquid and gaseous states of matter," and remains an important instrument to analyze the critical behavior of fluid systems. The transition to ferromagnetism has also been explained, since the beginning of the twentieth century, by a phenomenological theory proposed by Pierre Curie, and developed by Pierre Weiss, which is closely related to the van der Waals theory. These **classical theories of phase transitions** are still used to describe qualitative aspects of phase transitions in all sorts of systems.

The classical theories were subjected to a more serious process of analysis in the 1960s, with the appearance of some techniques to carry out much more refined experiments in the neighborhood of the so-called critical points. Several thermodynamic quantities (as the specific heat, the compressibility, and the magnetic susceptibility) present a quite peculiar behavior in the critical region, with asymptotic divergences that have been characterized by a collection of **critical exponents**. It was soon recognized that the critical behavior of equivalent thermodynamic quantities (as the compressibility of a fluid and the susceptibility of a ferromagnet) display a universal character, characterized by the same, well-defined, critical exponent. The classical theories for distinct systems also lead to the same set of (classical) critical exponents. In the 1960s, new experimental data, as well

as a number of theoretical calculations, indicated the existence of classes of critical universality, where just a few parameters (as the dimensionality of the systems under consideration) were sufficient to define a set of critical exponents, in general completely different from the set of classical values.

In this chapter, we present some of the classical theories, which are embraced by a very general phenomenology proposed by Landau in the 1930s. In the next chapter, we discuss some properties of the Ising model, including the exact calculation of the thermodynamic functions in one dimension, and the construction of an alternative model with long-range interactions. Owing to the universal character of critical phenomena, it becomes relevant to consider simplified but nontrivial statistical models, as the Ising model, with interactions involving many degrees of freedom.

From the phenomenological point of view, theories of homogeneity (or scaling) have been proposed to replace the classical phenomenology of Landau, without leading, however, to the determination of the critical exponents. Perturbative schemes, which are in general very useful to treat many-body systems, completely fail in the vicinity of critical points. Nowadays, it is known that this is directly related to the lack of a characteristic length at the critical point. Indeed, except at critical points, correlations between elements of a many-body system decay exponentially with distance. So, there is a well-defined correlation length, but it becomes larger and larger as the critical point is approached. At criticality, correlations decay much slower, with just a power law of distance, and no characteristic length. These ideas have been included in the renormalization-group theory, initially proposed by Kenneth Wilson in the 1970s, representing an advance of tremendous impact in this area. The renormalization-group theory offers a justification for the phenomenological scaling laws and explains the existence of the universality classes of critical behavior. There are many schemes to construct and improve the renormalization-group transformations, leading to (nonclassical) values of the critical exponents.

12.1 Simple fluids. Van der Waals equation

Consider the phase diagram of a simple fluid, in terms of thermodynamic fields, as pressure and temperature. In figure 12.1, solid lines indicate coexistence of phases, with different densities but the same values of the thermodynamic fields. If the system performs a thermodynamic path across a solid line, there will be a first-order transition, according to an old classification of Ehrenfest. At the triple point, T_t, p_t, there is a coexistence among three distinct phases, with different values of the thermodynamic densities (specific volume, entropy per mole). The triple point of pure water, located at $T_t = 273.16$ K, and $p_t = 4.58$ mmHg, is actually used as a standard to establish the Kelvin (absolute) scale of temperatures. The critical point,

FIGURE 12.1. Sketch of the phase diagram, in terms of pressure and temperature, for a simple fluid with a single component. The solid lines represent first-order transitions. At the triple point there is coexistence among three distinct phases.

T_c, p_c, represents the terminus of a line of coexistence of phases (for pure water, the critical temperature is very high, of the order of 650 K; for nitrogen or carbon monoxide, for example, it is considerably lower, below 150 K). In a thermodynamic path along the coexistence curve between liquid and gas phases in the $p - T$ phase diagram, as the critical point is approached, the difference between densities of liquid and gas decreases, and finally vanishes at the critical point, where both phases become identical. The critical transition is then called continuous (or second order). In the neighborhood of the critical point, the anomalous behavior of some thermodynamic derivatives, as the compressibility and the specific heat, gives rise to a new "critical state" of matter.

It may be more interesting to draw a $p - v$ diagram, where $v = V/N$ is specific volume (V is total volume, and N is total number of moles). In figure 12.2, there is a plateau, for $T < T_c$, which indicates the coexistence, at a given pressure, of a liquid phase (with specific volume v_L) and a gaseous phase (with specific volume v_G). As the temperature T increases, the difference $\psi = v_G - v_L$ decreases, and finally vanishes at the critical point (we could as well have chosen ψ as the difference between densities, $\rho_L - \rho_G$, where $\rho = 1/v$). Above the critical temperature, the pressure is a well-behaved function of specific volume.

Now we are prepared to introduce a critical exponent to characterize the asymptotic behavior of ψ as $T \to T_c$,

$$\psi \sim B \left(\frac{T_c - T}{T_c} \right)^\beta , \tag{12.1}$$

FIGURE 12.2. Isotherms for a simple fluid in the neighborhood of the critical point. The plateau indicates the coexistence between two distinct phases, with specific volumes v_L and v_G.

where the prefactor B and the critical temperature T_c do not have any universal character, but the exponent β is about $1/3$ for all fluids (and for analogous quantities in other physical systems). Figure 12.3 represents the data collected by Guggenheim in 1945 for ρ/ρ_c (where ρ_c is the density at the critical point) versus T/T_c along the coexistence curve of eight distinct fluid systems. As pointed out by Guggenheim, these data are fitted to a cubic equation with "high relative accuracy."

The coexistence, or first-order transition, line in the $p - T$ diagram is described by the Clausius–Clapeyron equation, given by comparing the expressions of the Gibbs free energy per mole in both phases. For example, along the coexistence curve between liquid and gas phases, we have

$$g_G\,(T,p) = g_L\,(T,p),\tag{12.2}$$

where g is the Gibbs free energy per mole. Therefore, from the differential form

$$dg = -sdT + vdp,\tag{12.3}$$

where s is the entropy per mole, we can write

$$-s_GdT + v_Gdp = -s_LdT + v_Ldp,\tag{12.4}$$

from which we have the differential form of the Clausius–Clapeyron equation,

$$\frac{dp}{dT} = \frac{s_L - s_G}{v_L - v_G} = \frac{L}{T\Delta v},\tag{12.5}$$

where Δv is the change of specific volume and L is the latent heat of transition.

FIGURE 12.3. Experimental data collected by Guggenheim in 1945 for the co-existence curve of eight different fluids (densities and temperatures are divided by the respective critical values). Note the fitting to a cubic equation.

The phenomenological model of van der Waals

At sufficiently high temperatures, the behavior of a diluted fluid system may be described by Boyle's law,

$$pv = RT, \tag{12.6}$$

where R is the universal gas constant. The first successful theory of the critical region, proposed by van der Waals, consists in the introduction of the following changes in Boyle's law,

$$\begin{aligned} v &\rightarrow v - b, \\ p &\rightarrow p + a/v^2, \end{aligned} \tag{12.7}$$

where b is an effective volume, supposed to account for the hard core of the repulsion interactions between molecules, and a/v^2 is a correction term associated with the attractive part of the pair interactions. Therefore, the van der Waals equation is given by

$$\left(p + \frac{a}{v^2}\right)(v - b) = RT. \tag{12.8}$$

In figure 12.4, we sketch a graph of p versus v for $T < T_c$. Instead of a flat plateau, as in the experimental data, there appear the so-called loops of van der Waals. Given the values of pressure and temperature, there may be up to three distinct values of v. Also, there is a thermodynamically unstable part of the curve, along which $(\partial p/\partial v)_T > 0$. This drawback has been corrected by Maxwell, by introducing the equal-area (or double tangent) construction, as discussed below. Above T_c, the van der Waals isotherms

FIGURE 12.4. Sketch of a van der Waals isotherm for $T < T_c$. The hatched region corresponds to Maxwell's construction of equal areas.

are one-to-one functions of p versus v, approaching the form of Boyle's law as the temperature increases.

To obtain the critical parameters in terms of the phenomenological parameters a and b, we write the van der Waals equation as a polynomial in v,

$$v^3 - \left(b + \frac{RT}{p}\right)v^2 + \frac{a}{p}v - \frac{ab}{p} = 0. \qquad (12.9)$$

At the critical point $(T = T_c; p = p_c)$, this polynomial has three distinct roots, and may be written as

$$(v - v_c)^3 = 0. \qquad (12.10)$$

Therefore, it is easy to see that

$$v_c = 3b, \quad T_c = \frac{8a}{27bR}, \quad \text{and} \quad p_c = \frac{a}{27b^2}. \qquad (12.11)$$

We may also write the van der Waals equation as

$$p = \frac{RT}{v - b} - \frac{a}{v^2}. \qquad (12.12)$$

Thus,

$$\left(\frac{\partial p}{\partial v}\right)_T = -\frac{RT}{(v - b)^2} + \frac{2a}{v^3}, \qquad (12.13)$$

from which we have

$$\left(\frac{\partial p}{\partial v}\right)_{T;v=v_c} = -\frac{RT}{4b^2} + \frac{2a}{27b^3} = -\frac{R}{4b^2}(T - T_c). \qquad (12.14)$$

Above the critical temperature, with $v = v_c$, there is a divergent asymptotic behavior of the isothermal compressibility,

$$\kappa_T = -\frac{1}{v}\left(\frac{\partial v}{\partial p}\right)_T \sim C\left(\frac{T_c - T}{T_c}\right)^{-\gamma}, \tag{12.15}$$

with the critical exponent $\gamma = 1$ (in disagreement with the experimental results, which give γ between 1.2 and 1.4 for fluid systems). With an additional algebraic effort (and using the Maxwell construction), it is not difficult to obtain the following results: (i) the same exponent $\gamma = 1$ still characterizes the divergent critical behavior of the compressibility for temperatures below T_c; (ii) the exponent $\beta = 1/2$ (in disagreement with the experimental result $\beta \approx 1/3$) characterizes the asymptotic behavior of $\psi = v_G - v_L$ (equivalently, of $\psi = \rho_L - \rho_G$) along the coexistence curve; (iii) the specific heat at constant volume (at the critical volume) displays just a discontinuity (which may be associated with a critical exponent $\alpha = 0$) across the critical temperature.

From equation (12.12), and considering the pressure as a function of T and v, we have

$$-\left(\frac{\partial f}{\partial v}\right) = \frac{RT}{v - b} - \frac{a}{v^2}. \tag{12.16}$$

Therefore, the Helmholtz free energy per mole of the van de Waals model is given by

$$f(T, v) = -RT\ln(v - b) - \frac{a}{v} + f_o(T), \tag{12.17}$$

where $f_o(T)$ is a well-behaved function of temperature. The first term of this expression is a convex function of v, but the second term is concave. The equal-area Maxwell construction is designed to recover the full convexity of the Helmholtz free energy (see figure 12.5). As indicated in the figure, we draw the double tangent to the curve of f versus v, and define the specific volumes v_L and v_G associated with the coexisting liquid and gas phases. Besides recovering the convexity of the Helmholtz free energy, the straight line between the points of tangency corresponds to the flat region in the $p - v$ diagram (see figure 12.4).

The Gibbs free energy per mol of the van der Waals model is obtained from the Legendre transformation

$$g(T, p) = \min_v \{f(T, v) + pv\}. \tag{12.18}$$

We should be careful to take the Legendre transformation of the properly convex function $f(T, v)$, duly corrected by the Maxwell construction. The graph of the concave function $g(T, p)$ versus pressure, at sufficiently low temperatures, displays a kink at $p = p_c$, with distinct derivatives at right and at left, corresponding to the specific volumes of the coexisting phases.

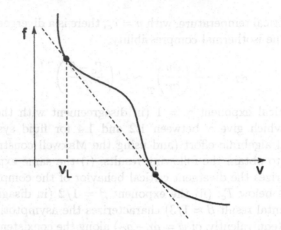

FIGURE 12.5. Sketch of the Helmholtz free energy per mole as a function of specific volume at a given temperature (below the critical temperature). Convexity of the curve is recovered by performing the double-tangent construction (which is equivalent to the equal-area construction of figure 12.4).

Expansion of the van der Waals free energy

To make contact with the Landau phenomenological theory of phase transitions, it is convenient to define

$$g(T, p; v) = f(T, v) + pv, \qquad (12.19)$$

where the function $f(T, v)$ is given by equation (12.17). Using the notation $\psi = v - v_c$, we write an expansion about the critical volume $v_c = 3b$,

$$g(T, p; v) = \sum_{n=0}^{\infty} g_n(T, p)\, \psi^n, \qquad (12.20)$$

where

$$g_o = -RT \ln(2b) - \frac{a}{3b} + f_o(T) + 3bp, \qquad (12.21)$$

$$g_1 = -\frac{RT}{2b} + \frac{a}{9b^2} + p = -\frac{4a}{27b^2}t + \frac{a}{27b^2}\Delta p, \qquad (12.22)$$

$$g_2 = \frac{1}{b^2}\left(\frac{RT}{8} - \frac{a}{27b}\right) = \frac{a}{27b^3}t, \qquad (12.23)$$

$$g_3 = -\frac{1}{3b^3}\left(\frac{RT}{8} - \frac{a}{27b}\right) = -\frac{a}{81b^4}t, \qquad (12.24)$$

$$g_4 = \frac{1}{b^4} \left(\frac{RT}{4 \cdot 2^4} - \frac{a}{3^5 b} \right) = \frac{a}{1944 b^5} + \frac{a}{216 b^5} t, \tag{12.25}$$

and so on, with

$$t = \frac{T - T_c}{T_c} \quad \text{and} \quad \Delta p = \frac{p - p_c}{p_c}. \tag{12.26}$$

At the critical point ($t = 0$; $\Delta p = 0$), the coefficients g_1, g_2, and g_3 vanish, but the coefficient g_4 is positive. As g_4 is positive, and as ψ is supposed to be small in the critical region, we still guarantee the existence of a minimum even if this expansion is truncated at the fourth-order term. In order to analyze the critical region, it is then enough to write

$$g(T, p; \psi) = g_o + g_1 \psi + g_2 \psi^2 + g_3 \psi^3 + g_4 \psi^4. \tag{12.27}$$

As the coefficients g_1, g_2, and g_3 vanish at the critical point, and as the choice of the order parameter ψ is rather arbitrary, we can also make a shift of ψ to eliminate the cubic term. If we write

$$\psi' = \psi - c, \tag{12.28}$$

where

$$c = -\frac{g_3}{4 g_4}, \tag{12.29}$$

we have the simplified expansion

$$g(T, p; \psi) = A_0 + A_1 \psi' + A_2 \left(\psi' \right)^2 + A_4 \left(\psi' \right)^4, \tag{12.30}$$

where

$$A_1 = g_1 - \frac{1}{2} \frac{g_2 g_3}{g_4} + \frac{1}{8} \frac{g_3^3}{g_4^2}, \tag{12.31}$$

$$A_2 = g_2 - \frac{3}{8} \frac{g_3^2}{g_4}, \tag{12.32}$$

and

$$A_4 = g_4. \tag{12.33}$$

At the critical point ($t = 0$; $\Delta p = 0$), it is clear that $A_1 = A_2 = 0$, with $A_4 > 0$, and that $\psi \rightarrow \psi' \rightarrow 0$, in agreement with the idea that there is room for choosing different forms of the parameter ψ to distinguish between the coexisting phases (as far as these choices lead to order parameters that

vanish at the critical point). To check that this is indeed the case, we note that

$$A_1 = -\frac{4a}{27b^2}t + \frac{a}{27b^2}\Delta p + O\left(t^2\right), \tag{12.34}$$

$$A_2 = \frac{a}{27b^3}t + O\left(t^2\right), \tag{12.35}$$

and

$$A_4 = \frac{a}{1944b^5} + O\left(t\right). \tag{12.36}$$

As we shall see later, it is not difficult to find the minima of equation (12.30) in the neighborhood of the critical point. Indeed, we shall see that expansions (12.27) or (12.30) already indicate that the van der Waals model fits into the more general scheme proposed by Landau to analyze the critical behavior.

In a simple fluid, critical points can only occur as isolated points in the $T - p$ phase diagram (since we have to fix two different coefficients, which are functions of two independent variables, T and p). In the case of a fluid with two distinct components, for example, we have three thermodynamic degrees of freedom, which may give rise to a line of critical points in a three-dimensional phase diagram (in terms of $T - p - \mu_1$, for example).

12.2 The simple uniaxial ferromagnet. The Curie–Weiss equation

Owing to the higher symmetry, the analysis of the phase diagram of a simple ferromagnmet is much easier than the corresponding treatment of a simple fluid. The phase diagram in terms of the external applied magnetic field, H, and the temperature (the analog of the $p - T$ phase diagram of simple fluids) is sketched in figure 12.6.

Along the coexistence line ($H = 0$, $T < T_c$), the ferro-1 ("spin up") and ferro-2 ("spin down") ordered phases have the same magnetic free energy per ion, $g(T, H = 0)$, and (spontaneous) magnetizations of the same intensity, but opposite signs (vis-à-vis the direction of the applied magnetic field). We then draw the graph of figure 12.7, where ψ is the spontaneous magnetization and T_c is the Curie temperature. The spontaneous magnetization vanishes above T_c. Although much more symmetric, it plays the same role as the difference $v_G - v_L$ (or $\rho_L - \rho_G$) for simple fluids. Thermodynamic quantities of this nature are called **order parameters**. For the uniaxial ferromagnet, the order parameter ψ is associated with the breaking

FIGURE 12.6. Sketch of the applied field versus temperature phase diagram of a simple uniaxial ferromagnetic system. The coexistence region is given by $H = 0$ with $T < T_c$.

FIGURE 12.7. Order parameter (spontaneous magnetization) as a function of temperature for a simple uniaxial ferromagnet.

FIGURE 12.8. Sketch of the isotherms of a simple uniaxial ferromagnet (magnetization versus applied field).

of symmetry of the system; at $H = 0$, above T_c, there is a more symmetric phase, $\psi = 0$; below T_c this symmetry is broken (since $\psi \neq 0$).

Landau introduced the concept of **order parameter** and gave several examples: the spontaneous magnetization of a ferromagnet; the difference $v_G - v_L$ along a coexistence curve of a fluid system; the difference between the densities of copper and zinc on a site of a crystalline lattice of the binary alloy of copper and zinc; the spontaneous polarization of a ferroelectric compound; the staggered magnetization of an antiferromagnet, etc.

We now look in more detail at the properties of the simple uniaxial ferromagnet. The typical $m - H$ diagram, where m is magnetization (taking into account eventual corrections to eliminate demagnetization effects associated with the shape of the crystal), is given by the sketch of figure 12.8. The situation is much more symmetric (and, thus, much simpler) than the previous case of fluids. Let us take advantage of this simplicity (and of the more intuitive language of ferromagnetism) to define some critical exponents:

(i) The order parameter associated with the transition behaves as

$$\psi = m\,(T, H \to 0+) \sim B\,(-t)^{\beta}\,, \qquad (12.37)$$

where $T < T_c$, and $t = (T - T_c)/T_c$. The prefactor B is not universal, but the exponent β is always close to $1/3$, as in the case of fluids. Also, we could have defined ψ from other quantities, such as $m\,(T, H \to 0-)$ or the difference $m\,(T, H \to 0+) - m\,(T, H \to 0-)$. We always use the sign \sim to indicate an asymptotic behavior in the neighborhood of the critical point, along a well-determined thermodynamic path on the phase diagram.

(ii) The magnetic susceptibility,

$$\chi\,(T, H) = \left(\frac{\partial m}{\partial H}\right)_T\,, \qquad (12.38)$$

corresponds to the derivative of the order parameter with respect to the canonically conjugated thermodynamic field. Its critical behavior is given by

$$\chi(T, H = 0) \sim \begin{cases} C_+ t^{-\gamma} & ; \quad (T > T_c), \\ C_- (-t)^{-\gamma} & ; \quad (T < T_c), \end{cases} \tag{12.39}$$

which leads to a critical exponent γ, whose values are between 1.2 and 1.4, as in the case of fluids. To be more precise, we should refer to distinct exponents, γ above T_c and γ' below T_c. This distinction, however, is usually irrelevant.

(iii) The specific heat in zero field (that is, along a locus parallel to the coexistence curve of the phase diagram in terms of the thermodynamic fields) behaves as

$$c_{H=0} \sim \begin{cases} A_+ t^{-\alpha} & ; \quad (T > T_c), \\ A_- (-t)^{-\alpha} & ; \quad (T < T_c), \end{cases} \tag{12.40}$$

with a positive, but very small, critical exponent α (which is thus compatible with a very weak divergence, maybe of a logarithmic character).

(iv) From the critical isotherm,

$$m(T = T_c, H) \sim H^{1/\delta}, \tag{12.41}$$

we define the exponent $\delta \approx 3$.

The Curie–Weiss phenomenological theory

Paramagnetism has been explained by Langevin on the basis of a classical model of localized and noninteracting magnetic moments. In the quantum version of Brillouin, with localized and noninteracting magnetic moments of spin 1/2 on the sites of a crystalline lattice, we describe paramagnetism with the spin Hamiltonian

$$\mathcal{H} = -H \sum_{i=1}^{N} \sigma_i, \tag{12.42}$$

where $\sigma_i = \pm 1$ for all sites of the lattice and H is the external magnetic field in suitable units (in terms of the Bohr magneton and the gyromagnetic constant). The (dimensionless) magnetization is given by the equation of state

$$m = \tanh\left(\frac{H}{k_B T}\right), \tag{12.43}$$

which is analogous to the Boyle law for an ideal fluid. It is obvious that $m = 0$ for $H = 0$, without any possibility of explaining ferromagnetism. If we use the standard notation of statistical mechanics, $\beta = 1/(k_B T)$, without any confusion with a critical exponent, the magnetic susceptibility is given by

$$\chi(T, H) = \left(\frac{\partial m}{\partial H}\right)_T = \frac{\beta}{\cosh^2(\beta H)}. \tag{12.44}$$

Thus, we have

$$\chi(T, H = 0) \equiv \chi_o(T) = \frac{1}{k_B T}, \tag{12.45}$$

which is the expression of the well-known Curie law of paramagnetism (note that the Curie law is usually written as $\chi_o = C/T$, where the constant C is proportional to the square of the magnetic moment).

To explain the occurrence of a spontaneous magnetization in ferromagnetic compounds, let us introduce the idea of an effective field (also known as **molecular** or **mean field**), as introduced by Pierre Weiss. In analogy with the phenomenological treatment of van der Waals for fluid systems, let us replace the external field H by the effective field

$$H_{eff} = H + \lambda m, \tag{12.46}$$

where m is the magnetization per spin and λ is a parameter that expresses the (average) effect on a particular lattice site of the field produced by all neighboring spins. We then have the equation of state

$$m = \tanh(\beta H + \beta \lambda m), \tag{12.47}$$

known as the **Curie–Weiss equation**. For $H = 0$, it is easy to search for the solutions of this equation. Considering figure 12.9, we see that $m = 0$ is always a solution. However, for $\beta \lambda > 1$, that is, for

$$T < T_c = \frac{\lambda}{k_B}, \tag{12.48}$$

there are two additional solutions, $\pm m \neq 0$, with equal intensities but opposite signs.

In the neighborhood of the critical point ($T \approx T_c$, $H \approx 0$), m is small, and we can expand equation (12.47),

$$\beta(H + \lambda m) = \tanh^{-1} m = m + \frac{1}{3}m^3 + \frac{1}{5}m^5 + \dots. \tag{12.49}$$

Now, we are able to deduce some asymptotic formulas in the neighborhood of the critical point:

FIGURE 12.9. Solutions of the Curie–Weiss equation in zero field.

(i) To calculate the spontaneous magnetization (that is, the magnetization with $H = 0$), we write

$$\beta \lambda m = m + \frac{1}{3} m^3 + \dots . \tag{12.50}$$

Therefore, $m^2 \sim -3t$, from which we have

$$m\,(T, H = 0) \sim \pm \sqrt{3}\,(-t)^{1/2} . \tag{12.51}$$

The exponent β associated with the order parameter is $1/2$, as in the van der Waals theory, but in open disagreement with the experimental data.

(ii) To calculate the magnetic susceptibility, let us take the derivative of both sides of equation (12.49) with respect to the magnetic field,

$$\beta \left(\frac{\partial H}{\partial m} + \lambda \right) = 1 + m^2 + m^4 + \dots . \tag{12.52}$$

For $H = 0$ and $T > T_c$, we have

$$\beta \left(\frac{1}{\chi_o} + \lambda \right) = 1. \tag{12.53}$$

Hence,

$$\chi_o = \chi\,(T, H = 0) \sim \frac{1}{k_B T_c} t^{-1}. \tag{12.54}$$

For $H = 0$ and $T < T_c$, we have

$$\beta \left(\frac{1}{\chi_o} + \lambda \right) = 1 - 3t + O\left(t^2\right). \tag{12.55}$$

Hence,

$$\chi_o \sim \frac{1}{2k_B T_c} (-t)^{-1}. \tag{12.56}$$

Therefore, $\gamma = 1$, and $C_+/C_- = 2$, as in the van der Waals theory for a simple fluid.

(iii) To obtain the critical isotherm, at $T = T_c$ (that is, $\beta\lambda = 1$), we write

$$\beta_c H + m = m + \frac{1}{3}m^3 + \dots . \tag{12.57}$$

We then have the asymptotic form

$$H \sim \frac{k_B T_c}{3} m^3, \tag{12.58}$$

from which $\delta = 3$, which is also the same value as calculated from the van der Waals theory. Later, we shall see that the specific heat at zero field is finite but displays a discontinuity at T_c. Therefore, it can be characterized by he critical exponent $\alpha = 0$.

Free energy of the Curie–Weiss theory

As in the van der Waals model, we can write an expansion for the free energy in terms of the order parameter (owing to the symmetry of the ferromagnetic system, this expansion will be much simpler). From the dimensionless magnetization, given by

$$m = m(T, H) = -\left(\frac{\partial g}{\partial H}\right)_T, \tag{12.59}$$

we write the magnetic Helmholtz free energy,

$$f(T, m) = g(T, H) + mH, \tag{12.60}$$

where

$$H = \left(\frac{\partial f}{\partial m}\right)_T. \tag{12.61}$$

Therefore, if we write the Curie–Weiss equation (12.47) in the form

$$H = \frac{1}{\beta} \tanh^{-1}m - \lambda m, \tag{12.62}$$

we have

$$f(T, m) = \frac{1}{\beta} \int \left(\tanh^{-1}m\right) dm - \frac{\lambda}{2}m^2 + f_o(T), \tag{12.63}$$

where $f_o(T)$ is a well-behaved function of temperature. Expanding $\tanh^{-1} m$ and integrating term by term, we have

$$f(T, m) = f_o(T) + \frac{1}{2\beta}(1 - \beta\lambda)m^2 + \frac{1}{\beta}\left(\frac{1}{12}m^4 + \frac{1}{30}m^6 + \dots\right).$$
(12.64)

In contrast with the case of fluids, since there are only even powers of the order parameter, the problem is drastically simplified (although, of course, there are similar problems of convexity). From the magnetic Gibbs free energy per particle,

$$g(T, H) = \min_m\{g(T, H; m)\} = \min_m\{f(T, m) - mH\},$$
(12.65)

we obtain the (Landau) expansion for ferromagnetism,

$$g(T, H; m) = f_o(T) - Hm + \frac{1}{2\beta}(1 - \beta\lambda)m^2 + \frac{1}{12\beta}m^4 + \frac{1}{30\beta}m^6 + \dots.$$
(12.66)

At the critical point, the coefficients of the linear and the quadratic terms should vanish, $H = 0$, and $\beta\lambda = 1$, respectively, while the coefficient of the quartic term remains positive (to guarantee the existence of a minimum). For $H = 0$ and $T > T_c$, the minimum of $g(T, H; m)$ is located at $m = 0$. For $H = 0$ and $T < T_c$, the solution $m = 0$ is a maximum. There are then two symmetric minima at $\pm m \neq 0$. It is not difficult to use this expansion to show that the specific heat in zero field remains finite, but exhibits a discontinuity, $\Delta c = (3k_B)/2$, at the critical temperature.

12.3 The Landau phenomenology

The Landau theory of continuous phase transitions is based on the expansion of the free energy in terms of the invariants of the order parameter. The free energy is thus supposed to be an analytic functions even in the neighborhood of the critical point.

In many cases, it is relatively easy to figure out a number of acceptable order parameters associated with a certain phase transition. The order parameter does not need to be a scalar (for more complex systems, it may be either a vector or even a tensor). In general, we have $\psi = 0$ in the more symmetric phase (usually, in the high-temperature, disordered phase), and $\psi \neq 0$ in the less symmetric (ordered) phase. We may give several examples: (i) the liquid–gas transition, in which ψ may be given either by $v_G - v_L$ or by $\rho_L - \rho_G$ (v is the specific volume and ρ is the particle density); (ii) the para-ferromagnetic transition, in which the order parameter ψ may be the magnetization vector (but becomes a scalar for

uniaxial systems) in zero applied field; (iii) the antiferromagnetic transition, in which ψ may be associated with the sublattice magnetization; (iv) the order–disorder transition in binary alloys of the type $Cu - Zn$, in which the order parameter may be chosen as the difference between the densities of atoms of either copper or zinc on the sites of one of the sublattices; (v) the structural transition in compounds as barium titanate, $BaTiO_3$, in which ψ is associated with the displacement of a sublattice of ions with respect to the other sublattice; (vi) the superfluid transition in liquid helium, in which ψ is associated with a complex wave function. In all these cases the free energy is expanded in terms of the invariants of the order parameter, which reflects the underlying symmetries of the physical system.

For a pure fluid, as the order parameter is a scalar, we write the expansion

$$
\begin{aligned}
g\left(T,p;\psi\right) = \; & g_o\left(T,p\right) + g_1\left(T,p\right)\psi + g_2\left(T,p\right)\psi^2 \\
& + g_3\left(T,p\right)\psi^3 + g_4\left(T,p\right)\psi^4 + \ldots ,
\end{aligned} \tag{12.67}
$$

where the coefficients g_n are functions of the thermodynamic fields (T and p). The van der Waals model is perfectly described by this phenomenology. Indeed, to account for the existence of a simple critical point, it is sufficient to have g_4 positive (to guarantee the existence of a minimum with respect to ψ). We thus ignore higher-order terms in this Landau expansion. Also, as there is some freedom to choose ψ, it is always possible to eliminate the cubic term of this expansion. Thus, without any loss of generality, in the neighborhood of a simple critical point, we can write the expansion

$$
g\left(T,p;\psi\right) = A_o\left(T,p\right) + A_1\left(T,p\right)\psi + A_2\left(T,p\right)\psi^2 + \psi^4, \tag{12.68}
$$

which has already been written for the van der Waals fluid. To have a stable minimum, the coefficients A_1 and A_2 should vanish at the critical point. However, it is important to point out that even the existence of this expansion cannot be taken for granted (the exact solution of the two-dimensional Ising model provides a known counterexample). The derivatives of $g\left(T,p;\psi\right)$ are given by

$$
\frac{\partial g}{\partial \psi} = A_1 + 2A_2\psi + 4\psi^3 = 0 \tag{12.69}
$$

and

$$
\frac{\partial^2 g}{\partial \psi^2} = 2A_2 + 12\psi^2. \tag{12.70}
$$

For $A_1 = 0$, we have either $\psi = 0$ or $\psi^2 = -A_2/2$. The solution $\psi = 0$ is stable for $A_2 > 0$. With $\psi \neq 0$, we have $\partial^2 g/\partial \psi^2 = -4A_2$, that is, the solution $\psi \neq 0$ is stable for $A_2 < 0$.

For a uniaxial ferromagnet, the Landau expansion is much simpler. Owing to symmetry, we can write the magnetic Helmholtz free energy

$$
f\left(T,m\right) = f_o\left(T\right) + A\left(T\right)m^2 + B\left(T\right)m^4 + \ldots . \tag{12.71}
$$

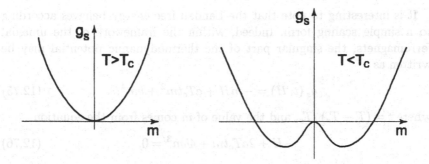

FIGURE 12.10. Singular part of the Landau thermodynamic potential of a simple uniaxial ferromagnet as a function of magnetization. Below the critical temperature, the paramagnetic minimum becomes a maximum (and there appear two symmetric minima).

Therefore,

$$g(T, H; m) = f_o(T) - Hm + A(T)m^2 + B(T)m^4 + \dots . \qquad (12.72)$$

At the critical point, we have $H = 0$ and $A(T) = 0$. In the neighborhood of the critical point, we then write

$$A(T) = a(T - T_c), \qquad (12.73)$$

with $a > 0$, $B(T) = b > 0$, and $f_o(T) \approx f_o(T_c)$. Therefore, we can write

$$g(T, H; m) = f_o(T_c) - Hm + a(T - T_c)m^2 + bm^4. \qquad (12.74)$$

For $H = 0$, with positive values of the constant a and b, we can draw the graphs sketched in figure 12.10, where $g_s = g - f_o$ represents the "singular part" of the thermodynamic potential. Now it is easy to calculate a number of thermodynamic quantities to show that $\beta = 1/2$, $\gamma = 1$, $C_+/C_- = 2$, and $\delta = 3$, regardless of the particular features of the systems under consideration. It is also easy to show that the specific heat in zero field exhibits a discontinuity $\Delta c = a^2 T_c/2b$ at the critical temperature. In zero field, above T_c, the minimum of g_s is a quadratic function of m (at the critical point, this minimum becomes a quartic function of m).

The Landau expansion associated with more-general cases may lead to much more involved results (of experimental relevance). For example, in the case of certain fluids, the cubic term may not be eliminated from a simple shift of the linear term, which leads to a first-order transition, with the coexistence of distinct phases. If there are additional independent variables, and A_4 may also vanish, we are forced to carry out the expansion to higher orders to be able to investigate the occurrence of multicritical phenomena (which are beyond the scope of this text).

It is interesting to note that the Landau free energy behaves according to a simple scaling form. Indeed, within the framework of the uniaxial ferromagnets, the singular part of the thermodynamic potential may be written as

$$g_s(t, H) = -mH + aT_c tm^2 + bm^4, \tag{12.75}$$

where $t = (T - T_c)/T_c$, and the value of m comes from the equation

$$-H + 2aT_c tm + 4bm^3 = 0. \tag{12.76}$$

Using these two equations, it is not difficult to see that we can write

$$g_s(t, H) = \lambda g_s(\lambda^x t, \lambda^y H), \tag{12.77}$$

with $x = -1/2$ and $y = -3/4$, for all values of λ. Therefore, with the choice $\lambda^x t = 1$, we have

$$g_s(t, H) = t^2 g_s\left(1, \frac{H}{t^{3/2}}\right) = t^2 F\left(\frac{H}{t^{3/2}}\right). \tag{12.78}$$

Later, we shall see that the scaling phenomenological hypotheses, which are much less restrictive than the Landau theory, just require that

$$g_s(t, H) = t^{2-\alpha} F\left(\frac{H}{t^{\beta\delta}}\right), \tag{12.79}$$

with exponents α and $\Delta = \delta\beta$ to be determined from the experimental data.

Exercises

1. Show that the van der Waals equation may be written as

$$\pi = \frac{4(1+t)}{1 + \frac{3}{2}\omega} - \frac{3}{(1+\omega)^2} - 1,$$

where $\pi = (p - p_c)/p_c$, $\omega = (v - v_c)/v_c$, and $t = (T - T_c)/T_c$.

Use this result to show that

$$\pi = 4t - 6t\omega - \frac{3}{2}\omega^3 + O\left(\omega^4, t\omega^2\right),$$

in the neighborhood of the critical point.

From the van der Waals isotherms, including the Maxwell construction, obtain the asymptotic form of the coexistence curve in the neighborhood of the critical point. In other words, use the equations

$$\pi = 4t - 6t\omega_1 - \frac{3}{2}\omega_1^3,$$

$$\pi = 4t - 6t\omega_2 - \frac{3}{2}\omega_2^3,$$

and

$$\int_{\omega_1}^{\omega_2} \left(4t - 6t\omega - \frac{3}{2}\omega^3\right) d\omega = \pi\left(\omega_2 - \omega_1\right),$$

which represent the "equal-area construction," to obtain asymptotic expressions for ω_1, ω_2, and π, in terms of t for $t \to 0-$. You may use truncations up to the order $\omega_1, \omega_2 \sim t^{1/2}$.

2. Use the results of the previous problem to show that the curve of coexistence and the "locus" of $v = v_c$ for the van der Waals model meet tangentially at the critical point.

3. Obtain the following asymptotic expressions for the isothermal compressibility of the van der Waals model:

$$\kappa_T\left(T, v = v_c\right) \sim Ct^{-\gamma} \quad (t \to 0+)$$

and

$$\kappa_T\left(T = T_c, p\right) \sim \overline{C}\pi^{-\gamma_p} \quad (p \to p_c+),$$

where

$$\kappa_T = -\frac{1}{v}\left(\frac{\partial v}{\partial p}\right)_T.$$

What are the values of the critical exponents and the critical prefactors of these quantities?

4. Consider the Berthelot gas, given by the equation of state

$$\left(p + \frac{a}{Tv^2}\right)(v - b) = RT.$$

(a) Obtain the critical parameters, v_c, T_c, and p_c.

(b) Obtain an expression for the Helmholtz free energy per mole, $f\left(T, v\right)$, except for an arbitrary function of temperature.

(c) The Gibbs free energy per mole may be written as

$$g\left(T, p\right) = \min_v \left\{g\left(T, p; v\right)\right\} = \min_v \left\{pv + f\left(T, v\right)\right\}.$$

Write an expansion for $g\left(T, p; v\right)$ in terms of $\psi = v - v_c$. Check the critical conditions, and make a shift of ψ in order to eliminate the cubic term of the expansion.

(d) What is the asymptotic expression of the curve of coexistence of phases in the immediate vicinity of the critical point?

(e) Use your results to obtain the critical exponents β, γ, δ, and α.

5. Consider the Curie–Weiss equation for ferromagnetism,

$$m = \tanh\left(\beta H + \beta \lambda m\right).$$

Obtain an asymptotic expression for the isothermal susceptibility, $\chi(T, H)$, at $T = T_c$ for $H \to 0$. Obtain asymptotic expressions for the spontaneous magnetization for $T \ll T_c$ (that is, for $T \to 0$) and $T \approx T_c$ (that is, for $t \to 0-$).

13
The Ising Model

Most of the experiments in the neighborhood of critical points indicate that critical exponents assume the same universal values, far from the predictions of the "classical theories" (as represented by Landau's phenomenology, for example). We now recognize that the universal values of the critical exponents depend on a just few ingredients:

(i) The dimension of physical systems. Usual three-dimensional systems are associated with a certain class of critical exponents. There are experimental realizations of two-dimensional systems, whose critical behavior is characterized by another class of distinct and well-defined critical exponents.

(ii) The dimension of the order parameter. For simple fluids and uniaxial ferromagnets, the order parameter is a scalar number. For an isotropic ferromagnet, the critical parameter is a three-dimensional vector.

(iii) The range of the microscopic interactions. For most systems of physical interest, the microscopic interactions are of short range. We will see that statistical systems with long-range microscopic interactions lead to the set of classical critical exponents.

Owing to the universal behavior of critical exponents, it is enough to analyze very simple (but nontrivial) models in order to construct a microscopic theory of the critical behavior. The Ising model, including short-range interactions between spin variables on the sites of a d-dimensional lattice, plays the role of a prototypical system. The Ising spin Hamiltonian is given

by

$$\mathcal{H} = -J \sum_{(ij)} \sigma_i \sigma_j - H \sum_{i=1}^{N} \sigma_i, \qquad (13.1)$$

where σ_i is a random variable assuming the values ± 1 on the sites $i = 1, 2, ..., N$ of a d-dimensional hypercubic lattice. The first term, where the sum is over pairs of nearest-neighbor sites, represents the interaction energies introduced to bring about an ordered ferromagnetic state (if $J > 0$). The second term, involving the interaction between the applied field H and the spin system, is of a purely paramagnetic character (as we have already seen in previous chapters of this book). Since it was proposed by Lenz and solved in one dimension by Ernst Ising in 1925, the Ising model has gone through a long history [see, for example, the paper by S. G. Brush, in Rev. Mod. Phys. **39**, 883 (1967)].

The Ising model can represent the main features of distinct physical systems. In the usual magnetic interpretation, the Ising spin variables are taken as spin components (that may be pointing either up or down, along the direction of the applied field) of crystalline magnetic ions. We may also consider a binary alloy of type AB. In this case, the spin variables indicate whether a certain site on the crystalline lattice is occupied by an atom of either type A or type B (neighbors of the same type contribute with an energy $-J$; neighbors of different types, contribute with $+J$). As another example, take the ± 1 spin variables to indicate either the presence $(+1)$ or the absence (-1) of a molecule in a certain cell of a "lattice gas" (which is a useful model for the critical behavior of a fluid system). This multiplicity of interpretations is compatible with the ability of the Ising model to represent the main features of the critical behavior of many different physical systems.

From the point of view of magnetism, the Ising Hamiltonian may be regarded as a kind of approximation for the Heisenberg Hamiltonian, associated with a highly anisotropic spin-1/2 magnetic insulator. The energy J is interpreted as the quantum exchange parameter of electrostatic origin. In this chapter, we take advantage of the more intuitive language of this magnetic analogy to derive some properties of the Ising model.

In order to solve the Ising problem, we have to obtain the canonical partition function

$$Z_N = Z(T, H, N) = \sum_{\{\sigma_i\}} \exp(-\beta \mathcal{H}), \qquad (13.2)$$

where the sum is over all configurations of spin variables, and the Hamiltonian is given by equation (13.1). From this partition function, we have the magnetic free energy per site,

$$g = g(T, H) = \lim_{N \to \infty} \left[-\frac{1}{\beta N} \ln Z_N \right]. \qquad (13.3)$$

In one dimension, it is relatively easy to obtain an expression for this free energy. We will use the technique of the transfer matrices, which can also be written in higher dimensions, to obtain a solution for the Ising chain. However, as shown by Ising in 1925, this one-dimensional solution is quite deceptive, since the free energy is an analytic function of T and H (except at the trivial point $T = H = 0$), which precludes the existence of a spontaneous magnetization (and of any phase transition).

Several approximate techniques have been developed to solve the Ising model in two and three dimensions. Some of them are quite simple and useful, and may lead to reasonable qualitative results for the phase diagrams (besides providing useful tools to investigate more complex model systems). However, as pointed out before, phase transitions are associated with a nonanalytic behavior of the free energy in the thermodynamic limit. As a consequence, we should be warned against any truncations or perturbative expansions around the critical point. Indeed, most of the approximate schemes can be written as a Landau expansion, leading to classical critical exponents.

In a mathematical "tour de force," Lars Onsager, in 1944, obtained an analytical solution for the Ising model on a square lattice, with nearest-neighbor interactions, in the absence of an external field. For $T \to T_c$, the specific heat diverges according to a logarithmic asymptotic form,

$$c_{H=0} \sim \ln |T - T_c|, \tag{13.4}$$

with a well-defined critical temperature, $k_B T_c / J = 2 / \ln \left(1 + \sqrt{2}\right)$. Therefore, the free energy is not analytic at T_c, and cannot be written as a Landau expansion. The Onsager solution has been reproduced and confirmed by different techniques on many planar lattices (with first-neighbor interactions). It represents a true milestone in the development of the modern theories of critical phenomena. By the first time, it was shown that a microscopic model leads to nonanalytic behavior within the framework of equilibrium statistical mechanics. The origins of this nonanalyticity were later explained, under much more general grounds, by the Yang and Lee theory of phase transitions (see Chapter 7), including the remarkable "circle theorem" about the zeros of the partition function in the thermodynamic limit. In the 1950s, C. N. Yang checked a result of Onsager for the spontaneous magnetization of the Ising ferromagnet on the square lattice to obtain the exponent $\beta = 1/8$, in sharp contrast with the classical value. Nowadays, although there are no exact solutions in a field, we may be sure that $\gamma = 7/4$ in two dimensions. All planar lattices, with short-range interactions, lead to the same set of critical exponents ($\alpha = 0$, $\beta = 1/8$, $\gamma = 7/4$), which are far from the experimental values for three-dimensional systems, and far as well from the classical Landau results.

The solution of the Ising model in three dimensions remains an open (and probably impossible) problem. However, we can use an argument due to

Peierls to prove the existence of spontaneous magnetization at sufficiently low temperatures. Also, since the 1960s there have been many efforts to obtain quite long series expansions (at high as well as low temperatures) for several thermodynamic quantities associated with the three-dimensional Ising model. From refined asymptotic analyses of these series, we obtain a range of values for the critical exponents in agreement with experimental measurements ($\beta \approx 5/16$, $\gamma \approx 5/4$, $\alpha \approx 1/8$). Also, more recent, and much more sophisticated, renormalization-group techniques lead to similar results. In the table below, we give the values of some usual thermodynamic critical exponents.

	Landau	Ising $(d=2)$	Ising $(d=3)$	Experiments
β	1/2	1/8	$\approx 5/16$	$0.3 - 0.35$
γ	1	7/4	$\approx 5/4$	$1.2 - 1.4$
δ	3	15	≈ 5	$4.2 - 4.8$
α	0	0 (log)	$\approx 1/8$	≈ 0

13.1 Exact solution in one dimension

In one dimension $(d = 1)$, the Ising Hamiltonian is written as

$$\mathcal{H} = -J \sum_{i=1}^{N} \sigma_i \sigma_{i+1} - H \sum_{i=1}^{N} \sigma_i. \tag{13.5}$$

The canonical partition function is given by

$$Z_N = \sum_{\{\sigma_i\}} \exp\left[K \sum_{i=1}^{N} \sigma_i \sigma_{i+1} + \frac{L}{2} \sum_{i=1}^{N} (\sigma_i + \sigma_{i+1}) \right], \tag{13.6}$$

where $K = \beta J$, $L = \beta H$, and the second term has been rearranged to take advantage of a more symmetric form. As a matter of convenience, we adopt periodic boundary conditions, $\sigma_{N+1} = \sigma_1$. Now it is interesting to write the partition function as

$$Z_N = \sum_{\sigma_1, \sigma_2, \cdots, \sigma_N} \prod_{i=1}^{N} T(\sigma_i, \sigma_{i+1}), \tag{13.7}$$

where

$$T(\sigma_i, \sigma_{i+1}) = \exp\left[K \sigma_i \sigma_{i+1} + \frac{L}{2} (\sigma_i + \sigma_{i+1}) \right]. \tag{13.8}$$

This last expression can also be written as a standard 2×2 matrix, whose indices are the spin variables, $\sigma_i = \pm 1$ and $\sigma_{i+1} = \pm 1$. We then define a

transfer matrix,

$$\mathbf{T} = \begin{pmatrix} T(+,+) & T(+,-) \\ T(-,+) & T(-,-) \end{pmatrix} = \begin{pmatrix} \exp(K+L) & \exp(-K) \\ \exp(-K) & \exp(K-L) \end{pmatrix},$$

(13.9)

and use the matrix formalism to see that equation (13.7) for the canonical partition function is a trace of a product of N identical transfer matrices,

$$Z_N = \mathrm{Tr}\,(\mathbf{T})^N.$$

(13.10)

Furthermore, the transfer matrix (13.9) is symmetric, and can thus be diagonalized by a unitary transformation,

$$\mathbf{U}\mathbf{T}\mathbf{U}^{-1} = \mathbf{D}, \quad \text{with} \quad \mathbf{U}^{-1} = \mathbf{U}^\dagger,$$

(13.11)

where \mathbf{D} is a diagonal matrix. Therefore, the canonical partition function can be written in terms of the eigenvalues of the transfer matrix,

$$Z_N = \mathrm{Tr}\,\left(\mathbf{U}^{-1}\mathbf{D}\mathbf{U}\right)^N = \mathrm{Tr}\,(\mathbf{D})^N = \lambda_1^N + \lambda_2^N,$$

(13.12)

where

$$\lambda_{1,2} = e^K \cosh L \pm \left[e^{2K}\cosh^2 L - 2\sinh(2K)\right]^{1/2},$$

(13.13)

are given by the roots of the secular equation, $\det(\mathbf{T} - \lambda\mathbf{I}) = 0$. It is easy to see that these eigenvalues are always positive, and that $\lambda_1 > \lambda_2$ (except at the trivial point $T = H = 0$). In zero field, we have,

$$\lambda_1 = 2\cosh K \geq \lambda_2 = 2\sinh K,$$

(13.14)

with a degeneracy ($\lambda_1 = \lambda_2$) in the limit $K \to \infty$ (that is, for $T \to 0$).

To obtain the free energy in the thermodynamic limit, it is convenient to write

$$Z_N = \lambda_1^N \left[1 + \left(\frac{\lambda_2}{\lambda_1}\right)^N\right].$$

(13.15)

Since $\lambda_2 < \lambda_1$, we have the limit

$$g(T,H) = \lim_{N\to\infty}\left[-\frac{1}{\beta N}\ln Z_N\right] = -\frac{1}{\beta}\ln\lambda_1,$$

(13.16)

that is,

$$g(T,H) = -\frac{1}{\beta}\ln\left\{e^{\beta J}\cosh(\beta H) + \left[e^{2\beta J}\cosh^2(\beta H) - 2\sinh(2\beta J)\right]^{1/2}\right\},$$

(13.17)

FIGURE 13.1. Ising chain with six sites and two different domains (at left and at right of a point wall).

which is an analytic function of T and H, from which we derive all the thermodynamic properties of the one-dimensional system.

The magnetization per spin is given by

$$m\left(T, H\right) = -\left(\frac{\partial g}{\partial H}\right)_T = \frac{\sinh\left(\beta H\right)}{\left[\sinh^2\left(\beta H\right) + \exp\left(-4\beta J\right)\right]^{1/2}}. \tag{13.18}$$

We then see that, as $m\left(T, H = 0\right) = 0$, this model is unable to explain ferromagnetism. From the entropy per spin, $s = s\left(T, H\right) = -\left(\partial g/\partial T\right)_H$, we can calculate the specific heat at constant field. In zero field, we have

$$c_{H=0} = \frac{J^2}{k_B T^2}\left[\operatorname{sech}\left(\frac{J}{k_B T}\right)\right]^2, \tag{13.19}$$

which is a well-behaved function, displaying just a broad maximum as a function of temperature.

According to an argument attributed to Landau, we can show that there is no ordered state (therefore, no phase transition) in a one-dimensional system with short-range interactions. Consider the ground state of the Ising chain, in the absence of an external field, with all spins pointing up. To create two distinct domains, it is enough to reverse the sign of just a single spin (see figure 13.1). This costs an amount of energy $\Delta U = 2J > 0$. However, there is an enormous increase of entropy, $\Delta S = k_B \ln N$, since there are N distinct positions to locate the separating wall between domains (for a chain with $N + 1$ sites). At finite temperatures, the free energy of this one-dimensional model undergoes a change

$$\Delta G = 2J - k_B T \ln N,$$

which becomes negative for sufficiently large values of N. Therefore, as the free energy decreases, there is a tendency to create more and more domains, which precludes the stability of any ordered phase. It is not difficult to check that similar arguments do not work in two dimensions, since the domain walls are not so simple, and both ΔU and ΔS are much more complicated.

Using the technique of the transfer matrices, we can calculate the spin–spin correlations,

$$\langle\sigma_k\sigma_l\rangle_N = \frac{1}{Z}\sum_{\{\sigma_i\}}\sigma_k\sigma_l\exp\left(-\beta\mathcal{H}\right). \tag{13.20}$$

For $l > k$, and a fair amount of algebra, it is possible to show that

$$\langle \sigma_k \sigma_l \rangle_N = \frac{\lambda_1^{N-(l-k)} \lambda_2^{(l-k)} + \lambda_1^{(l-k)} \lambda_2^{N-(l-k)}}{\lambda_1^N + \lambda_2^N}. \tag{13.21}$$

Thus, in the thermodynamic limit, we have

$$\langle \sigma_k \sigma_l \rangle = \lim_{N \to \infty} \langle \sigma_k \sigma_l \rangle_N = \left(\frac{\lambda_2}{\lambda_1} \right)^{(l-k)}, \tag{13.22}$$

which still works for $l < k$, if we replace the difference $(l - k)$ by its absolute value, $|l - k|$. In zero field, we write the pair correlation,

$$\langle \sigma_k \sigma_l \rangle_{H=0} = (\tanh K)^r, \tag{13.23}$$

where $r = |l - k|$ is the distance between sites k and l. This expression can also be written as

$$\langle \sigma_k \sigma_l \rangle_{H=0} = \exp \{ r \ln (\tanh K) \} = \exp \left(-\frac{r}{\xi} \right), \tag{13.24}$$

from which we define the **correlation length**,

$$\xi = \frac{1}{|\ln (\tanh K)|}. \tag{13.25}$$

Now, we see that ξ diverges for $K \to \infty$ (that is, at the trivial critical point, $T = 0$). For $T \neq 0$, correlations decay exponentially, with the characteristic length ξ. For the Ising model in two dimensions, at $T \neq T_c$, and for large enough distances, it can be exactly shown that correlations decay exponentially, with a correlation length of the form $\xi \sim |t|^{-\nu}$, where $\nu = 1$ and $t = (T - T_c)/T_c \to 0$. At the critical point ($T_c = H = 0$), spin–spin correlations decay asymptotically as a power law,

$$\langle \sigma_k \sigma_l \rangle_{cr} \sim \frac{1}{r^{d-2+\eta}},$$

where $\eta = 1/4$, for $d = 2$ and $r \to \infty$. From the classical Ornstein and Zernike theory for the decay of the critical correlations, we obtain the (classical) critical exponents $\nu = 1/2$ and $\eta = 0$.

13.2 Mean-field approximation for the Ising model

The standard mean-field approximation, also known as the Bragg–Williams method, can be obtained in the canonical formalism if we suppose that, besides the constraints of fixed temperature and external magnetic field,

there is an additional internal constraint that fixes the magnetization per spin.

For a spin-1/2 model, we write

$$N_+ + N_- = N \tag{13.26}$$

and

$$N_+ - N_- = Nm, \tag{13.27}$$

where N_+ (N_-) is the number of spins up (down), N is the total number of spins, and m is the dimensionless magnetization per spin. Given N_+ and N_- (that is, N and m), we can write the total entropy,

$$S = k_B \ln \frac{N!}{N_+! N_-!} = k_B \ln \frac{N!}{\left(\frac{N+Nm}{2}\right)! \left(\frac{N-Nm}{2}\right)!}. \tag{13.28}$$

Now, if we take into account the translational symmetry of the Hamiltonian, the internal energy of a nearest-neighbor Ising model on a d-dimensional hypercubic lattice is given by

$$U = \langle \mathcal{H} \rangle = -JdN \langle \sigma_i \sigma_j \rangle - HNm. \tag{13.29}$$

Therefore, with the additional constraint of fixed magnetization, the magnetic free energy per spin is given by

$$g(T, H; m) = \frac{1}{N}(U - TS) = -Jd \langle \sigma_i \sigma_j \rangle - Hm$$
$$- \frac{k_B T}{N} \ln \frac{N!}{\left(\frac{N+Nm}{2}\right)! \left(\frac{N-Nm}{2}\right)!}. \tag{13.30}$$

Up to this point there are no approximations. The difficult problem is the calculation of the pair correlations in terms of T, H, and m.

The Bragg–Williams approximation consists in neglecting fluctuations in the correlation functions. We then assume the approximation

$$\langle \sigma_i \sigma_j \rangle \approx \langle \sigma_i \rangle \langle \sigma_j \rangle = m^2. \tag{13.31}$$

Introducing a Stirling expansion to take care of the factorials, using the approximate form of the spin–spin correlations, and taking the thermodynamic limit, we can write the following Bragg–Williams free energy per spin,

$$g_{BW}(T, H; m) = -Jdm^2 - Hm - \frac{1}{\beta} \ln 2$$
$$+ \frac{1}{2\beta}[(1+m)\ln(1+m) + (1-m)\ln(1-m)]. \tag{13.32}$$

To remove the internal constraint of fixed magnetization, we minimize g_{BW} with respect to m. Hence, we obtain

$$\frac{\partial g_{BW}}{\partial m} = -2Jdm - H + \frac{1}{2\beta} \ln \frac{1+m}{1-m} = 0, \qquad (13.33)$$

from which the Curie–Weiss equation follows,

$$m = \tanh \left(\beta 2Jdm + \beta H \right), \qquad (13.34)$$

where the phenomenological parameter λ is identified as the product $2dJ$. In this approximation, the critical temperature is given by $k_B T_c = 2dJ$, and there is a transition even in one dimension. Although this result is completely wrong, especially at low dimensions, we anticipate that mean-field approximations become much better as the dimension increases.

The Bragg–Williams free energy, given by equation (13.32), can also be written as

$$g_{BW} \left(T, H; m \right) = -Jdm^2 - Hm - \frac{1}{\beta} \ln 2 + \frac{1}{\beta} \int \left(\tanh^{-1} m \right) dm, \quad (13.35)$$

which leads to an identification with the function $g \left(T, H; m \right)$, as obtained in the last chapter from the phenomenological equation of Curie–Weiss. We thus recover all of the classical results for the critical behavior.

The mean-field approximation can also be obtained from an elegant variational formalism based on the Peierls–Bogoliubov inequality, coming from convexity arguments, already known by Gibbs himself [see, for example, H. Falk, Am. J. Phys. **38**, 858 (1970)]. For all classical systems (in fact, also for quantum systems), we can write the inequality

$$G \left(\mathcal{H} \right) \leq G_o \left(\mathcal{H}_o \right) + \langle \mathcal{H} - \mathcal{H}_o \rangle_o = \Phi, \qquad (13.36)$$

where $G \left(\mathcal{H} \right)$ and $G_o \left(\mathcal{H}_o \right)$ are free energies associated with two different systems given by the Hamiltonians \mathcal{H} and \mathcal{H}_o, respectively, and the thermal average is taken with respect to a canonical distribution associated with \mathcal{H}_o. If we choose a noninteracting (trial) Hamiltonian,

$$\mathcal{H}_o = -\eta \sum_{i=1}^{N} \sigma_i, \qquad (13.37)$$

where η is a parameter, we have

$$Z_o = \sum_{\{\sigma_i\}} \exp \left(-\beta \mathcal{H}_o \right) = \left(2 \cosh \beta \eta \right)^N. \qquad (13.38)$$

Thus

$$G_o = -\frac{N}{\beta} \ln \left[2 \cosh \beta \eta \right] \qquad (13.39)$$

and

$$\left\langle \mathcal{H} - \mathcal{H}_o \right\rangle_o = -JdN \left\langle \sigma_i \sigma_j \right\rangle_o - HN \left\langle \sigma_i \right\rangle_o + \eta N \left\langle \sigma_i \right\rangle_o, \tag{13.40}$$

with

$$\left\langle \sigma_i \sigma_j \right\rangle_o = \left\langle \sigma_i \right\rangle_o^2 = \left(\tanh \beta \eta \right)^2 \text{ and } \left\langle \sigma_i \right\rangle_o = \tanh \beta \eta. \tag{13.41}$$

Hence, we have

$$\frac{1}{N} \Phi = \frac{1}{N} \Phi \left(T, H, N; \eta \right) = -\frac{1}{\beta} \ln 2 - \frac{1}{\beta} \ln \left(\cosh \beta \eta \right)$$
$$- Jd \left(\tanh \beta \eta \right)^2 - H \tanh \beta \eta + \eta \tanh \beta \eta. \tag{13.42}$$

This expression is just an upper bound for the free energy of the Ising system under consideration. In the (mean-field) approximation, the free energy per spin will be given by the minimum of $\Phi \left(T, H, N; \eta \right)$ with respect to the field parameter η,

$$g_{MF} = \frac{1}{N} \min_{\eta} \Phi \left(T, H, N; \eta \right), \tag{13.43}$$

which corresponds to the smaller upper bound that comes from Bogoliubov's inequality with a free trial Hamiltonian. It should be noted that η depends on m through the relation $m = \tanh \beta \eta$, from which we recover the previous results of the Bragg–Williams approximation.

13.3 The Curie–Weiss model

Instead of working with an approximate solution on a Bravais lattice, it may be interesting to introduce a (simplifying) modification in the very definition of the statistical model. With a suitable modification, some physical features are not lost, and the new problem can be exactly solved. According to this strategy, a deformation of the interaction term of the nearest-neighbor Ising Hamiltonian leads to the **Curie–Weiss model**,

$$\mathcal{H}_{CW} = -\frac{J}{2N} \sum_{i=1}^{N} \sum_{j=1}^{N} \sigma_i \sigma_j - H \sum_{i=1}^{N} \sigma_i, \tag{13.44}$$

in which each spin interacts with all neighbors. The interactions are long ranged (indeed, of infinite range), but very weak, of the order $1/N$, to preserve the existence of the thermodynamic limit. In zero field, the ground-state energy per spin of this Curie–Weiss model is given by $U_{CW}/N = -J/2$, which should be compared with the corresponding result for a nearest-neighbor Ising ferromagnet on a hypercubic d-dimensional lattice, $U/N = -Jd$.

The canonical partition function associated with the Curie–Weiss model is given by

$$Z = \sum_{\{\sigma i\}} \exp \left[\frac{\beta J}{2N} \left(\sum_{i=1}^{N} \sigma_i \right)^2 + \beta H \sum_{i=1}^{N} \sigma_i \right]. \qquad (13.45)$$

Now we use the Gaussian identity,

$$\int_{-\infty}^{+\infty} \exp \left(-x^2 + 2ax \right) dx = \sqrt{\pi} \exp \left(a^2 \right), \qquad (13.46)$$

to calculate the sum over the spin variables in equation (13.45),

$$\begin{aligned} Z &= \sum_{\{\sigma i\}} \frac{1}{\sqrt{\pi}} \int_{-\infty}^{+\infty} dx \exp \left[-x^2 + 2 \left(\frac{\beta J}{2N} \right)^{1/2} x \sum_{i=1}^{N} \sigma_i + \beta H \sum_{i=1}^{N} \sigma_i \right] \\ &= \frac{1}{\sqrt{\pi}} \int_{-\infty}^{+\infty} dx \exp \left(-x^2 \right) \left\{ 2 \cosh \left[2 \left(\frac{\beta J}{2N} \right)^{1/2} x + \beta H \right] \right\}^N. \end{aligned} \qquad (13.47)$$

Introducing the change of variables

$$2 \left(\frac{\beta J}{2N} \right)^{1/2} x = \beta J m, \qquad (13.48)$$

we have

$$Z = \left(\frac{N}{2\pi \beta J} \right)^{1/2} \int_{-\infty}^{+\infty} dm \exp \left[-N\beta g \left(T, H; m \right) \right], \qquad (13.49)$$

where

$$g \left(T, H; m \right) = \frac{1}{2} J m^2 - \frac{1}{\beta} \ln \left[2 \cosh \left(\beta J m + \beta H \right) \right]. \qquad (13.50)$$

In order to obtain the free energy per spin in the thermodynamic limit, we use Laplace's method to calculate the asymptotic form, as $N \to \infty$, of the integral (13.49). We thus have

$$g \left(T, H \right) = \lim_{N \to \infty} \left\{ -\frac{1}{\beta N} \ln Z \right\} = \min_{m} \left\{ g \left(T, H; m \right) \right\}. \qquad (13.51)$$

Hence,

$$\frac{\partial g \left(T, H; m \right)}{\partial m} = J m - J \tanh \left(\beta J m + \beta H \right) = 0, \qquad (13.52)$$

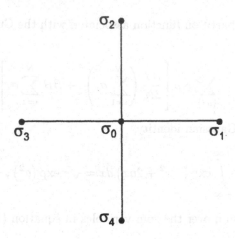

FIGURE 13.2. Spin cluster with a central site and four surrounding sites.

from which we have the (Curie–Weiss) equation of state,

$$m = \tanh\left(\beta Jm + \beta H\right), \qquad (13.53)$$

which is the trademark of the mean-field approximation for the Ising model. It is easy to check that this model with infinite-range interactions leads to the same classical results of the Bragg-Williams approximation for the Ising model on a Bravais lattice. As in the phenomenological treatment of Landau, the expansion of the "functional" $g\left(T, H; m\right)$ in powers of m gives rise to a (rigorous) analysis of the transition in the Curie-Weiss model. Although the coefficients of the various powers of m are different from the corresponding terms in the expansion of the Bragg–Williams "functional," all the critical parameters are exactly the same [see, for example, C. E. I. Carneiro, V. B. Henriques, and S. R. Salinas, Physica A162, 88 (1989)].

13.4 The Bethe–Peierls approximation

There are some self-consistent approximations for the Ising model on a Bravais lattice, usually correct in one dimension, which do take into account some short-range fluctuations, and are thus capable of displaying some features of the phase diagrams that are not shown by the standard mean-field approximations (although critical exponents keep their classical values). The Bethe–Peierls approximation is very representative of these self-consistent calculations.

Consider a cluster of a central spin σ_o, in an external field H, and q surrounding spins, in an effective field H_e, which is supposed to mimic the effects of the remaining crystalline lattice (see figure 13.2). The spin

Hamiltonian of this cluster is given by

$$\mathcal{H}_c = -J\sigma_o\left(\sigma_1 + \ldots + \sigma_q\right) - H\sigma_o - H_e\left(\sigma_1 + \ldots + \sigma_q\right). \tag{13.54}$$

Thus, we write the canonical partition function of the cluster,

$$Z_c = \sum_{\{\sigma_i\}} \exp\left(-\beta\mathcal{H}_c\right) = \sum_{\sigma_o} \exp\left(\beta H\sigma_o\right)\left[2\cosh\left(\beta J\sigma_o + \beta H_e\right)\right]^q. \tag{13.55}$$

From this expression, we calculate the spin magnetization of the central site,

$$m_o = \frac{1}{\beta}\frac{\partial}{\partial H}\ln Z_c, \tag{13.56}$$

and the spin magnetization of one of the surrounding sites,

$$m_p = \frac{1}{\beta q}\frac{\partial}{\partial H_e}\ln Z_c. \tag{13.57}$$

After some algebraic manipulations, equation (13.56) may be written as

$$m_o = \tanh\left[\beta H + q\tanh^{-1}\left(\tanh\beta J\tanh\beta H_e\right)\right]. \tag{13.58}$$

Also, it is not difficult to write equation (13.57) in the more convenient form

$$\begin{aligned} m_p &= \frac{\left(1 - \tanh^2\beta J\right)\tanh\beta H_e + m_o\left(1 - \tanh^2\beta H_e\right)\tanh\beta J}{1 - \left(\tanh\beta J\tanh\beta H_e\right)^2} \\ &= \frac{\tanh\left(\beta H_e + \beta J\right) + \tanh\left(\beta H_e - \beta J\right)\exp\left(-2\tanh^{-1}m_o\right)}{1 + \exp\left(-2\tanh^{-1}m_o\right)}. \end{aligned}$$
$$\tag{13.59}$$

The self-consistent condition, from which we eliminate the effective field, is given by

$$m_o = m_p = m, \tag{13.60}$$

which leads to the **equation of state** of the Bethe–Peierls approximation, $m = m(T, H)$.

In zero field ($H = 0$), and in the neighborhood of the critical temperature, m and H_e are very small. Therefore, we can write expansions for equations (13.58) and (13.59),

$$m = q\left(\tanh\beta J\right)\tanh\beta H_e + \ldots, \tag{13.61}$$

$$m = \frac{1}{\cosh^2\beta J}\beta H_e + m\tanh\beta J + \ldots. \tag{13.62}$$

From these expansions, it is easy to check that the critical temperature, in this Bethe–Peierls approximation, will be given by

$$\frac{k_B T_c}{J} = 2 \left[\ln \frac{q}{q-2} \right]^{-1}. \tag{13.63}$$

In one dimension (that is, for $q = 2$), there is no phase transition ($T_c = 0$). For $q = 4$ (which corresponds to a square lattice), $k_B T_c / J = 2/\ln 2 = 2.885...$, smaller than the critical temperature from the Bragg–Williams approximation, $k_B T_c / J = 4$, but still larger than the exact Onsager value, $k_B T_c / J = 2/\ln\left(1 + \sqrt{2}\right) = 2.269...$.

Approximations of the Bethe–Peierls type, based on self-consistent calculations for a small cluster of spins, give an equation of state, but do not lead to an expression for the free energy of the system (in general, it is inconsistent to write the free energy from the canonical partition function Z_c of the cluster). In order to obtain a consistent free energy from the equation of state, we write

$$g = -\int m\left(T, H\right) dH + g_o\left(T\right), \tag{13.64}$$

where $g_o\left(T\right)$ is a well-behaved function of temperature (which may be found, for example, from a comparison with the high-temperature limit of the exact free energy). For the Ising ferromagnet, we can go through some algebraic manipulations to calculate this integral. Initially, note that equation (13.58) may be written as

$$m = \frac{\exp\left(2\beta H\right) - x^q}{\exp\left(2\beta H\right) + x^q}, \tag{13.65}$$

where

$$x = \frac{1 - \tanh \beta J \tanh \beta H_e}{1 + \tanh \beta J \tanh \beta H_e}. \tag{13.66}$$

We can now use equation (13.59) to write

$$\exp\left(2\beta H\right) = x^{q-1} \frac{\exp\left(2\beta J\right) - x}{x \exp\left(2\beta J\right) - 1}. \tag{13.67}$$

Therefore, the magnetization m and the field H can be expressed in terms of the new variable x. Hence, from equation (13.64), we have

$$g = -\int m \frac{\partial H}{\partial x} dx + g_o\left(T\right), \tag{13.68}$$

where

$$m = \frac{-zx^2 + z}{zx^2 - 2x + z} \tag{13.69}$$

and

$$2\beta \frac{\partial H}{\partial x} = \frac{q-1}{x} - \frac{1}{z-x} - \frac{z}{zx-1},\qquad(13.70)$$

with $z = \exp(2\beta J)$. Now, we perform some additional algebraic manipulations to show that

$$g = -\frac{q-1}{2\beta}\ln x + \frac{1}{2\beta}\ln(z-x) + \frac{1}{2\beta}\ln(zx-1)$$

$$+\frac{q-2}{2\beta}\ln\left(zx^2 - 2x + z\right) + g_o(T),\qquad(13.71)$$

which is the free energy associated with the Bethe–Peierls approximation.

Finally, it is interesting to remark that the results of the Bethe–Peierls approximation can be recovered from an exact solution of the Ising model on a Cayley tree. This graph is a peculiar layered structure, in which the spins belonging to a certain generation interact with q other spins belonging to the next generation, such that there are no closed cycles. Indeed, the correspondence between the approximate and exact solutions works in the central part of the limit of a large Cayley tree (which is then called a Bethe lattice). The interested reader may check the works of C. J. Thompson, J. Stat. Phys. **27**, 441 (1982), and of M. J. Oliveira and S. R. Salinas, Rev. Bras. Fis. **15**, 189 (1985).

13.5 Exact results on the square lattice

The discussion of the Onsager solution (and of its several alternatives) is certainly beyond the scope of this book. The interested reader, with plenty of spare time, should check the work of T. D. Schultz, D. C. Mattis, and E. H. Lieb, Rev. Mod. Phys. **36**, 856 (1964), where the technique of the transfer matrix is used to reduce the calculation of the eigenvalues to the problem of diagonalizing the Hamiltonian of a system of free fermions. The transfer matrix is written in terms of Pauli spin operators, which are then changed into fermions through the ingenious Jordan–Wigner transformation. We shall limit our considerations to a mere listing of some of the Onsager results.

In the **thermodynamic limit**, the free energy of the Ising model (on a square lattice, with nearest-neighbor interactions, in zero field) may be written as a double integral,

$$-\beta g(T) = \ln 2 + \frac{1}{2\pi^2}\int_0^\pi\int_0^\pi \ln\left[\cosh^2 2K - \sinh 2K\left(\cos\theta_1 + \cos\theta_2\right)\right]d\theta_1 d\theta_2,$$

$$(13.72)$$

where $K = \beta J$ (in the original solution, Onsager already considered different interactions along the two directions of the square lattice). Therefore, we have the internal energy

$$
u = -\frac{J}{\tanh K}\left[1 + \frac{\sinh^2 2K - 1}{\pi^2}\right.
$$

$$
\left. \times \int_0^\pi \int_0^\pi \frac{d\theta_1 d\theta_2}{\cosh^2 2K - \sinh 2K\,(\cos\theta_1 + \cos\theta_2)}\right]. \qquad (13.73)
$$

The integral in this expression logarithmically diverges for

$$
\cosh^2 2K = 2\sinh 2K, \qquad (13.74)
$$

that is,

$$
\sinh 2K = 1, \qquad (13.75)
$$

which gives the Onsager critical temperature,

$$
K_c^{-1} = \frac{k_B T_c}{J} = \frac{2}{\ln\left(1 + \sqrt{2}\right)} = 2.269.... \qquad (13.76)
$$

In the neighborhood of the critical temperature, it is convenient to introduce a (small) parameter,

$$
\delta = (\sinh 2K - 1)^2. \qquad (13.77)
$$

For $\delta \to 0$, we write

$$
\int_0^\pi \int_0^\pi \frac{d\theta_1 d\theta_2}{\cosh^2 2K - \sinh 2K\,(\cos\theta_1 + \cos\theta_2)} \sim \int_0^\pi \int_0^\pi \frac{d\theta_1 d\theta_2}{\delta + \frac{1}{2}\sinh 2K\,(\theta_1^2 + \theta_2^2)}
$$

$$
= 2\pi \int_0 \frac{r\,dr}{\delta + \frac{1}{2}r^2 \sinh 2K} \sim -2\pi\ln\delta, \qquad (13.78)
$$

where we have kept the leading term only (and used polar coordinates to simplify the integral in the intermediate step). From this asymptotic form, we have the internal energy in the neighborhood of the critical temperature,

$$
u \sim -\frac{J}{\tanh K_c}\left[1 + A\,(K - K_c)\ln|K - K_c|\right], \qquad (13.79)
$$

where A is a constant. Taking the derivative with respect to temperature, we have the famous asymptotic formula for the zero-field specific heat,

$$
c_{H=0} \sim B\ln|K - K_c|, \qquad (13.80)
$$

where B is a constant and $K \to K_c$.

From equation (13.73), we can write an analytic expression for the internal energy in terms of an elliptic integral of the first kind,

$$u = -\frac{J}{\tanh K} \left[1 + (2 \tanh^2 2K - 1) \frac{2}{\pi} K(k_1) \right], \qquad (13.81)$$

where

$$k_1 = \frac{2 \sinh (2K)}{\cosh^2 2K}, \qquad (13.82)$$

and $K(k_1)$ is a complete elliptic integral of the first kind,

$$K(k_1) = \int_0^{\pi/2} \left[1 - k_1^2 \sin^2 \theta \right]^{-1/2} d\theta. \qquad (13.83)$$

The specific heat can also be written in terms of complete elliptic integrals (of first and second kind). Unfortunately, however, we do not have generalizations of these results for either three dimensions or in the presence of an external field!

Exercises

1. Consider a one-dimensional spin-1 model, given by the Hamiltonian

$$\mathcal{H} = -J \sum_{i=1}^{N} S_i S_{i+1} + D \sum_{i=1}^{N} S_i^2,$$

 where $J > 0$, $D > 0$, and $S_i = -1, 0, +1$, for all lattice sites.

 (a) Assuming periodic boundary conditions, calculate the eigenvalues of the transfer matrix.

 (b) Obtain expressions for the internal energy and the entropy per spin.

 (c) What is the ground state of this model ($T = 0$) as a function of the parameter $d = D/J$? Obtain the asymptotic form of the eigenvalues of the transfer matrix, for $T \to 0$, in the characteristic regimes of the parameter d.

2. The one-dimensional Ising ferromagnet is given by the Hamiltonian

$$\mathcal{H} = -J \sum_{i=1}^{N} \sigma_i \sigma_{i+1} - H \sum_{i=1}^{N} \sigma_i,$$

 with $J > 0$ and $\sigma_i = \pm 1$ for all lattice sites.

 (a) In zero field $(H = 0)$, show that

$$\langle \sigma_k \sigma_l \rangle = (\tanh \beta J)^{|k-l|}.$$

 (b) Consider the fluctuations of the magnetization in the canonical ensemble to show that

$$\chi(T, H) \equiv \left(\frac{\partial m}{\partial H}\right)_\beta = \frac{\beta}{N}\left\langle \sum_{i,j=1}^{N} \sigma_i \sigma_j \right\rangle_N$$

$$-\frac{\beta}{N}\left\langle \sum_{i=1}^{N} \sigma_i \right\rangle_N \left\langle \sum_{j=1}^{N} \sigma_j \right\rangle_N.$$

 (c) Use the previous results to obtain an expression for the magnetic susceptibility in zero field, $\chi_0 = \chi(T, h \to 0)$. Sketch a graph of χ_o versus temperature.

 (d) Obtain an expression, in zero field, for the four-spin correlation function, $\langle \sigma_i \sigma_j \sigma_k \sigma_l \rangle$, with $1 \leq i \leq j \leq k \leq l \leq N$.

 (e) Show that the specific heat in zero field may be written as a sum over four-spin correlation functions.

3. For a well-behaved convex function $\phi(x)$ of a random variable x, show that

$$\phi(x) \geq \phi(\langle x \rangle) + (x - \langle x \rangle) \phi'(x).$$

Taking the average with respect to a positive measure, show that

$$\langle \phi(x) \rangle \geq \phi(\langle x \rangle),$$

which is known, for $\phi(x) = \exp(x)$, as Jensen's inequality. Show that we obtain the classical version of the Peierls–Bogoliubov inequality if we assume that

$$\langle \cdots \rangle = \frac{1}{Z_o} \text{Tr}[\exp(-\beta \mathcal{H})(\cdots)]$$

and that

$$\phi(x) = \exp[\beta(\mathcal{H}_o - \mathcal{H})].$$

4. The Blume–Capel model on a lattice of coordination q is given by the spin Hamiltonian

$$\mathcal{H} = -J\sum_{(ij)} S_i S_j + D\sum_{i=1}^{N} S_i^2 - H\sum_{i=1}^{N} S_i,$$

where the first sum is over nearest-neighbor sites, all parameters are positive, and $S_i = -1, 0, +1$ for $i = 1, 2, ..., N$. Use the Bogoliubov–Peierls variational principle, with a trial Hamiltonian of the form

$$\mathcal{H}_o = +D \sum_{i=1}^{N} S_i^2 - \eta \sum_{i=1}^{N} S_i,$$

where η is a variational parameter, to obtain an approximate solution for the free energy of this system,

$$g_{approx}(T, H) = \min_{\eta} \{ g(T, H; m) \},$$

where m is a function of η that corresponds to the magnetization per spin. In zero field ($H = 0$), obtain the coefficients of the expansion

$$g(T, H = 0; m) = A + Bm^2 + Cm^4 + Dm^6 + \dots .$$

Now consider the phase diagram in zero field (in the $d = D/J$ versus $t = k_B T/J$ plane, for $H = 0$). Obtain an expression for the line of second-order transitions (λ line), given by $B = 0$ with $C > 0$, and locate the tricritical point ($B = C = 0$, with $D > 0$). What happens in the region of the phase diagram where $C < 0$? What happens at $T = 0$?

5. Sketch a graph of the specific heat in zero field as a function of temperature for the Ising ferromagnet in the Curie–Weiss version. Obtain an expression for the magnetic susceptibility $\chi(T, H)$. Sketch the qualitative form of χ versus the magnetic field H for three typical values of temperature ($T_1 < T_c$, $T_2 = T_c$, and $T_3 > T_c$).

6. The Curie–Weiss version of the Blume–Capel model, with a ferromagnetic ground state, is given by the spin Hamiltonian

$$\mathcal{H} = -\frac{J}{2N} \left(\sum_{i=1}^{N} S_i \right)^2 + D \sum_{i=1}^{N} S_i^2 - H \sum_{i=1}^{N} S_i,$$

where the parameters are positive, and $S_i = +1, 0, -1$ for all sites.

(a) Show that the free energy per spin may be written as

$$g(t, d, h) = \min_{y} \{ g(t, d, h; y) \},$$

where $t = k_B T/J$, $d = D/J$, and $h = H/J$. Obtain an expression for $g(t, d, h; y)$.

(b) From an expansion of $g(t, d, h; y)$ in powers of y, obtain expressions for the critical line and the tricritical point (compare with the results of problem 3).

(c) Sketch the phase diagram in the $d - t$ plane (for $h = 0$).

(d) Sketch graphs of the spontaneous magnetization versus d for some characteristic values of t.

7. Use the Bethe–Peierls approximation for the ferromagnetic Ising model in the absence of an external field to obtain an expression for the expected value $\langle \sigma_o \sigma_1 \rangle$, where σ_o is the central spin of the cluster and σ_1 is the spin on a surrounding site. Sketch a qualitative graph of $\langle \sigma_o \sigma_1 \rangle$ versus temperature. Sketch a graph of the specific heat in zero field versus temperature (compare with the result for the Curie–Weiss version of the Ising model).

14

Scaling Theories and the Renormalization Group

It is too strong to assume that the free energy of a model system in the critical region can be expanded as a power series of the order parameter. The divergent specific heat of the Onsager exact solution for the two-dimensional Ising model precludes an expansion whose coefficients are analytic functions of temperature. In the 1960s, there appeared a number of (weaker) scaling hypotheses, based on some general assumptions about the form of the thermodynamic potentials. Although these scaling hypotheses do not lead to a microscopic treatment of critical phenomena, they do provide a way of going beyond the phenomenological equations of van der Waals and Curie–Weiss. The microscopic justification of these ideas, as well as a real possibility of calculating values for the critical exponents to compare with experimental data and theoretical predictions, were provided by the advent of the modern renormalization-group techniques.

14.1 Scaling theory of the thermodynamic potentials

In the neighborhood of a critical point, we assume that the free energy per spin $g(T, H)$ of a simple uniaxial ferromagnet can be written as the sum of a regular, and less interesting part, $g_o(T, H)$, and a singular part, $g_s(T, H)$, which contains all of the anomalies of the critical behavior. It is convenient to write the singular part of this thermodynamic potential in terms of the reduced variables $t = (T - T_c)/T_c$ and H, which vanish at the

critical point. We then have

$$g(T, H) = g_o(T, H) + g_s(t, H). \tag{14.1}$$

The **scaling** or **homogeneity** hypothesis consists in the assumption that g_s is a **generalized homogeneous function** of the variables t and H,

$$g_s(t, H) = \lambda g_s\left(\lambda^a t, \lambda^b H\right), \tag{14.2}$$

for all values of λ, and where a and b are two critical exponents. As we have already shown, it is interesting to remark that Landau's phenomenological theory also leads to a generalized homogeneous thermodynamic potential, but with well-defined (classical) exponents. As the parameter λ is arbitrary, we can choose $\lambda^a t = 1$, such that $\lambda = t^{-1/a}$. Thus, we can write

$$g_s(t, H) = t^{-1/a} g_s\left(1, \frac{H}{t^{b/a}}\right) = t^{-1/a} F\left(\frac{H}{t^{b/a}}\right). \tag{14.3}$$

Assuming that $F(x)$ is well behaved, we can use the standard tools of thermodynamics to write the critical exponents $(\alpha, \beta, \gamma, ...)$ in terms of a and b. Considering the differential form of the thermodynamic potential,

$$dg = -sdT - mdH, \tag{14.4}$$

we have the singular parts of the entropy and the magnetization per spin,

$$s = -\frac{\partial g_s}{\partial T} = -\frac{1}{T_c}\left\{-\frac{1}{a}t^{-1/a-1}F\left(\frac{H}{t^{b/a}}\right) - \frac{b}{a}t^{-1/a-b/a-1}HF'\left(\frac{H}{t^{b/a}}\right)\right\} \tag{14.5}$$

and

$$m = -\frac{\partial g_s}{\partial H} = -t^{-1/a-b/a}F'\left(\frac{H}{t^{b/a}}\right). \tag{14.6}$$

In zero field, we write

$$s(t, H = 0) = \frac{1}{aT_c}t^{-1/a-1}F(0) \tag{14.7}$$

and

$$m(t, H = 0) = -t^{-1/a-b/a}F'(0). \tag{14.8}$$

From this last expression, we have the critical exponent associated with the spontaneous magnetization,

$$\beta = -\frac{1}{a} - \frac{b}{a}. \tag{14.9}$$

In order to obtain the critical exponent α associated with the specific heat, we take the derivative of $s(t, H)$ with respect to temperature. Hence, we have

$$c(t, H = 0) = T \left(\frac{\partial s}{\partial T} \right)_{H=0} \sim -\frac{1}{aT_c} \left(\frac{1}{a} + 1 \right) t^{-1/a-2} F(0). \quad (14.10)$$

Thus,

$$\alpha = 2 + \frac{1}{a}. \quad (14.11)$$

To obtain the critical exponent γ, we write the magnetic susceptibility,

$$\chi(T, H) = \left(\frac{\partial m}{\partial H} \right)_T = -t^{-1/a-2b/a} F'' \left(\frac{H}{t^{b/a}} \right). \quad (14.12)$$

In zero field, we have

$$\chi_o(T) = \chi(T, H = 0) = -t^{-1/a-2b/a} F''(0), \quad (14.13)$$

from which we have the exponent

$$\gamma = \frac{2b}{a} + \frac{1}{a}. \quad (14.14)$$

From equations (14.9), (14.11), and (14.14), we see that only two exponents are sufficient to define all other thermodynamic exponents, and we then write the (scaling) relation

$$\alpha + 2\beta + \gamma = 2. \quad (14.15)$$

Scaling relations of this same nature have been widely confirmed by the analysis of both experimental data and exact as well as numerical results (from high-temperature series expansions, for example). The classical exponents ($\alpha = 0$, $\beta = 1/2$, $\gamma = 1$) also obey this scaling relation. It should be pointed out that a relation of this sort was anticipated by Rushbrooke, as the inequality $\alpha + 2\beta + \gamma \geq 2$, which can be derived from purely thermodynamic arguments. There are several examples of inequalities involving the critical exponents, derived on the basis different arguments, which end up being fulfilled as simple equalities by experimental and theoretical values.

In the context of this phenomenological theory, the magnetization can be written as

$$m(t, H) = t^\beta Y \left(\frac{H}{t^\Delta} \right), \quad (14.16)$$

where $Y(x)$ is a well-behaved function, and $\Delta = b/a = 2 - \alpha - \beta = \beta + \gamma$. We then write

$$\frac{m}{t^\beta} = Y \left(\frac{H}{t^\Delta} \right). \quad (14.17)$$

FIGURE 14.1. Properly scaled values of the applied magnetic field versus the magnetization for the compound $CrBr_3$. Note that there are two distinct scaling functions, below and above the critical temperature.

The plot of the experimental data for m/t^β, that is, for the magnetization in a suitable scale given by t^β, versus H/t^Δ, where t^Δ is the scale associated with the external field H, leads to the universal function $Y(x)$. Graphs in terms of these suitably scaled variables, in agreement with the homogeneity ideas, have been plotted for a wide variety of systems. From these experimental plots, we obtain estimates for the critical exponents and the critical temperature. For example, in figure 14.1, we reproduce an analysis of Ho and Litster for the ferromagnetic compound $CrBr_3$ (note that there are two distinct scaling functions, above and below the critical temperature).

We have already mentioned that the Landau free energy associated with a simple uniaxial ferromagnet, given by

$$g_s(t, H) = -mH + tm^2 + m^4, \tag{14.18}$$

with

$$-H + 2tm + 4m^3 = 0, \tag{14.19}$$

obeys the scaling form

$$g_s(t, H) = \lambda g_s(\lambda^x t, \lambda^y H), \tag{14.20}$$

where $x = -1/2$ and $y = -3/4$. Therefore, we can write the singular part of the Landau phenomenological free energy in the scaling form

$$g_s(t, H) = t^2 F\left(\frac{H}{t^{3/2}}\right), \qquad (14.21)$$

with the classical exponents $\alpha = 0$ and $\Delta = \beta + \gamma = 3/2$.

14.2 Scaling of the critical correlations

The critical point has been defined as the "terminus" of a line of first-order transitions, where the order parameter finally vanishes. As an alternative definition, of a microscopic character, we assume that the correlation length associated with the order parameter diverges at the critical point. Of course, we expect that these definitions are entirely equivalent.

To discuss this point of view, let us use the language of the Ising model, given by the Hamiltonian

$$\mathcal{H} = -J\sum_{(ij)} \sigma_i \sigma_j - H\sum_{i=1}^{N} \sigma_i, \qquad (14.22)$$

where the first sum is over pairs of nearest-neighbor sites, and $\sigma_i = \pm 1$, for all sites. The dimensionless thermodynamic magnetization per site is given by

$$m(\beta, H) = \lim_{N\to\infty} m_N(\beta, H) = \lim_{N\to\infty} \left\langle \frac{1}{N}\sum_{i=1}^{N} \sigma_i \right\rangle_N, \qquad (14.23)$$

and the spontaneous magnetization is written as

$$m_o(\beta) = \lim_{H\to 0^+} m(\beta, H). \qquad (14.24)$$

Taking into account the translational symmetry of the system, the spin–spin correlation function is given by

$$\Gamma_N(k, l) = \langle \sigma_k \sigma_l \rangle_N - \langle \sigma_k \rangle_N \langle \sigma_l \rangle_N = \langle \sigma_k \sigma_l \rangle_N - m_N^2. \qquad (14.25)$$

In the thermodynamic limit, we have

$$\Gamma(k.l) = \lim_{N\to\infty} \Gamma_N(k, l) = \langle \sigma_k \sigma_l \rangle - m^2. \qquad (14.26)$$

In zero field, $m_N = 0$ due to the symmetry of the Hamiltonian with respect to reversing the signs of the spin variables. Therefore, in zero field, the existence of **long-range order** comes from the relation

$$\lim_{|\vec{r}_k - \vec{r}_l|\to\infty} \langle \sigma_k \sigma_l \rangle \to m_o^2 \neq 0, \qquad (14.27)$$

from which Onsager, and later Yang, have obtained the spontaneous magnetization of the Ising ferromagnet on the square lattice (without having to calculate a field-dependent thermodynamic potential).

At the critical point ($T = T_c$ and $H = 0$), the long-distance spin–spin correlation functions display the asymptotic behavior

$$\Gamma(\vec{r}) \to B_d r^{-(d-2+\eta)}, \tag{14.28}$$

where B_d is a nonuniversal factor and $\eta = 1/4$ for all lattices in $d = 2$ dimensions. For $H = 0$ and $T \to T_c$, we have the long-distance asymptotic behavior,

$$\Gamma(\vec{r}) \to B_d(\vec{r}) \exp\left(-\frac{r}{\xi}\right), \tag{14.29}$$

where the correlation length is given by

$$\xi = \begin{cases} D_+ t^{-\nu} & ; \quad T \to T_c+, \\ D_- (-t)^{-\nu} & ; \quad T \to T_c-. \end{cases} \tag{14.30}$$

For the two-dimensional Ising model, $\nu = 1$. According to the classical theory of Ornstein and Zernike, $\nu = 1/2$ and $\eta = 0$. The experimental results (for three-dimensional systems) lead to values of ν between 0.6 and 0.7, and η less than 0.1. Numerical analysis of high-temperature series for the three-dimensional Ising model give $\nu \approx 0.643$ and $\eta \approx 0.05$, with the error in the last digit (although there should be even better results in the very recent literature).

Now we write the homogeneous form of the pair correlation function,

$$\Gamma(r,t,H) = \lambda\Gamma\left(\lambda^a r, \lambda^b t, \lambda^c H\right) = t^{-1/b} F\left(\frac{r}{t^{a/b}}, \frac{H}{t^{c/b}}\right), \tag{14.31}$$

in terms of new exponents a, b, and c. With a convenient choice of these exponents, we have the scaling form

$$\Gamma(r,t,H) = t^{\nu(d-2+\eta)} F\left(\frac{r}{t^{-\nu}}, \frac{H}{t^{\Delta}}\right). \tag{14.32}$$

Again, let us use the Ising model. It is easy to show that

$$\chi_N(T,H) \equiv \left(\frac{\partial m_N}{\partial H}\right)_\beta = \frac{\beta}{N}\left\langle \sum_{i,j=1}^{N} \sigma_i \sigma_j \right\rangle_N - \frac{\beta}{N}\left\langle \sum_{i=1}^{N} \sigma_i \right\rangle_N \left\langle \sum_{j=1}^{N} \sigma_j \right\rangle_N, \tag{14.33}$$

in other words, that

$$\chi_N(T,H) = \frac{\beta}{N} \sum_{i,j=1}^{N} \Gamma_N(i,j) = \beta \sum_{j=1}^{N} \Gamma_N(1,j), \tag{14.34}$$

where we have taken advantage of the translational symmetry of the system. Therefore, in the thermodynamic limit, $N \to \infty$, we have a static form of the **fluctuation-dissipation** theorem,

$$\chi(T, H) = \beta \sum_{\vec{r}} \Gamma(\vec{r}), \qquad (14.35)$$

which shows that the susceptibility is related to the fluctuations of the magnetization.

Inserting the asymptotic expressions for the susceptibility and the correlations into equation (14.35), we have

$$t^{-\gamma} \sim \beta \int d^d \vec{r} \frac{B}{r^{d-2+\eta}} \exp\left(-\frac{r}{t^{-\nu}}\right), \qquad (14.36)$$

from which we obtain Fisher's relation,

$$\gamma = \nu(2 - \eta), \qquad (14.37)$$

which is fulfilled by the scaling equation (14.32). There is also a (hyperscaling) relation,

$$2 - \alpha = d\nu, \qquad (14.38)$$

whose deduction is based on rather qualitative arguments, but that seems to be confirmed by available experimental and numerical data (from the classical point of view, this relation holds for $\alpha = 0$ and $d = 4$, which points out the special role of the four-dimensional lattice). Therefore, the new critical exponents associated with the correlation functions are entirely determined by just two thermodynamic critical exponents!

14.3 The Kadanoff construction

What is the microscopic justification of the homogeneity assumption and the scaling relations? How to calculate the critical exponents a and b (and all other thermodynamic critical exponents)? Answers to these questions were given in the 1970s with the advent of the renormalization-group ideas.

To introduce the Kadanoff construction, consider again the Ising ferromagnet on a d-dimensional hypercubic lattice, given by the Hamiltonian of equation (14.22),

$$\mathcal{H} = -J \sum_{(ij)} \sigma_i \sigma_j - H \sum_{i=1}^{N} \sigma_i, \qquad (14.39)$$

where the first sum is over pairs of nearest-neighbor sites and $\sigma_i = \pm 1$ for all sites. Close to the critical temperature, the correlation length becomes

FIGURE 14.2. In the neighborhood of the critical temperature, the correlation length ξ is very large. Therefore, the four sites belonging to the cell of dimension $b = 2$ should be very strongly correlated.

very large as compared with the interatomic distances of the crystalline lattice. As we get closer to the critical point, there appear bigger and bigger clusters of highly correlated spins (see the square lattice in figure 14.2).

In the neighborhood of the critical point, as indicated in figure 14.2, we consider cells of dimension b ($b << \xi$) with b^d highly correlated spins. According to an argument introduced by Kadanoff (and by Patashinski and Pokrovski, under the name of similarity hypothesis), let us associate with each one of these cells a new spin variable, θ_α, with $\alpha = 1, 2, ..., N/b^d$, so that $\theta_\alpha = \pm 1$ for all α, and let us assume that the Hamiltonian of the system may be expressed by a **similar** expression as before, given by

$$\mathcal{H}' = -J' \sum_{(\alpha\beta)} \theta_\alpha \theta_\beta - H' \sum_\alpha \theta_\alpha, \qquad (14.40)$$

but with new parameters, J' and H' [as J defines the critical temperature, the transformation of J into J' in the partition function is equivalent to a change from $t = (T - T_c)/T_c$ to t'].

What is the relation between (t', H') and (t, H)? As both Hamiltonians have the same form, we can write

$$g(t', H') = b^d g(t, H), \qquad (14.41)$$

where g is the free energy per spin. The similarity hypothesis consists in the assumption that

$$t' = b^{\lambda_1} t \quad \text{and} \quad H' = b^{\lambda_2} H, \qquad (14.42)$$

where λ_1 and λ_2 are two critical exponents. We then see that this assumption preserves the physical aspects of the problem and gives a justification for the scaling laws. Indeed, the free energy is written as a generalized homogeneous function,

$$g\left(b^{\lambda_1}t, b^{\lambda_2}H\right) = b^d g\left(t, H\right). \tag{14.43}$$

We should now search for particular ways to implement these ideas for specific statistical models.

14.4 Renormalization of the ferromagnetic Ising chain

In one dimension, the Ising Hamiltonian is written as

$$\mathcal{H} = -J\sum_{i=1}^{N} \sigma_i \sigma_{i+1} - H\sum_{i=1}^{N}\sigma_i, \tag{14.44}$$

with the canonical partition function given by

$$Z = \text{Tr} \exp\left[K\sum_{i=1}^{N}\sigma_i\sigma_{i+1} + L\sum_{i=1}^{N}\sigma_i\right], \tag{14.45}$$

where $K = \beta J$, $L = \beta H$, and the trace means a sum over all configurations of the spin variables.

If we consider a cell of length $b = 2$ (two lattice spacings), we can eliminate half of the degrees of freedom by summing over spin variables on the even sites of the chain (according to a procedure called decimation). For example, we have independent sums of the form

$$\sum_{\sigma_4} \exp\left[K\sigma_3\sigma_4 + K\sigma_4\sigma_5 + L\sigma_4\right] = 2\cosh\left(K\sigma_3 + K\sigma_5 + L\right)$$

$$= A\exp\left[B\sigma_3\sigma_5 + C\left(\sigma_3 + \sigma_5\right)\right], \tag{14.46}$$

where

$$B = \frac{1}{4}\ln\left[\cosh\left(2K + L\right)\cosh\left(2K - L\right)\right] - \frac{1}{2}\ln\cosh L \tag{14.47}$$

and

$$C = \frac{1}{4}\ln\frac{\cosh\left(2K + L\right)}{\cosh\left(2K - L\right)}. \tag{14.48}$$

FIGURE 14.3. Flow lines associated with the recursion relation $K' = K'(K)$ for the one-dimensional Ising model in zero field.

The canonical partition function is then written as

$$Z = A^{N/2} \sum_{\{\theta_\alpha\}} \exp\left[K' \sum_\alpha \theta_\alpha \theta_{\alpha+1} + L' \sum_\alpha \theta_\alpha \right], \qquad (14.49)$$

where $\{\alpha\}$ is the set of the remaining odd sites of the lattice, and $\theta_\alpha = \sigma_i$, if the original site i is odd. The new parameters, K' and L', are given by

$$K' = \frac{1}{4} \ln\left[\cosh\left(2K + L\right)\cosh\left(2K - L\right)\right] - \frac{1}{2}\ln\cosh L \qquad (14.50)$$

and

$$L' = L + \frac{1}{2}\ln\frac{\cosh\left(2K + L\right)}{\cosh\left(2K - L\right)}, \qquad (14.51)$$

which define the (nonlinear) recursion relations in the space of parameters associated with this system.

In zero field ($H = 0$), we have just one single recursion relation,

$$K' = \frac{1}{2}\ln\left[\cosh\left(2K\right)\right]. \qquad (14.52)$$

The fixed points ($K' = K = K^*$) of this recursion relation are given by $K^* = 0$ (corresponding to infinite temperature) and $K^* = \infty$ (corresponding to zero temperature). It is easy to see that, for any initial condition $K > 0$, the successive application of this recursion relation leads to a flow line toward the trivial fixed point at $K^* = 0$. The fixed point of physical interest, associated with the zero-temperature phase transition, is fully unstable (the arrows in figure 14.3 indicate the flow direction).

As the transition in the one-dimensional model occurs at zero temperature, it is convenient to introduce new parameter variables,

$$x = \exp\left(-4K\right) \quad \text{and} \quad y = \exp\left(-2L\right), \qquad (14.53)$$

in terms of which the recursion relations (14.50) and (14.51) are written as

$$x' = \frac{x\left(1 + y\right)^2}{x\left(1 + y^2\right) + y\left(1 + x^2\right)} \qquad (14.54)$$

and

$$y' = y\left(\frac{x + y}{1 + xy}\right). \qquad (14.55)$$

FIGURE 14.4. Flow lines associated with the recursion relations for the one-dimensional Ising model. Note the line of fixed points at $x = 1$. The "physical" fixed point, located at $x^* = 0$, $y^* = 1$, is fully unstable.

In zero field ($H = 0$), we have $y = 1$ and

$$x' = \frac{4x}{(1+x)^2}, \tag{14.56}$$

with fixed points at $x^* = 0$ (zero temperature) and $x^* = 1$ (infinite temperature). The linearization of this recursion relation around the physical fixed point ($x^* = 0$) leads to the linear form

$$x' \approx 4x = 2^2 x. \tag{14.57}$$

Therefore, we already have a critical scaling exponent, $\lambda_t = 2$, known as the thermal critical exponent. As $\lambda_t > 0$, fixed point $x^* = 0$ is unstable along the x direction.

Now we consider the complete system of recursion relations (14.54) and (14.55), with $H \neq 0$. It is not difficult to see that there are two isolated fixed points, along the $x = 0$ axis, and a line of fixed points at $x = 1$ (see figure 14.4, where the arrows indicate the directions of the flow lines). The linearization of the recursion relations around the physical fixed point ($x^* = 0$; $y^* = 1$) leads to the equations

$$x' \approx 2^2 x \quad \text{and} \quad \Delta y' \approx 2\Delta y, \tag{14.58}$$

where $\Delta y = y - 1$. Therefore, the critical exponents are given by $\lambda_t = 2 > 0$ and $\lambda_H = 1 > 0$, which show that this fixed point is fully unstable. Unfortunately, the phase transition in the Ising chain is rather artificial, as it occurs at zero temperature, which makes it difficult to give a precise and unambiguous description of the asymptotic behavior of thermodynamic

FIGURE 14.5. Simple square lattice divided into two sublattices.

quantities in terms of critical exponents. A more complete renormalization-group analysis of the Ising chain may be found in the paper by D. R. Nelson and M. E. Fisher, Ann. Phys. (NY) **91**, 226 (1975). In the next section, we try to perform a similar analysis for the Ising ferromagnet on the square lattice.

14.5 Renormalization of the Ising model on the square lattice

A simple square lattice may be divided into two distinct square sublattices, as indicated in figure 14.5. In analogy with the decimation technique for the Ising chain, let us fix the spin variables belonging to one of the sublattices, and sum over the spin variables associated with the other sublattice.

Considering figure 14.5, we have

$$\sum_{\sigma_o} \exp \left[K\sigma_o \left(\sigma_1 + \sigma_2 + \sigma_3 + \sigma_4 \right) \right] = 2 \cosh \left[K \left(\sigma_1 + \sigma_2 + \sigma_3 + \sigma_4 \right) \right]$$

$$= A \exp \left[B \left(\sigma_1\sigma_2 + \sigma_2\sigma_3 + \sigma_3\sigma_4 + \sigma_1\sigma_4 \right) \right.$$

$$\left. + C \left(\sigma_1\sigma_3 + \sigma_2\sigma_4 \right) + D\sigma_1\sigma_2\sigma_3\sigma_4 \right], \tag{14.59}$$

where

$$A = 2 \left(\cosh 2K \right)^{1/2} \left(\cosh 4K \right)^{1/8}, \tag{14.60}$$

$$B = C = \frac{1}{8} \ln \cosh 4K, \qquad (14.61)$$

and

$$D = \frac{1}{8} \ln \cosh 4K - \frac{1}{2} \ln \cosh 2K. \qquad (14.62)$$

Therefore, after eliminating the spin variables associated with one of the sublattices, we do obtain a new spin model on the remaining sublattice, **but it involves second-neighbor and four-spin interaction terms**, besides the standard nearest-neighbor interactions. If we perform a second step of the transformation (to produce flow lines in parameter space), the problem involves even more complicated interaction terms. Thus, in contrast with the case of the Ising chain, it is now impossible to preserve the form of the Hamiltonian as we keep applying the renormalization-group transformations. However, at the beginning of the iterations, we can assume a more complex Hamiltonian, including distant-neighbor and multi-spin interaction terms, to follow the flux lines in a multiparameter space. Indeed, this has been tried by Kenneth Wilson [see Rev. Mod. Phys. **47**, 773 (1975)], who reports (approximate) calculations for a parameter space with 200 components. This kind of calculation, however, is too cumbersome to produce useful results!

Now, in order to give an idea of a more realistic renormalization-group analysis, we use the techniques of last paragraph to obtain a set of (highly simplified) recursion relations. After performing the first decimation of spins associated with one of the sublattices, we have

$$K_1' = 2B = \frac{1}{4} \ln \cosh 4K_1, \qquad (14.63)$$

for the new interactions between first neighbors (where $K_1 \equiv K \equiv \beta J$ represents the first-neighbor interactions in the initial Ising Hamiltonian),

$$K_2' = B = \frac{1}{8} \ln \cosh 4K_1, \qquad (14.64)$$

for the second-neighbor interactions, and

$$U' = D = \frac{1}{8} \ln \cosh 4K_1 - \frac{1}{2} \ln \cosh 2K_1, \qquad (14.65)$$

for the four-spin interaction term. To (drastically) simplify the problem, and at the same time produce an interesting toy model, let us take into account only the leading term of a high-temperature expansion (for small $K = \beta J$) of these expressions. Thus, we have

$$K_1' \approx 2K_1^2, \qquad (14.66)$$

$$K_2' \approx K_1^2, \tag{14.67}$$

and

$$U' \approx O\left(K_1^4\right). \tag{14.68}$$

We then discard the higher-order four-spin parameter, which already represents an enormous simplification, and keep first- and second-neighbor interactions only. Furthermore, as second-neighbor interactions are generated anyway, it is interesting to include second-neighbor interaction terms in the initial Hamiltonian. So, we perform the decimations for an initial Ising Hamiltonian with first- and second-neighbor interactions, and then take the high-temperature limit. Keeping terms up to order K_1^2 and K_2^2, it is a simple matter of algebra to show that the recursion relations are given by

$$K_1' = 2K_1^2 + K_2,$$
$$K_2' = K_1^2. \tag{14.69}$$

In spite of these uncontrolled approximations, this artificial toy model is pedagogically useful to illustrate the analysis of the fixed points and the flow lines in parameter space. Of course, we cannot hope to use this kind of approximation to obtain quantitative results for the critical behavior of the Ising ferromagnet on a square lattice. But we can use this toy model to learn some techniques associated with the proposals of the renormalization-group scheme.

The fixed points of the toy model are given by $O \equiv (K_1^* = 0; K_2^* = 0)$ and $P \equiv (K_1^* = 1/3; K_2^* = 1/9)$. The linearization of the recursion relations around the nontrivial fixed point P leads to the matrix equation

$$\begin{pmatrix} \Delta K_1' \\ \Delta K_2' \end{pmatrix} = \begin{pmatrix} 4/3 & 1 \\ 2/3 & 0 \end{pmatrix} \begin{pmatrix} \Delta K_1 \\ \Delta K_2 \end{pmatrix}, \tag{14.70}$$

where $\Delta K_1 = K_1 - K_1^*$ and $\Delta K_2 = K_2 - K_2^*$. The eigenvalues of the linear transformation matrix are given by

$$\Lambda_1 = \frac{2 + \sqrt{10}}{3} = 1.7207... > 1 \tag{14.71}$$

and

$$|\Lambda_2| = \left| \frac{2 - \sqrt{10}}{3} \right| = 0.3874... < 1. \tag{14.72}$$

Therefore, in the two-dimensional $K_1 - K_2$ space, fixed point P is **unstable** along one direction (associated with the eigenvalue Λ_1) and linearly

FIGURE 14.6. Fixed points and flow lines corresponding to the recursion relations associated with the "high-temperature approximation" for the two-dimensional Ising model. Parameters K_1 and K_2 refer to first and second neighbors, respectively. The physical fixed point ($K_1^* = 1/3$; $K_2^* = 1/9$) can be reached if we begin the iterations from the initial condition $K_1 = K_c = \beta_c J$ and $K_2 = 0$ (where K_c is associated with the critical temperature of a ferromagnetic Ising model with nearest-neighbor interactions on the simple square lattice).

stable along another direction (associated with Λ_2). The characteristic flow lines of figure 14.6 can be checked numerically. From the initial condition $K_1 = K_c = \beta_c J$, $K_2 = 0$, we can reach the nontrivial fixed point P. The numerically calculated value $K_c = 0.3921...$ should correspond to the critical temperature of the two-dimensional Ising model with interactions restricted to nearest neighbors (Onsager's solution gives $K_c = 0.4407...$). Although this numerical result is rather poor, it is important to see that it points out the universal character of the critical behavior. Even if there are second-neighbor interactions, in which case we have to start the iterations from an initial state belonging to the separatrix of the flow lines, we do reach the nontrivial fixed point (whose eigenvalues define the universal thermodynamic critical exponents).

Now it is interesting to present some considerations on the general scheme of application of the renormalization-group transformations. Later, we will refer again to the (universal) critical behavior associated with the nontrivial fixed point of this Ising toy model.

14.6 General scheme of application of the renormalization group

In the vicinity of the critical point the correlation length ξ becomes very large, much larger than the distance between sites of the crystalline lattice.

Then, according to Kadanoff, we can eliminate a certain number of degrees of freedom. In a spin system, the spin variables within a block of dimension b, with $b << \xi$, behave in a coherent way, and thus may be replaced by an effective block-spin variable. This scale transformation, involving a factor b in all lengths and another factor c in the intensity of the spin variables, leads to a new Hamiltonian \mathcal{H}'. If we begin with a sufficiently general initial Hamiltonian \mathcal{H}, we then obtain a set of (presumably analytic) recursion relations, involving the parameters of \mathcal{H} and \mathcal{H}'. At the critical point, and after a certain number of iterations of the recursion relations, we eliminate the irrelevant features of the initial Hamiltonian, and reach a fixed point \mathcal{H}^* of the transformation. This is expected to happen because the correlation length becomes infinite, which guarantees the invariance of the physical behavior under any change of length scale. At the critical point, there is a symmetry of dilation, which is also a characteristic feature of the fractal figures. Everything works as if we were looking at the system with a set magnifying lenses, with increasing resolution; changing the lenses, we still see the same overall features of the system.

In a slightly more precise way, the main steps of the renormalization-group scheme are formulated as follows:

(i) Given a certain Hamiltonian, written as $-\beta \mathcal{H} \{\sigma\} = \mathcal{K} \{\sigma\}$, we obtain $\mathcal{K}' \{\sigma'\}$ from the transformation

$$\mathcal{K}' \{\sigma'\} = \mathbf{R}(\mathcal{K} \{\sigma\}). \tag{14.73}$$

To preserve the partition function, we require that

$$Z_N(\mathcal{K}) = \operatorname{Tr} \exp (\mathcal{K}) = \operatorname{Tr}' \exp (\mathcal{K}') = Z_{N'} (\mathcal{K}'), \tag{14.74}$$

where

$$N' = \frac{N}{b^d}. \tag{14.75}$$

The transformation \mathbf{R} includes a change of length scale ($\vec{r} \to \vec{r}' = \vec{r}/b$), and may also include a renormalization of the spin variables ($\sigma \to \sigma' = \sigma/c$).

(ii) We then write new expressions for the free energy per spin and for the pair correlation function,

$$g (\mathcal{K}') = b^d g (\mathcal{K}) \tag{14.76}$$

and

$$\Gamma (\vec{r}; \mathcal{K}) = c^2 \Gamma \left(\frac{1}{b} \vec{r}; \mathcal{K}' \right), \tag{14.77}$$

respectively.

(iii) Now the transformation should be repeated many times,

$$\mathcal{K}' = \mathbf{R}\mathcal{K}, \quad \mathcal{K}'' = \mathbf{R}\mathcal{K}', \dots \tag{14.78}$$

At criticality, we reach a fixed point of the transformation,

$$\mathcal{K}^* = \mathbf{R}\mathcal{K}^*. \tag{14.79}$$

Close to the critical point, we linearize the transformation about \mathcal{K}^*. We then write

$$\mathbf{R}\mathcal{K} = \mathbf{R}\left(\mathcal{K}^* + \mathcal{Q}\right) = \mathcal{K}^* + \mathbf{L}\mathcal{Q} + \dots, \tag{14.80}$$

where \mathbf{L} is a linear operator. The operator \mathcal{Q} may be written in a diagonal form, in terms of the eigenvectors of \mathbf{L},

$$\mathcal{Q} = \sum_j h_j \mathcal{Q}_j, \tag{14.81}$$

so that

$$\mathbf{L}\mathcal{Q}_j = \Lambda_j \mathcal{Q}_j, \tag{14.82}$$

where $\{\mathcal{Q}_j\}$ is a complete set of basis operators, and the set $\{h_j\}$ is used to parametrize the initial Hamiltonian. Thus we have

$$\mathcal{K}' = \mathbf{R}\mathcal{K} = \mathcal{K}^* + \sum_j h_j \Lambda_j \mathcal{Q}_j, \tag{14.83}$$

where we can write

$$\Lambda_j = b^{\lambda_j} \tag{14.84}$$

to guarantee the semigroup properties of the transformation \mathbf{R}.

Now we have a conceptual basis to discuss the ideas of universality and to formulate the theories of scaling. The set of values $\{h_j\}$ parametrizes the initial Hamiltonian, while the successive applications of the recurrence relations leads to the flow lines in parameter space. After n iterations of the recursion relations, we have

$$\mathcal{K}^{(n)} = \mathcal{K}^* + \sum_j h_j b^{n\lambda_j} \mathcal{Q}_j. \tag{14.85}$$

If $\lambda_j > 0$, operator \mathcal{Q}_j is called **relevant**. If $\lambda_j < 0$, operator \mathcal{Q}_j is **irrelevant**. Of course, after a large number of iterations, only the relevant operators still survive, which leads the Hamiltonian far away from the fixed point. Therefore, at the critical point, we require that $h_j = 0$ for all parameters h_j associated with a relevant operator \mathcal{Q}_j. The marginal case, $\lambda_j = 0$,

is treated separately. Different fixed points correspond to different **critical universality classes**. In parameter space, there are distinct regions (basins of attraction) associated with fluxes to each one of these different fixed points.

Since the critical point is located by giving the values of two thermodynamic fields, there should be only two relevant operators associated with the critical behavior of simple systems. For the simple uniaxial ferromagnet, the relevant operators may be chosen as the energy and the magnetization. We thus define $h_1 = t = (T - T_c)/T_c$ and $h_2 = H$. At the critical point, we have $h_1 = h_2 = 0$, that is, $t = 0$ and $H = 0$.

The scaling theory comes naturally from this formalism. After n iterations, the free energy, given by equation (14.76), may be written as

$$
\begin{aligned}
g(h_1, h_2, h_3, ...) &= b^{-dn} g\left(b^{\lambda_1 n} h_1, b^{\lambda_2 n} h_2, b^{\lambda_3 n} h_3, ...\right) \\
&= h_1^{d/\lambda_1} g\left(1, \frac{h_2}{h_1^{\lambda_2/\lambda_1}}, \frac{h_3}{h_1^{\lambda_3/\lambda_1}},\right).
\end{aligned}
\tag{14.86}
$$

With the choice $h_1 = t$ and $h_2 = H$, we have

$$
\frac{d}{\lambda_1} = 2 - \alpha, \text{ that is, } \lambda_1 > 0,
\tag{14.87}
$$

$$
\frac{\lambda_2}{\lambda_1} = \Delta = \beta + \gamma > 0,
\tag{14.88}
$$

and

$$
\frac{\lambda_3}{\lambda_1} = \phi,
\tag{14.89}
$$

which is known as a crossover exponent. At a standard critical point, $\lambda_3 < 0$, since there should be only two relevant operators. Thus, $\phi < 0$, which leads, even if $h_3 \neq 0$, to the existence of an asymptotic scaling expression for the free energy,

$$
g(h_1, h_2, h_3, ...) \approx t^{2-\alpha} g\left(1, \frac{H}{t^{\Delta}}, 0, 0, ...\right).
\tag{14.90}
$$

We can also construct a scaling theory for the critical correlations,

$$
\begin{aligned}
\Gamma(\vec{r}; t, H, h_3, ...) &= b^{-(d-2+\eta)} \Gamma\left(\frac{1}{b}\vec{r}; b^{\lambda_1} t, b^{\lambda_2} H, b^{\lambda_3} h_3, ...\right) \\
&= t^{(d-2+\eta)/\lambda_1} \Gamma\left(\frac{1}{t^{-1/\lambda_1}}\vec{r}; 1, \frac{H}{t^{\lambda_2/\lambda_1}}, \frac{h_3}{t^{\lambda_3/\lambda_1}}, ...\right),
\end{aligned}
\tag{14.91}
$$

where $1/\lambda_1 = \nu$, that is, $2 - \alpha = d$, which automatically fulfills the hyper-scaling relation.

We now use this general framework to complete the analysis of the critical behavior of the toy model associated with the Ising ferromagnet on the square lattice. Consider again the recursion relations given by equation (14.59), whose flow lines and fixed points in a two-dimensional parameter space are sketched in figure 14.6. From equation (14.60), the linearization about the nontrivial fixed point leads to the matrix relation

$$\begin{pmatrix} \Delta K_1' \\ \Delta K_2' \end{pmatrix} = \begin{pmatrix} 4/3 & 1 \\ 2/3 & 0 \end{pmatrix} \begin{pmatrix} \Delta K_1 \\ \Delta K_2 \end{pmatrix}. \tag{14.92}$$

Diagonalizing the matrix, we obtain the eigenvalues Λ_1 and Λ_2, given by equations (14.71) and (14.72). As the original lattice spacing is reduced by the factor $b = \sqrt{2}$ under an iteration of the recursion relations, we write

$$\Lambda_1 = \left(\sqrt{2}\right)^{\lambda_1} \quad \text{and} \quad |\Lambda_2| = \left(\sqrt{2}\right)^{\lambda_2}. \tag{14.93}$$

Therefore,

$$\lambda_1 = \frac{2\ln\Lambda_1}{\ln 2} = 1.56609... \tag{14.94}$$

and

$$\lambda_2 = \frac{2\ln|\Lambda_2|}{\ln 2} = -2.7360.... \tag{14.95}$$

To derive the scaling forms in zero field, it is enough to introduce the obvious definitions, $h_1 = t = \delta K_1$, $h_2 = H = 0$ (zero field), and $h_3 = \delta K_2$, where the new parameters, δK_1 and δK_2, are linear combinations of ΔK_1 and ΔK_2 produced by the diagonalization of the transformation matrix in equation (14.92). The contribution of the second-neighbor interactions is irrelevant ($\lambda_2 < 0$), which confirms the ideas of universality. As $\lambda_1 > 0$, the energy operator is relevant. The specific heat critical exponent, given by

$$\alpha = 2 - \frac{d}{\lambda_1} = 2 - \frac{2}{1.566...} = 0.722..., \tag{14.96}$$

is very far from the exact result for the two-dimensional Ising model!

14.7 Renormalization group for the Ising ferromagnet on the triangular lattice

In the case of a few special models, as the Ising chain, we can do exact calculations to eliminate part of the degrees of freedom in order to write

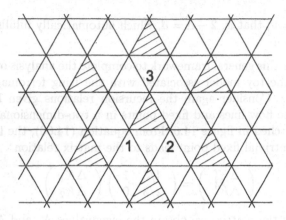

FIGURE 14.7. Sketch of a triangular lattice. The hatched cells define a new triangular lattice.

(exact) renormalization-group recursion relations, but in general we are forced to resort to some approximations. In this chapter, for example, we have used a very crude approximation to analyze the critical behavior of the Ising ferromagnet on the square lattice. There are, however, a number of more systematic, and sometimes even controlled, approximate schemes that preserve the main physical features of the critical behavior of statistical systems. The experience of using these techniques has shown that approximations for the recursion relations are not so damaging, as far as we look at the critical behavior, as compared with mean-field or self-consistent approximations for the thermodynamic potentials. The more successful approximate scheme is a perturbative expansion in momentum space, around the solution of the model in the trivial $d = 4$ dimension, in terms of the small parameter $\epsilon = 4 - d$. The development of this expansion, however, requires field-theoretical techniques, which demand a more thorough introduction, and are certainly beyond the scope of this book. We then restrict the analysis to a much simpler perturbative expansion, in the real space of the crystalline lattice, whose lowest-order terms already lead to some interesting results.

According to a pioneering work of Niemeyer and van Leeuwen, consider the Ising ferromagnet, in zero field, on a triangular lattice, and perform a partial sum over the spin variables belonging to the hatched cells in figure 14.7. If we suppose that the original triangular lattice has a unit lattice parameter, the transformed lattice is still triangular, with a lattice parameter $b = \sqrt{3}$. Now we replace the spin variables of the hatched cells by a new set of spin variables, $\{\theta\}$, so that, for each cell,

$$\theta = \mathrm{sgn}\,(\sigma_1 + \sigma_2 + \sigma_3). \tag{14.97}$$

This is the so-called **majority rule**, according to which the sign of the new spin variable follows the predominant sign of the cell (for a square lattice, we can still write a slightly more complicated rule).

There is a projection operator that provides a way to sum over the initial spin variables, with fixed values of θ for each cell. For a particular cell I, we define

$$P_I = \frac{1}{2}\left[1 + \theta_I \operatorname{sgn}\left(\sigma_I^1 + \sigma_I^2 + \sigma_I^3\right)\right]. \tag{14.98}$$

If the overall sign is positive, we have

$$P_I = \frac{1}{2}\left[1 + \theta_I\right], \tag{14.99}$$

so that $P_I = +1$ for $\theta_I = +1$ and $P_I = 0$ for $\theta_I = -1$. If the overall sign is negative, we have

$$P_I = \frac{1}{2}\left[1 - \theta_I\right], \tag{14.100}$$

so that $P_I = 0$ for $\theta_I = +1$ and $P_I = +1$ for $\theta_I = -1$. It is also easy to see that

$$\sum_{\theta_I} P_I = 1. \tag{14.101}$$

Therefore, we can define a projection operator P, given by

$$P = \prod_I P_I, \tag{14.102}$$

where the subscript I runs over all hatched cells of the figure. Thus, we have

$$Z = \sum_{\{\sigma\}} \exp\left(-\beta\mathcal{H}\right) = \sum_{\{\sigma\}}\left(\sum_{\{\theta\}} P\right) \exp\left(-\beta\mathcal{H}\right) = \sum_{\{\sigma\}}\sum_{\{\theta\}} P \exp\left(-\beta\mathcal{H}\right). \tag{14.103}$$

For the Ising model in zero field, we write

$$-\beta\mathcal{H} = H_o + V, \tag{14.104}$$

where

$$H_o = K \sum_I \left(\sigma_I^1\sigma_I^2 + \sigma_I^2\sigma_I^3 + \sigma_I^3\sigma_I^1\right), \tag{14.105}$$

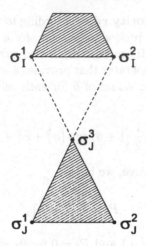

FIGURE 14.8. Two distinct and neighboring hatched cells (I and J) on the triangular lattice.

and V contains interaction terms of the form $K\left(\sigma_I^3\sigma_J^1 + \sigma_I^3\sigma_J^2\right)$, in which I and J are two distinct, but neighboring, hatched cells, as indicated in figure 14.8. We then have

$$Z = \sum_{\{\sigma\}}\sum_{\{\theta\}} P \exp\left(H_o + V\right). \tag{14.106}$$

Introducing the definitions

$$Z_o = \sum_{\{\sigma\}} P \exp\left(H_o\right), \tag{14.107}$$

which is a factorized sum, and

$$\langle\cdots\rangle_o = \frac{1}{Z_o}\sum_{\{\sigma\}}(\cdots)\, P \exp\left(H_o\right), \tag{14.108}$$

we write the canonical partition function,

$$Z = \sum_{\{\theta\}} Z_o \langle\exp\left(V\right)\rangle_o. \tag{14.109}$$

The calculation of Z_o is trivial. Since equation (14.107) is a product of sums for each cell, it is enough to consider a single cell. Hence, we write

$$\sum_{\sigma_1,\sigma_2,\sigma_3} P_I \exp\left(H_o^I\right) = \sum_{\sigma_1,\sigma_2,\sigma_3} \{\tfrac{1}{2}\left[1 + \theta\,\mathrm{sgn}\left(\sigma_1 + \sigma_2 + \sigma_3\right)\right]$$

$$\times \exp\left[K\left(\sigma_1\sigma_2 + \sigma_2\sigma_3 + \sigma_3\sigma_1\right)\right]\} = \exp\left(3K\right) + 3\exp\left(-K\right). \quad (14.110)$$

which does not depend on the value of θ. Thus, we have

$$Z_o = \left[\exp\left(3K\right) + 3\exp\left(-K\right)\right]^{N'}, \quad (14.111)$$

where

$$N' = \frac{N}{\left(\sqrt{3}\right)^2} = \frac{N}{3}. \quad (14.112)$$

Now we consider the expansions,

$$\langle\exp\left(V\right)\rangle_o = 1 + \langle V\rangle_o + \frac{1}{2!}\langle V^2\rangle_o + \dots \quad (14.113)$$

and

$$\exp\left(\langle V\rangle_o\right) = 1 + \langle V\rangle_o + \frac{1}{2!}\langle V\rangle_o^2 + \dots \quad (14.114)$$

Thus, up to first-order terms in the interaction V, we have

$$\langle\exp\left(V\right)\rangle_o = \exp\left(\langle V\rangle_o\right), \quad (14.115)$$

which will be sufficient to illustrate the performance of this method. It should be noted that we can write $\langle\exp\left(V\right)\rangle_o$ as the exponential of a systematic expansion in terms of some quantities, known as the cumulants of V.

Keeping first-order terms only, we now calculate terms of the form

$$\sum_{\{\sigma\}} V P \exp\left(\mathcal{H}_o\right). \quad (14.116)$$

Since V is a sum that involves pairs of spins belonging to distinct cells, this expression factorizes. A typical term is given by

$$\left(e^{3K} + 3e^{-K}\right)^{N'-2} K$$

$$\times\left\{\sum_{\sigma_1,\sigma_2,\sigma_3} \sigma_3\frac{1}{2}\left[1 + \theta_I \operatorname{sgn}\left(\sigma_1 + \sigma_2 + \sigma_3\right)\right]\exp\left[K\left(\sigma_1\sigma_2 + \sigma_2\sigma_3 + \sigma_3\sigma_1\right)\right]\right\}$$

$$\times\left\{\sum_{\sigma_1,\sigma_2,\sigma_3} \sigma_1\frac{1}{2}\left[1 + \theta_J \operatorname{sgn}\left(\sigma_1 + \sigma_2 + \sigma_3\right)\right]\exp\left[K\left(\sigma_1\sigma_2 + \sigma_2\sigma_3 + \sigma_3\sigma_1\right)\right]\right\},$$

$$(14.117)$$

where the first (second) sum refers to spins belonging to cell I (J). It is then enough to show that

$$\sum_{\sigma_1,\sigma_2,\sigma_3} \sigma_1 \frac{1}{2} \left[1 + \theta \operatorname{sgn}\left(\sigma_1 + \sigma_2 + \sigma_3\right)\right] \exp\left[K\left(\sigma_1\sigma_2 + \sigma_2\sigma_3 + \sigma_3\sigma_1\right)\right]$$

$$= \theta\left(e^{3K} + e^{-K}\right). \tag{14.118}$$

Therefore, the contribution of the typical term is given by

$$\left(e^{3K} + 3e^{-K}\right)^{N'-2} K\left(e^{3K} + e^{-K}\right)^2 \theta_I\theta_J. \tag{14.119}$$

Recalling that there are always two terms of this form for each neighboring cell, we finally write the recursion relation (up to terms of first order in V),

$$K' = 2K\left(\frac{e^{3K} + e^{-K}}{e^{3K} + 3e^{-K}}\right)^2. \tag{14.120}$$

There are two trivial fixed points, $K^* = 0$ and $K^* = \infty$. However, there is also a nontrivial unstable fixed point,

$$K^* = \frac{1}{4}\ln\frac{3 - \sqrt{2}}{\sqrt{2} - 1} = 0.335..., \tag{14.121}$$

with eigenvalue

$$\Lambda_1 = \left(\frac{\partial K'}{\partial K}\right)_{K=K^*} = 1.62... > 1. \tag{14.122}$$

The linear form of the recursion relation about this unstable fixed point is given by

$$\Delta K' \approx 1.62\Delta K = \left(\sqrt{3}\right)^{\lambda_1} \Delta K, \tag{14.123}$$

where

$$\lambda_1 = 0.887..., \tag{14.124}$$

which leads to the specific heat critical exponent $\alpha = -0.28...$, quite far from the exact Onsager result!

It is not difficult to introduce a magnetic field to obtain a second critical exponent. Also, it is interesting to carry out the calculations up to at least second-order terms, in which case we are forced to include second-neighbor interactions in the initial spin Hamiltonian. In zero field, and keeping terms up to second order, we have $K_c \approx 0.251$ (which should be compared with the exact value for the triangular lattice, $K_c = 0.27...$), and the critical exponent $\alpha \approx 0.002$ (see T. Niemeyer and J. M. J. van Leeuwen, in *Phase Transitions and Critical Phenomena*, edited by C. Domb and M. S. Green, volume 6, Academic Press, New York, 1976).

Exercises

1. Use the scaling form of the singular part of the free energy associated with a simple ferromagnet to show the scaling relation

$$\gamma = \beta (\delta - 1),$$

where δ is the exponent associated with the critical isotherm, $m \sim H^{1/\delta}$, at $T = T_c$.

2. In order to show the experimental inadequacy of the classical exponent values, J. S. Kouvel and M. E. Fisher [Phys. Rev. **136**, A1626 (1964)] performed a detailed analysis of the old data of P. Weiss and R. Forrer [Ann. Phys. (Paris) **5**, 133 (1926)] for the magnetic equation of state of nickel. In the table below, we reproduce some of the experimental data for the magnetization m (in suitable experimental units) at different values of temperature (in degrees Celsius) and magnetic field (in Gauss). Note that the field values should be corrected for demagnetization effects (for each value of the magnetization m, Weiss and Forrer obtain the thermodynamic applied field by subtracting the factor $39.3m$ from the field values in the first column of the table).

	348.78 °C	350.66 °C	352.53 °C	358.18 °C
430 G	11.361	10.452	8.114	–
890 G	13.787	–	–	–
1355 G	14.307	12.646	10.626	4.235
3230 G	15.651	14.322	12.752	7.567
6015 G	16.938	15.780	14.519	10.312
10070 G	18.209	17.271	16.198	12.775
14210 G	19.212	18.348	17.395	14.385
17775 G	19.976	19.141	18.293	15.504

According to an analysis of J. S. Kouvel and J. B. Comley [Phys. Rev. Lett. **20**, 1237 (1968)], the critical parameters of nickel are given by $T_c = 627.4\,\mathrm{K}$, $\gamma = 1.34\,(1)$, $\beta = 0.378\,(4)$, and $\delta = 4.58\,(5)$, where the presumed uncertainties are indicated in parenthesis. Check that these values are compatible with the scaling relation $\gamma = \beta (\delta - 1)$. Use the data of the table to show that the magnetization obeys the scaling form

$$m = |t|^{\beta} Y_{\pm} \left(\frac{H}{|t|^{\Delta}} \right),$$

where H is the applied magnetic field (after corrections for demagnetization), $t = (T - T_c)/T_c$, $\Delta = \beta + \gamma$, and there might be two distinct functions, Y_{\pm}, above and below the critical temperature.

3. The spontaneous magnetization of the Ising ferromagnet on the square lattice is given by the Onsager expression,

$$m = \left[1 - (\sinh 2K)^{-4}\right]^{1/8},$$

where $K = J/(k_B T)$. In the neighborhood of the critical point, we have the asymptotic form $m \sim B(-t)^\beta$, where $t = (T - T_c)/T_c \to 0$. What are the values of the prefactor B and the critical exponent β?

4. Consider the Ising ferromagnet, given by the spin Hamiltonian

$$\mathcal{H} = -J \sum_{(ij)} \sigma_i \sigma_j - H \sum_{i=1}^{N} \sigma_i,$$

where $J > 0$, $\sigma_i = \pm 1$, and the first sum is over nearest neighbors on a simple triangular lattice. Use the real-space renormalization-group scheme of Niemeyer and van Leeuwen, up to first-order terms, as discussed in the last section, to obtain the thermal and magnetic critical exponents. Consider the Ising ferromagnet, given by the Hamiltonian

$$\mathcal{H} = -J \sum_{(ij)} \sigma_i \sigma_j - H \sum_{i=1}^{N} \sigma_i,$$

where $J > 0$, $\sigma_i = \pm 1$, and the first sum is over nearest-neighbor sites on a square lattice. Choose a unit cell consisting of five spins, as indicated in figure 14.9. Use the real-space renormalization-group scheme of Niemeyer and van Leeuwen to write a set of recursion relations up to first-order terms in the interactions. Obtain the values of the thermal and magnetic critical exponents.

5. Use the Migdal–Kadanoff real-space renormalization-group scheme, as illustrated in figure 14.10, to analyze the critical behavior of the Ising ferromagnet, in zero field, on the simple square lattice. According to the scheme of Migdal–Kadanoff (see the illustration), we eliminate the \times sites, double the remaining interactions, and finally perform a decimation of the intermediate spins. Show that the recursion relation is given by

$$K' = \frac{1}{2} \ln \cosh (4K),$$

where $K = \beta J$. Obtain the fixed points and the thermal exponent associated with the physical fixed point. What is the estimate for the critical temperature?

6. The real-space renormalization-group is carried out ...
... with a certain Ising ferromagnet, with specified magnetic interactions ...

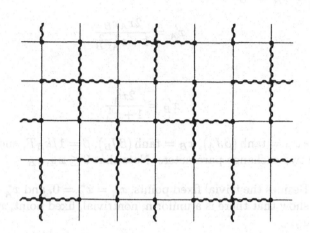

FIGURE 14.9. Unit cells with five sites on the square lattice.

FIGURE 14.10. Some steps of the Migdal–Kadanoff renormalization-group scheme for a spin system on a square lattice.

6. The real-space renormalization-group recursion relations associated
 with a certain Ising ferromagnet with aperiodic exchange interactions
 [see S. T. R. Pinho, T. A. S. Haddad, and S. R. Salinas, Physica A**257**,
 515 (1998)] are given by

$$x'_A = \frac{2x_A x_B}{1 + x_A^2 x_B^2}$$

and

$$x'_B = \frac{2x_A^2}{1 + x_A^4},$$

where $x_A = \tanh{(\beta J_A)}$, $x_B = \tanh{(\beta J_B)}$, $\beta = 1/k_B T$, and $J_A, J_B >$
0 are two exchange parameters. Note that $0 \leq x_A, x_B \leq 1$.

 (a) Besides the trivial fixed points, $x_A^* = x_B^* = 0$, and $x_A^* = x_B^* = 1$,
 show that there is a uniform, nontrivial, fixed point, $x_A^* = x_B^* = x_o$.

 (b) Perform a linear analysis of stability in the neighborhood of
 the nontrivial fixed point. Find the eigenvalues and discuss the
 nature of the stability of this fixed point.

 (c) Given the ratio $r = J_B/J_A = 2$, use numerical methods to find
 the critical temperature, $k_B T_c/J_A$. Do the critical exponents
 depend on the ratio r?

7. Consider again the nearest-neighbor Ising ferromagnet on the trian-
 gular lattice in zero field ($H = 0$). Obtain recursion relations up
 to second order in the interaction terms (note that we have to in-
 clude second-neighbor interactions in the initial Hamiltonian). Find
 the fixed points of these recursion relations and sketch the flow lines
 in parameter space. Calculate the critical temperature and obtain the
 thermal critical exponent. If you feel that the problem is too difficult,
 check the work of Niemeyer and van Leeuwen.

15
Nonequilibrium Phenomena: I. Kinetic Methods

In the previous chapters of this book we have always referred to phenomena in macroscopic equilibrium, which can be treated and well understood by the formalism of Gibbs ensembles. In this chapter, we revisit the old problems of Ludwig Boltzmann, who directed most of his efforts to the investigation of the time evolution of systems toward equilibrium. Using a kinetic method, we initially derive the famous **Boltzmann equation** for a diluted fluid, which still remains an important theoretical technique to deal with transport problems. We then prove the H-theorem, which comes from the Boltzmann equation and provides a description of the route to equilibrium. The historical debates about the meaning of the H-theorem, including objections and answers by Boltzmann, very much up to date, contributed a great deal to clarify the character of statistical physics.

There is no satisfactory Gibbsian description of statistical physics out of equilibrium. From the classical point of view, the time evolution of a fluid system is described by Liouville's equation for the density ρ of microscopic states in phase space. Up to this point, however, there is no progress with respect to the original Hamilton equations of the system. In order to go beyond Hamilton's equation, it is interesting to define reduced densities that obey a set of coupled equations, known as the Bogoliubov, Born, Green, Kirkwood, and Yvon (BBGKY) hierarchy. Although there are several approximations to deal with these coupled equations, including the deduction of Boltzmann's transport equation for a sufficiently diluted fluid, it has not been possible to devise a really satisfactory way of decoupling and solving this hierarchy.

In order to treat non-equilibrium phenomena in dense fluids, we can resort to alternative approaches, coming from the early investigations of Brownian motion. Taking into account the stochastic nature of the phenomena, it is interesting to write from the beginning a **master equation**, based on probabilistic arguments, for the time evolution of the probability of finding a system in a determined microscopic state. For **Markovian processes** (in which the time evolution does not depend on the previous history of the system), the master equation assumes a tractable form, depending on **transition probabilities** between successive microscopic states. At least in principle, these transition probabilities can be obtained from the mechanical equations of motion. In general, however, this problem is as difficult as the attempts to decouple the BBGKY hierarchy. In some cases, if we keep the Gibbsian character of the final equilibrium state, there are plausible choices for the expressions of the transition probabilities.

In this chapter, we limit our considerations to the kinetic methods (with just an exception for a presentation of the beautiful Ehrenfest **urn model**). In the following chapter, we will present some examples of stochastic methods to deal with non-equilibrium phenomena.

15.1 Boltzmann's kinetic method

Consider a classical gas of N particles (monatomic molecules, for example), inside a container of volume V, at a given temperature T. We may introduce a molecular distribution function, $f(\vec{r}, \vec{p}, t)$, defined in the space μ of molecular positions and momenta, such that $f(\vec{r}, \vec{p}, t) d^3\vec{r} d^3\vec{p}$ is the number of molecules, at time t, with coordinates belonging to the elementary volume $d^3\vec{r} d^3\vec{p}$. In equilibrium, which corresponds to the limit $t \to \infty$, we should recover the classical Maxwell–Boltzmann molecular distribution.

In the absence of molecular collisions, a molecule at position (\vec{r}, \vec{p}) in μ space, at time t, changes to position $(\vec{r} + \vec{v}\delta t, \vec{p} + \vec{F}\delta t)$ at time $t + \delta t$, where $\vec{v} = \vec{p}/m$ and \vec{F} is an external force. In the absence of interactions, the molecular distribution function does not change with time. On the other hand, in the presence of molecular interactions (binary collisions, for a sufficiently diluted gas), we should write

$$f\left(\vec{r} + \frac{1}{m}\vec{p}\delta t, \vec{p} + \vec{F}\delta t, t + \delta t\right) = f(\vec{r}, \vec{p}, t) + \left(\frac{\partial f}{\partial t}\right)_{coll} \delta t, \quad (15.1)$$

in other words,

$$\left(\frac{\partial}{\partial t} + \frac{1}{m}\vec{p} \cdot \vec{\nabla}_r + \vec{F} \cdot \vec{\nabla}_p\right) f(\vec{r}, \vec{p}, t) = \left(\frac{\partial f}{\partial t}\right)_{coll}, \quad (15.2)$$

where $\vec{\nabla}_r$ and $\vec{\nabla}_p$ are gradient operators with respect to position and momentum, respectively. In order to attribute a meaning to this equation, we

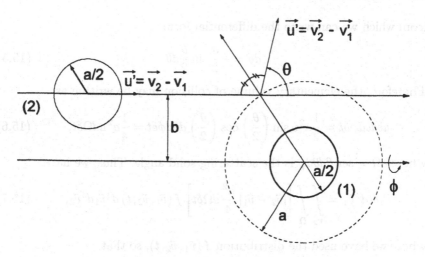

FIGURE 15.1. Schematic representation of the collision between two rigid spheres of diameter a. In the scheme, we indicate the relative velocities, \vec{u} and \vec{u}', before and after the collision, the impact parameter b, and the scattering angle θ.

have to calculate the contribution of the collision term. Taking into account binary collisions only, which should be highly dominating for a diluted fluid, we may write

$$\left(\frac{\partial f}{\partial t}\right)_{col} \delta t = (R_f - R_i)\,\delta t, \qquad (15.3)$$

where $R_i d^3\vec{r} d^3\vec{p}\delta t$ is the number of (binary) collisions taking place between times t and $t+\delta t$, so that one of the **initial** molecules is "within the volume" $d^3\vec{r} d^3\vec{p}$, while $R_f d^3\vec{r} d^3\vec{p}\delta t$ is the number of collisions in the interval δt so that one of the **final** molecules is within the volume $d^3\vec{r} d^3\vec{p}$.

Now we carry out a simple mechanical calculation for a gas of uniformly distributed rigid spheres of diameter a (see the sketch in figure 15.1), in the absence of external forces. As the system is uniform, the distribution function f is independent of position \vec{r}. Consider, for example, a certain molecule (1), with velocity \vec{v}_1 "within the volume" $d^3\vec{v}_1$ in the space of velocities, and a second molecule (2), with velocity \vec{v}_2 within $d^3\vec{v}_2$. In figure 15.1, \vec{v}_1' and \vec{v}_2' are velocities after the collision, b is the impact parameter of collision, and θ is the scattering angle. The elementary volume of the collision cylinder is given by $(u\delta t)\,(db)\,(bd\phi)$, where ϕ is an angular cylindrical coordinate and $u = |\vec{v}_2 - \vec{v}_1|$ is the intensity of the relative velocity. As the center of molecule (2) is inside this elementary volume, it will certainly collide with molecule (1) during the time interval δt. Looking at the figure, we have

$$\sin\frac{\pi - \theta}{2} = \frac{b}{a}, \qquad (15.4)$$

from which we can write the differential form

$$db = -\frac{a}{2} \sin \frac{\theta}{2} d\theta. \tag{15.5}$$

Therefore, the elementary volume of collision may be written as

$$ubdbd\phi\delta t = \frac{1}{2}a^2 u \sin\left(\frac{\theta}{2}\right) \cos\left(\frac{\theta}{2}\right) d\theta d\phi \delta t = \frac{1}{4}a^2 u d\Omega \delta t, \tag{15.6}$$

where $d\Omega = \text{sen}\,\theta d\theta d\phi$ is the scattering solid angle. Thus, we have

$$R_i \delta t d^3 \vec{v}_1 = \int\limits_{v_2} \int\limits_{\Omega} \left[|\vec{v}_2 - \vec{v}_1| \frac{a^2}{4} d\Omega \delta t \right] f\left(\vec{v}_1, \vec{v}_2, t\right) d^3 \vec{v}_1 d^3 \vec{v}_2, \tag{15.7}$$

where we have used the distribution $f\left(\vec{v}_1, \vec{v}_2, t\right)$, so that

$$f\left(\vec{v}_1, \vec{v}_2, t\right) d^3\vec{v}_1 d^3\vec{v}_2 d^3\vec{r}_1 d^3\vec{r}_2$$

gives the number of pairs of molecules, at time t, with coordinates within the elementary volumes $d^3\vec{v}_1 d^3\vec{r}_1$ and $d^3\vec{v}_2 d^3\vec{r}_2$, respectively. In order to proceed, we have to introduce a hypothesis of statistical independence, known as **"Stosszahlansatz"** or **molecular chaos**, which is the more delicate step of Boltzmann's method. According to this "Stosszahlansatz," which cannot be justified under a purely mechanical basis, we assume that molecular velocities are not correlated, that is, we write

$$f\left(\vec{v}_1, \vec{v}_2, t\right) = f\left(\vec{v}_1, t\right) f\left(\vec{v}_2, t\right). \tag{15.8}$$

Therefore, equation (15.7) is reduced to

$$R_i \delta t d^3 \vec{v}_1 = \int\limits_{\Omega} \frac{1}{4}a^2 d\Omega \int\limits_{v_2} |\vec{v}_2 - \vec{v}_1| f\left(\vec{v}_1, t\right) f\left(\vec{v}_2, t\right) d^3\vec{v}_1 d^3\vec{v}_2. \tag{15.9}$$

To obtain R_f it is enough to consider the reverse collision, $(\vec{v}_1'; \vec{v}_2') \rightarrow (\vec{v}_1; \vec{v}_2)$. We then have

$$R_f \delta t d^3 \vec{v}_1 = \int\limits_{\Omega} \frac{1}{4}a^2 d\Omega \int\limits_{v_2} |\vec{v}_2' - \vec{v}_1'| f\left(\vec{v}_1', t\right) f\left(\vec{v}_2', t\right) d^3\vec{v}_1' d^3\vec{v}_2'. \tag{15.10}$$

From the mechanical conservation of kinetic energy and momentum of the colliding particles, we have

$$\vec{v}_1 + \vec{v}_2 = \vec{v}_1' + \vec{v}_2' \tag{15.11}$$

and

$$|\vec{v}_1|^2 + |\vec{v}_2|^2 = |\vec{v}_1'|^2 + |\vec{v}_2'|^2, \tag{15.12}$$

from which we can show that

$$|\vec{v}_2 - \vec{v}_1| = |\vec{v}_2' - \vec{v}_1'| \tag{15.13}$$

and

$$d^3\vec{v}_1 d^3\vec{v}_2 = d^3\vec{v}_1' d^3\vec{v}_2'. \tag{15.14}$$

Therefore, we write the collision term

$$\left(\frac{\partial f_1}{\partial t}\right)_{coll} = \int_\Omega \frac{1}{4}a^2 d\Omega \int_{v_2} d^3\vec{v}_2 \, |\vec{v}_2 - \vec{v}_1| \, (f_1' f_2' - f_1 f_2), \tag{15.15}$$

where we have introduced the notation $f_1 = f(\vec{v}_1, t)$, $f_1' = f(\vec{v}_1', t)$, and so on. Thus, for spatially uniform systems, in the absence of external forces, the Boltzmann transport equation for the molecular distributions is given by

$$\frac{\partial f_1}{\partial t} = \int_\Omega \frac{1}{4}a^2 d\Omega \int_{v_2} d^3\vec{v}_2 \, |\vec{v}_2 - \vec{v}_1| \, (f_1' f_2' - f_1 f_2), \tag{15.16}$$

supplemented by the conservation laws, (15.11) and (15.12).

Now we consider a molecular distribution of the form $f = f(\vec{r}, \vec{p}, t)$ and write the Boltzmann transport equation in a more general case, in the presence of an external force \vec{F} and of binary collisions associated with a velocity-independent cross section $\sigma(\Omega)$. We then have the more general form

$$\left(\frac{\partial}{\partial t} + \frac{1}{m}\vec{p}_1 \cdot \vec{\nabla}_r + \vec{F} \cdot \vec{\nabla}_{p_1}\right) f(\vec{r}, \vec{p}_1, t) = \left(\frac{\partial f}{\partial t}\right)_{coll}$$

$$= \int_\Omega \sigma(\Omega) \, d\Omega \int_{p_2} d^3\vec{p}_2 \, |\vec{p}_2 - \vec{p}_1| \, (f_1' f_2' - f_1 f_2), \tag{15.17}$$

with the notation $f_1 = f(\vec{r}, \vec{p}_1, t)$, $f_1' = f(\vec{r}, \vec{p}_1', t)$, and so on.

This nonlinear integrodifferential equation represents a quite formidable mathematical problem. There is a long history of the developments of approximate methods to obtain the distribution function in order to calculate the time evolution of quantities of physical interest. The reader may check the texts by S. Chapman and T. G. Cowling, *The mathematical theory of non-uniform gases*, Cambridge University Press, New York, 1953, and by C. Cercignani, *Mathematical methods in kinetic theory*, Plenum Press, New York, 1969. In the remainder of this section, we examine some more immediate consequences of this equation. An approximation (of zero order) that is already able to produce some interesting physical results is left as an exercise.

The H-theorem

Consider a simple diluted gas in the absence of external forces. In equilibrium, we should have

$$f(\vec{v}, t) \rightarrow f_o(\vec{v}),$$ (15.18)

that is,

$$\int \sigma(\Omega) \, d\Omega \int_{v_2} d^3 \vec{v}_2 \, |\vec{v}_2 - \vec{v}_1| \, [f_o(\vec{v}_2') f_o(\vec{v}_1') - f_o(\vec{v}_2) f_o(\vec{v}_1)] = 0.$$ (15.19)

Therefore, we have a **sufficient** condition for equilibrium,

$$f_o(\vec{v}_2') f_o(\vec{v}_1') - f_o(\vec{v}_2) f_o(\vec{v}_1) = 0.$$ (15.20)

From the H-theorem, we can show that this is also a necessary condition for equilibrium.

The H-theorem can be stated as follows. If $f(\vec{v}, t)$ is a solution of the Boltzmann equation, then the function $H(t)$, given by

$$H(t) = \int d^3 \vec{v} f(\vec{v}, t) \ln f(\vec{v}, t),$$ (15.21)

obeys the inequality

$$\frac{dH}{dt} \leq 0.$$ (15.22)

In contrast with mechanical quantities, which do not distinguish between past and future, the function $H(t)$ has a well-defined direction in time, which opens up the possibility of making a connection with the second law of thermodynamics (in fact, in equilibrium, one can show that $-H$ is proportional to the entropy of an ideal gas).

In order to demonstrate the theorem, we write

$$\frac{dH}{dt} = \int d^3 \vec{v} \, [\ln f + 1] \frac{\partial f}{\partial t}.$$ (15.23)

Therefore,

$$\frac{\partial f}{\partial t} = 0 \rightarrow \frac{dH}{dt} = 0,$$ (15.24)

in other words, $dH/dt = 0$ is a **necessary** condition for equilibrium. From Boltzmann's equation, we can also write

$$\frac{dH}{dt} = \int d^3 \vec{v}_1 \int \sigma(\Omega) \, d\Omega \int d^3 \vec{v}_2 \, |\vec{v}_2 - \vec{v}_1| \, (f_1' f_2' - f_1 f_2) \, [\ln f_1 + 1].$$ (15.25)

Exchanging the integration variables, we have

$$\frac{dH}{dt} = \int d^3\vec{v}_2 \int \sigma\left(\Omega\right) d\Omega \int d^3\vec{v}_1 \left|\vec{v}_2 - \vec{v}_1\right| \left(f_1' f_2' - f_1 f_2\right) \left[\ln f_2 + 1\right].$$
(15.26)

Thus, we write

$$\frac{dH}{dt} = \frac{1}{2} \int d^3\vec{v}_1 \int \sigma\left(\Omega\right) d\Omega \int d^3\vec{v}_2 \left|\vec{v}_2 - \vec{v}_1\right| \left(f_1' f_2' - f_1 f_2\right) \left[2 + \ln\left(f_1 f_2\right)\right].$$
(15.27)

Changing again the integration variables, and taking into account equations (15.13) and (15.14), we also have

$$\frac{dH}{dt} = \frac{1}{2} \int d^3\vec{v}_1 \int \sigma\left(\Omega\right) d\Omega \int d^3\vec{v}_2 \left|\vec{v}_2 - \vec{v}_1\right| \left(f_1 f_2 - f_1' f_2'\right) \left[2 + \ln\left(f_1' f_2'\right)\right].$$
(15.28)

Thus, we can write

$$\frac{dH}{dt} = -\frac{1}{4} \int d^3\vec{v}_1 \int d^3\vec{v}_2 \int \sigma\left(\Omega\right) d\Omega \left|\vec{v}_2 - \vec{v}_1\right| \left(f_1 f_2 - f_1' f_2'\right)$$

$$\times \left[\ln\left(f_1 f_2\right) - \ln\left(f_1' f_2'\right)\right].$$
(15.29)

To complete the demonstration of the theorem, we use the inequality

$$\left(y - x\right)\left(\ln y - \ln x\right) \geq 0,$$
(15.30)

which holds for all positive values of x and y (and which is based on the concavity of the logarithm function). The equality, which corresponds $dH/dt = 0$, that is, to the necessary condition of equilibrium, only holds for $x = y$, that is, for $f_1 f_2 - f_1' f_2'$, which shows that equation (15.20) is also a **necessary condition** of equilibrium. Although there may be some mathematical problems with this proof, which cannot be rigorously done except for a gas of hard spheres (for which it has been proved that solutions exist and are unique, and that all integrals are duly convergent), the H-theorem remains one the more solid results of Boltzmann's program.

To obtain the distribution of molecular velocities in equilibrium, it is sufficient to write equation (15.20) in the form

$$f_o\left(\vec{v}_2'\right) f_o\left(\vec{v}_1'\right) = f_o\left(\vec{v}_2\right) f_o\left(\vec{v}_1\right),$$
(15.31)

from which we have

$$\ln f_o\left(\vec{v}_2'\right) + \ln f_o\left(\vec{v}_1'\right) = \ln f_o\left(\vec{v}_2\right) + \ln f_o\left(\vec{v}_1\right).$$
(15.32)

Since this last expression displays the form of a conservation law, we can write $\ln f_o(\vec{v})$ in terms of conserved quantities (momentum and kinetic energy). Thus, we have

$$\ln f_o(\vec{v}) = A + B(\vec{v} - \vec{v}_o)^2, \tag{15.33}$$

where A, B, and \vec{v}_o are three parameters to be determined from the constraints of the problem. Assuming that the average velocity vanishes, $\langle \vec{v} \rangle = 0$, that the mean kinetic energy is given by

$$\left\langle \frac{1}{2}m\vec{v}^2 \right\rangle = \frac{3}{2}k_B T, \tag{15.34}$$

and that the distribution is normalized,

$$\int d^3\vec{r}\, d^3\vec{v}\, f_o(\vec{v}) = N, \tag{15.35}$$

we recover the equilibrium Maxwell–Boltzmann distribution as written in previous chapters,

$$f_o(\vec{v}) = \frac{N}{V}\left(\frac{m}{2\pi k_B T}\right)^{3/2} \exp\left(-\frac{mv^2}{2k_B T}\right). \tag{15.36}$$

Except for the prefactor N/V, which is inserted to fulfill the normalization condition given by equation (15.35), we have the same expression as obtained in the context of Gibbs ensembles for an ideal monatomic gas.

Now we discuss some historical objections to the H-theorem.

Historical objections to the H-theorem

At the end of the nineteenth century, Boltzmann worked in a cultural environment in which the energeticists (Ostwald, Mach, Duhem) are proposing that macroscopic thermodynamics (of Clausius and Kelvin) is the paradigmatic model for all scientific theories. Thermodynamics organizes in a systematic way the phenomenological knowledge on the thermal behavior of physical systems. It does not require any microscopic hypothesis on the constituents of material bodies. Boltzmann, in contrast, is a realist thinker, who insists that a gas should be treated as a system of microscopic particles that behave according to the laws of mechanics.

The first objection to the H-theorem was raised in 1876 by Loschmidt, a colleague of Boltzmann in Vienna and a firm supporter of atomism. Loschmidt claimed that the laws of mechanics are reversible in time, and thus cannot lead to a function as H (or entropy) that does have a privileged direction in time. Each physical process where the entropy increases should correspond to an analogous process, with reversed particle velocities,

where the entropy decreases. Therefore, either the increase or the decrease of entropy depend on the initial conditions of the system only.

In order to answer this objection, Boltzmann was forced to emphasize the statistical character of his theory. He pointed out that he uses statistical distributions and assumes that particle velocities are not correlated (hypothesis of molecular chaos). Furthermore, Boltzmann concedes that there may be some initial conditions that lead to a decrease of entropy. In phase space, however, these initial conditions should be associated with much smaller regions than the regions corresponding to initial conditions that lead to an increase of entropy. According to Gibbs, as quoted by Boltzmann in the preface to his "Lessons on the theory of gases," the impossibility of an uncompensated decrease of entropy seems to be reduced to an improbability. The well-known model of a gas of particles with discrete energies (see Chapters 4 and 5), which has probably been a source of inspiration for Max Planck, was motivated by the need to clarify the statistical character of the second law of thermodynamics.

In the following years after the debate with Loschmidt, Boltzmann had many discussions, mainly with his English colleagues, supporters of atomism and of the kinetic theory of gases, in order to clarify the meaning of the H-theorem. He even suggested a graph of the function H versus time. In general, H should be very close to its minimum value, but there should be some well-pronounced peaks. To have a glimpse of this graph of H versus time, we can imagine a succession of reversed trees, most of them very short, with horizontal, or almost horizontal, branches. Occasionally, there appears a much taller tree (maybe more inclined), but the probability of these occurrences enormously decreases with the height of the tree. See a sketch of H versus time t in figure 15.2.

In 1896, Zermelo, a former student of Planck in Berlin, proposed one the more serious objections against Boltzmann's formulations. Zermelo's objection is based on a mechanical theorem that had recently been demonstrated by Poincaré. According to this theorem, any system of particles, with position-dependent interaction forces, returns, after a certain time τ_R, to an arbitrarily close neighborhood of its initial position. The only restrictive conditions on the system are finite bounds for the coordinates of position and momentum (that is, particles should be kept inside a container, and with a finite energy). It should not matter that the theorem may not hold for a very small set (of zero measure) of initial conditions. It is obvious that it excludes the existence of any mechanical function with a privileged direction in time!

The answer to Zermelo was immediate. Again, Boltzmann emphasized that the H-theorem is not purely mechanical, and that the statistical laws are fundamental to explain the behavior of a system with a large number of particles. Also, Boltzmann pointed out that the period τ_R for a gas of particles is very large, maybe of the order of the age of the universe. Although there are objections to this calculation, more-recent numerical

FIGURE 15.2. Sketch of a typical curve of the function H versus time. The dashed line represents an average value (as obtained from the Boltzmann equation).

experiments for hard discs and hard spheres seem to give full support to the claims of an extremely long Poincaré recurrence time [see, for example, A. Bellemans and J. Orban, Phys. Lett. **24A**, 620 (1967); A. Aharony, Phys. Lett. **37A**, 143 (1971)].

By the end of the nineteenth century, in spite of the support of his English colleagues, Boltzmann was in a weak defensive position. On the European continent, the energeticists had gained overwhelming influence, and all fundamentals of the kinetic theory of gases were under severe criticism (despite the success for calculating transport coefficients). There were also real problems with kinetic theory, as the applications of the equipartition theorem to obtain the specific heat of diatomic gases (although available experimental spectra already indicated the existence of much more complicated internal molecular structures). Under these circumstances, for the first time, Boltzmann gave up a realistic approach toward physical phenomena, and introduced a cosmological hypothesis, of subjective character, to give a justification of the H-theorem. Boltzmann suggested that the entropy of the universe is maximum, but that there are fluctuations, and occasionally deep valleys occur in which the entropy is a minimum. Life should be possible at the bottom of these deep valleys only, going up toward a maximum of entropy. Therefore, entropy increases for the existing civilizations, which determines a highly subjective direction of time, depending on our location in the universe. Hypotheses of this sort on the arrow of time continuously reappeared during the last 100 years. The main arguments of Boltzmann, however, are still up-to-date and extremely relevant for the modern understanding of irreversible phenomena [see, for example, the paper by Joel Lebowitz, "Boltzmann's entropy and time's arrow," Physics Today, September 1993, page 32].

Ehrenfest's urn model

One of the most brilliant explanations of the nature of the H-theorem (which is also one of the first examples of application of stochastic methods) has been proposed by Paul and Tatiana Ehrenfest in the beginning of the twentieth century. According to Ehrenfest's **urn model**, consider two boxes, 1 and 2, and $2R$ distinct balls, numbered from 1 to $2R$. At the initial time, $t = 0$, almost all the balls are in box 1. Let us then choose a random number (using a roulette wheel, for example), between 1 and $2R$, and move to the other box (urn) the ball associated with this number. We keep repeating this procedure, at each τ seconds, always moving, from one box to the other, the ball associated with the chosen number. At each step of this process, we have $N_1 + N_2 = 2R$, where N_1 (N_2) is the number of balls in box 1 (2). The equilibrium situation, after a long time (many spins of the roulette wheel), should correspond to $N_1 = N_2 = R$. However, there should be fluctuations about equilibrium, which lead to a distribution of occupation numbers, as in a usual problem of equilibrium statistical physics. From the dynamical point of view, it is interesting to investigate how these fluctuations decay with time from a certain initial state.

Given R and the initial conditions, it is easy to use a random number generator and a microcomputer to obtain a graph of $|N_1 - N_2|$ as a function of the number of steps s (time is given by $t = s\tau$). There are peaks, and very high values of $|N_1 - N_2|$, but after each peak the curve has a tendency to fall down again, which roughly agrees with the picture of "reversed trees" proposed by Boltzmann for the shape of the H-function (see figure 15.2).

We can carry out some analytic probabilistic calculations for this relatively simple urn model. Let us call $P(N_1, s; N_o)$ the probability of finding N_1 balls in box 1, after a set of s steps (that is, at time $t = s\tau$), if we begin with $N_1 = N_o$ balls in box 1, at initial time $t = 0$. A (macroscopic) state with N_1 balls in box 1 at time $(s+1)\tau$ can only be reached from states with either $N_1 - 1$ or $N_1 + 1$ balls in box 1 at the immediately previous time, $s\tau$. Thus, we can write

$$P(N_1, s + 1; N_o) = w(N_1 - 1 \rightarrow N_1) P(N_1 - 1, s; N_o)$$

$$+ w(N_1 + 1 \rightarrow N_1) P(N_1 + 1, s; N_o), \tag{15.37}$$

where it is plausible to assume that the **transition probabilities** are given by

$$w(N_1 - 1 \rightarrow N_1) = \frac{2R - (N_1 - 1)}{2R} \tag{15.38}$$

and

$$w(N_1 + 1 \rightarrow N_1) = \frac{N_1 + 1}{2R}. \tag{15.39}$$

It should be noted that $w(N_1 - 1 \rightarrow N_1)$ corresponds to the probability, at time $t = s\tau$, of choosing a ball from box 2 [which contains $2R - (N_1 - 1)$ balls), while $w(N_1 + 1 \rightarrow N_1)$ corresponds to the probability, at this same time, of choosing a ball from box 1 (which contains $N_1 + 1$ balls]. Therefore, the **master equation** (15.37) can be written as

$$P(N_1, s+1; N_o) = \frac{2R - (N_1 - 1)}{2R} P(N_1 - 1, s; N_o)$$

$$+ \frac{N_1 + 1}{2R} P(N_1 + 1, s; N_o). \tag{15.40}$$

Now, it is convenient to adopt a more-symmetric notation, by introducing the new variable $n = N_1 - R$, such that $N_1 = R + n$, $N_2 = R - n$, with the equilibrium situation given by $n = 0$. In terms of this new variable, we have

$$P(n, s+1; n_o) = \frac{R - (n+1)}{2R} P(n-1, s; n_o) + \frac{R + (n+1)}{2R} P(n+1, s; n_o), \tag{15.41}$$

with the initial condition $P(n, 0; n_o) = \delta_{n_o, n}$, where $n_o = N_o - R$. This **master equation** refers to the time evolution of a probability distribution, and plays a similar role as the Boltzmann transport equation for a gas of particles.

In the stationary situation, which corresponds to equilibrium, the probability $P(n, s; n_o)$ should not depend on time s (neither on the initial conditions!). Therefore, the equilibrium probability should come from the equation

$$P(n) = \frac{R - (n+1)}{2R} P(n-1) + \frac{R + (n+1)}{2R} P(n+1), \tag{15.42}$$

whose solution is the binomial form

$$P(n) = \frac{(2R)!}{(R+n)!(R-n)!} \left(\frac{1}{2}\right)^{2R}, \tag{15.43}$$

as it should have been anticipated from an application of the hypothesis of equal *a priori* probabilities at the very beginning. The expected value and the quadratic deviation with respect to the mean are given by $\langle n \rangle = 0$ and $\langle n^2 \rangle = R/2$, respectively. For $R \rightarrow \infty$, this binomial distribution tends to a Gaussian form,

$$P(n) \rightarrow p_G(n) = (\pi R)^{-1/2} \exp\left(-\frac{n^2}{R}\right). \tag{15.44}$$

Now it is interesting to calculate the time evolution of the expected value,

$$\langle n(s) \rangle = \sum_n n P(n, s; n_o). \tag{15.45}$$

From equation (15.41), it is easy to see that

$$\langle n(s+1) \rangle = \left(1 - \frac{1}{R}\right) \langle n(s) \rangle. \tag{15.46}$$

Thus, if $\langle n(s=0) \rangle = n_o$, we can write

$$\langle n(s) \rangle = n_o \left(1 - \frac{1}{R}\right)^s, \tag{15.47}$$

which leads to an exponential decay as R and s are big enough. Indeed, at the limit $s \to \infty$ (long enough time), $R \to \infty$ (sufficiently large number of balls), and $\tau \to 0$ (time intervals very short), with $t = s\tau$ and $1/\gamma = R\tau$ fixed, we have

$$\langle n(s) \rangle \to n_o \exp(-\gamma t), \tag{15.48}$$

which indicates the exponential decay of the expected values toward its vanishing equilibrium value. The system displays, as an average, an irreversible decay to equilibrium (the hatched curve in figure 15.2 represents this average behavior).

We may also use the urn model to discuss Poincaré recurrence times. The interested reader should check the work by Mark Kac, reprinted in the book *Selected papers in noise and stochastic processes*, edited by N. Wax (Dover, New York, 1954). Let us call $P'(n, s; m)$ the probability of finding the system in a state n, at time $t = s\tau$, after coming from an initial state m at time $t = 0$. Thus, it is possible to show that

$$\sum_{s=1}^{\infty} P'(n, s; n) = 1, \tag{15.49}$$

which means that each state of the system should occur again with probability 1 (the analog of Poincaré's recurrence theorem for the urn model). Also, it can be shown that

$$\tau_R = \sum_{s=1}^{\infty} s\tau P'(n, s; n) = \tau \frac{(R+n)!(R-n)!}{(2R)!} 2^{2R}, \tag{15.50}$$

which is the analog of Poincaré's recurrence time. For example, for $R = 10,000$, $n = 10,000$, and $\tau = 1\,\text{s}$, we have

$$\tau_R = 2^{20000} \approx 10^{6000} \text{ years}, \tag{15.51}$$

which is an exaggeratedly long time! On the other hand, if $R+n$ and $R-n$ are comparable, then τ_R is small. Of course, it only makes sense to refer to irreversibility for sufficiently large recurrence times (in other words, as the initial conditions are sufficiently far from the final equilibrium situation). If τ_R is small, there are only fluctuations about the equilibrium solution.

15.2 The BBGKY hierarchy

The time evolution in phase space of a classical gas of particles can be described by Liouville's equation. However, instead of dealing with the full complexity of the distribution ρ of representative points of the system in phase space, it is more convenient to define reduced distributions of one particle (f_1), two particles (f_2), and so on. The set $\{f_s\}$ is equivalent to knowing the distribution function ρ, from which we can calculate the average values of statistical mechanics. Let us show that the equation for the time evolution of f_1 can be written in terms of f_2, that the equation for the time evolution of f_2 can be written in terms of f_3, and so on, which defines a hierarchical set of (nonlinear) coupled equations. The main difficulty of this approach consists in finding a reasonable way of truncating this hierarchical set to find the reduced distributions.

The Liouville theorem

Consider the phase space of a classical gas of N particles, with $3N$ position coordinates, $q_1, q_2, ...,$ and $3N$ momentum coordinates, $p_1, p_2,$ The number of representative points of the system, at time t, within the hypervolume

$$d^{3N}q\,d^{3N}p = dq_1 \cdots dq_{3N}\,dp_1 \cdots dp_{3N}, \tag{15.52}$$

is given by $\rho(q, p, t)\,d^{3N}q\,d^{3N}p$, where

$$\rho(q, p, t) = \rho(q_1, ..., q_{3N}, p_1, ..., p_{3N}, t). \tag{15.53}$$

For a conservative system, associated with the (time-independent) Hamiltonian

$$\mathcal{H}(q, p) = \mathcal{H}(q_1, ..., q_{3N}, p_1, ..., p_{3N}), \tag{15.54}$$

the trajectories of the representative points of the system in phase space are given by the Hamilton equations,

$$\dot{p}_i = -\frac{\partial \mathcal{H}}{\partial q_i} \quad \text{and} \quad \dot{q}_i = \frac{\partial \mathcal{H}}{\partial p_i}, \tag{15.55}$$

where $i = 1, ...3N$. Owing to the uniqueness of the solution of these differential equations, the trajectories do not intersect, which guarantees the conservation of the number of representative points in phase space. Therefore, as we have done in Chapter 2, we can write a continuity equation,

$$-\frac{d}{dt} \int_\Gamma \rho(q, p, t) \, d\Gamma = \oint_S \vec{J} \cdot d\vec{S}, \qquad (15.56)$$

where $d\Gamma = d^{3N}q \, d^{3N}p$ is an element of volume in phase space Γ, and $d\vec{S}$ is an elementary vector area normal to the hypersurface S. The generalized flux is given by

$$\vec{J} = \rho\vec{v} = \rho(\dot{q}_1, ..., \dot{q}_{3N}, \dot{p}_1, ..., \dot{p}_{3N}), \qquad (15.57)$$

where \vec{v} is a generalized velocity.

Using the Gauss theorem, we can write equation (15.56) as

$$\int_\Gamma -\frac{\partial \rho}{\partial t} \, d\Gamma = \int_\Gamma \left(\vec{\nabla} \cdot \vec{J}\right) d\Gamma, \qquad (15.58)$$

from which we have the continuity equation,

$$\vec{\nabla} \cdot (\rho\vec{v}) + \frac{\partial \rho}{\partial t} = 0. \qquad (15.59)$$

From equation (15.57), the generalized divergent of the flux is given by

$$\begin{aligned}
\vec{\nabla} \cdot (\rho\vec{v}) &= \sum_{i=1}^{3N} \left[\frac{\partial}{\partial q_i}(\rho\dot{q}_i) + \frac{\partial}{\partial p_i}(\rho\dot{p}_i)\right] \\
&= \rho \sum_{i=1}^{3N} \left[\frac{\partial \dot{q}_i}{\partial q_i} + \frac{\partial \dot{p}_i}{\partial p_i}\right] + \sum_{i=1}^{3N} \left[\dot{q}_i \frac{\partial \rho}{\partial q_i} + \dot{p}_i \frac{\partial \rho}{\partial p_i}\right]. \qquad (15.60)
\end{aligned}$$

Using the Hamilton equations (15.55), we can write

$$\frac{\partial \dot{q}_i}{\partial q_i} = -\frac{\partial \dot{p}_i}{\partial p_i} = \frac{\partial^2 \mathcal{H}}{\partial q_i \partial p_i}. \qquad (15.61)$$

Thus, the generalized divergent of the flux is given by

$$\vec{\nabla} \cdot (\rho\vec{v}) = \sum_{i=1}^{3N} \left[\frac{\partial \mathcal{H}}{\partial p_i} \frac{\partial \rho}{\partial q_i} - \frac{\partial \mathcal{H}}{\partial q_i} \frac{\partial \rho}{\partial p_i}\right] = \{\rho, \mathcal{H}\}, \qquad (15.62)$$

where $\{\rho, \mathcal{H}\}$ are the Poisson parentheses of the density ρ with the Hamiltonian \mathcal{H}. Therefore, Liouville's theorem may be expressed by the relation

$$\{\rho, \mathcal{H}\} + \frac{\partial \rho}{\partial t} = 0, \qquad (15.63)$$

which is equivalent to stating that ρ is a constant of motion (the total derivative of ρ with respect to time is zero).

In analogy with Schrödinger's equation, Liouville's theorem can also be written as

$$i\frac{\partial \rho}{\partial t} = \mathcal{L}\rho, \tag{15.64}$$

where $i = \sqrt{-1}$ and \mathcal{L} is the Liouvillian operator. Thus, we write the formal solution

$$\rho = \rho(q, p, t) = \rho(q, p, 0) \exp[-i\mathcal{L}t], \tag{15.65}$$

which is quite elegant, but does not provide any practical means to proceed with the calculations.

So far, we have rederived the same results already obtained in Chapter 2. Now, let us consider Cartesian coordinates, with the notation $z_i = (\vec{r}_i, \vec{p}_i)$ and $dz_i = d^3\vec{r}_i d^3\vec{p}_i$. The distribution function in Γ space will be written as

$$\rho = \rho(z_1, z_2, ..., z_N, t) = \rho(1, 2, ..., N, t). \tag{15.66}$$

Also, let us consider normalized distributions, so that

$$\int \rho(1, 2, ..., N, t) \, dz_1 dz_2 \cdots dz_N = 1. \tag{15.67}$$

Thus, the mean value of a physical quantity $O(z_1, ..., z_N)$ can be written as

$$\langle O \rangle = \int \cdots \int O(1, 2, ..., N) \rho(1, 2, ..., N, t) \, dz_1 dz_2 \cdots dz_N. \tag{15.68}$$

Liouville's theorem is then given by

$$\frac{\partial \rho}{\partial t} = -\{\rho, \mathcal{H}\} = \sum_{i=1}^{N} \{(\nabla_{p_i}\rho) \cdot (\nabla_{r_i}\mathcal{H}) - (\nabla_{r_i}\rho) \cdot (\nabla_{p_i}\mathcal{H})\}, \tag{15.69}$$

where ∇_{r_i} and ∇_{p_i} are gradient operators with respect to \vec{r}_i and \vec{p}_i, respectively.

For a gas of monatomic particles, we have the Hamiltonian

$$\mathcal{H} = \sum_{i=1}^{N} \frac{1}{2m}\vec{p}_i^2 + \sum_{i<j} v_{ij}, \tag{15.70}$$

where the pair interactions are given by a spherically symmetric potential,

$$v_{ij} = v_{ji} = v(|\vec{r}_i - \vec{r}_j|). \tag{15.71}$$

Thus, we write the Hamilton equations,

$$\nabla_{p_i}\mathcal{H} = \frac{1}{m}\vec{p}_i \quad \text{and} \quad \nabla_{r_i}\mathcal{H} = -\sum_{j=1(j\neq i)}^{N}\vec{K}_{ij}. \tag{15.72}$$

where

$$\vec{K}_{ij} = -\nabla_{r_i}v\left(|\vec{r}_i - \vec{r}_j|\right). \tag{15.73}$$

Hence, Liouville's theorem can also be written as

$$\frac{\partial \rho}{\partial t} = \left\{\sum_{i=1}^{N}\left[-\frac{1}{m}\vec{p}_i\cdot\nabla_{r_i} - \sum_{j=1(j\neq i)}^{N}\vec{K}_{ij}\cdot\nabla_{p_i}\right]\right\}\rho, \tag{15.74}$$

that is,

$$\left[\frac{\partial}{\partial t} + h_N\left(1,2,...,N\right)\right]\rho\left(1,2,...,N,t\right) = 0, \tag{15.75}$$

where

$$h_N = \sum_{i=1}^{N}\frac{1}{m}\vec{p}_i\cdot\nabla_{r_i} + \frac{1}{2}\sum_{j=1(j\neq i)}^{N}\vec{K}_{ij}\cdot\left(\nabla_{p_i} - \nabla_{p_j}\right). \tag{15.76}$$

At this point, we are prepared to deduce the BBGKY hierarchical equations for the reduced densities. We closely follow the treatment of Kerson Huang (Section 3.5 of *Statistical Mechanics*, second edition, John Wiley and Sons, New York, 1987). The interested reader may also check N. N. Bogoliubov, in *Studies in Statistical Mechanics*, J. de Boer and G. Uhlenbeck, editors, volume 1, North-Holland, Amsterdam, 1962.

Reduced distribution functions. Deduction of the BBGKY hierarchy

The single-particle distribution function is given by

$$\begin{aligned}
f_1\left(\vec{r},\vec{p},t\right) &= \left\langle\sum_{i=1}^{N}\delta^3\left(\vec{p}-\vec{p}_i\right)\delta^3\left(\vec{r}-\vec{r}_i\right)\right\rangle \\
&= N\int\cdots\int dz_2\cdots dz_N\rho\left(1,2,...,N,t\right), \tag{15.77}
\end{aligned}$$

where we assume the normalization

$$\int f_1\left(\vec{r},\vec{p},t\right)d^3\vec{r}d^3\vec{p} = N. \tag{15.78}$$

We analogously define the distribution function of s particles,

$$f_s(1, 2, ..., z_s, t) = \frac{N!}{(N-s)!} \int dz_{s+1} \cdots \int dz_N \rho(1, 2, ..., N, t), \quad (15.79)$$

where $s = 1, 2, ..., N$ (the combinatorial prefactor indicates that we do not distinguish which individual particles have coordinates z_1, which particles have coordinates z_2, and so on). Now we establish a time evolution equation for f_1 in terms of the function f_2, another equation for the evolution of f_2 in terms of f_3, and so on.

From Liouville's theorem, as written in equation (15.75), we have

$$\frac{\partial f_s}{\partial t} = \frac{N!}{(N-s)!} \int dz_{s+1} \cdots \int dz_N \frac{\partial \rho}{\partial t} = -\frac{N!}{(N-s)!} \int dz_{s+1} \cdots \int dz_N h_N \rho,$$
$$(15.80)$$

where h_N, given by equation (15.76), can be written as

$$h_N = \sum_{i=1}^{N} S_i + \frac{1}{2} \sum_{i,j=1(i \neq j)}^{N} P_{ij}, \quad (15.81)$$

with

$$S_i = \frac{1}{m} \vec{p}_i \cdot \nabla_{r_i} \quad (15.82)$$

and

$$P_{ij} = \vec{K}_{ij} \cdot (\nabla_{p_i} - \nabla_{p_j}) = P_{ji}. \quad (15.83)$$

Taking into account terms that depend only on the coordinates $z_1, z_2, ..., z_s$, we can write

$$h_N(1, 2, ..., N) = \sum_{i=1}^{s} S_i + \sum_{i=s+1}^{N} S_i + \frac{1}{2} \sum_{i,j=1(i \neq j)}^{s} P_{ij} + \frac{1}{2} \sum_{i,j=s+1(i \neq j)}^{N} P_{ij}$$

$$+ \sum_{i=1}^{s} \sum_{j=s+1}^{N} P_{ij} = h_s(1, 2, ..., s) + h_{N-s}(s+1, s+2, ..., N) + \sum_{i=1}^{s} \sum_{j=s+1}^{N} P_{ij}.$$
$$(15.84)$$

Now, it is easy to see that

$$\int \cdots \int dz_{s+1} \cdots dz_N h_{N-s}(s+1, s+2, ..., N) \rho(1, 2, ..., N) = 0, \quad (15.85)$$

since (i) the coefficients of terms depending on \vec{p} do not depend themselves on \vec{p}; (ii) the coefficients of terms depending on \vec{r} do not depend themselves

on \vec{r}; (iii) function ρ should vanish at the external surface enclosing the phase space accessible to the system. Thus, we have

$$\left(\frac{\partial}{\partial t} + h_s\right) f_s = -\frac{N!}{(N-s)!} \int dz_{s+1} \cdots \int dz_N \sum_{i=1}^{s} \sum_{j=s+1}^{N} P_{ij}\rho\left(1, 2, ..., N, t\right)$$

$$= -\frac{N!}{(N-s+1)!} \int dz_{s+1} \cdots \int dz_N \sum_{i=1}^{s} P_{i,s+1}\rho\left(1, 2, ..., N, t\right), \quad (15.86)$$

since the sum over j gives $N - s$ identical terms. Form this last equation, we can also write

$$\left(\frac{\partial}{\partial t} + h_s\right) f_s = -\sum_{i=1}^{s} \int dz_{s+1} P_{i,s+1} \frac{N!}{(N-s+1)!}$$

$$\times \int dz_{s+2} \cdots \int dz_N \rho\left(1, 2, ..., N, t\right)$$

$$= -\sum_{i=1}^{s} \int dz_{s+1} P_{i,s+1} f_{s+1}\left(1, 2, ..., s+1\right). (15.87)$$

Inserting into this equation the expression for $P_{i,s+1}$, given by equation (15.83), we have

$$\left(\frac{\partial}{\partial t} + h_s\right) f_s = -\sum_{i=1}^{s} \int dz_{s+1} \vec{K}_{i,s+1} \cdot \left(\nabla_{p_i} - \nabla_{p_{s+1}}\right) f_{s+1}\left(1, 2, ..., s+1\right).$$

$$(15.88)$$

If we note that there is no contribution from the gradient with respect to \vec{p}_{s+1}, since it leads to a term that vanishes at the surface, then we finally write the BBGKY hierarchy,

$$\left(\frac{\partial}{\partial t} + h_s\right) f_s = -\sum_{i=1}^{s} \int dz_{s+1} \left(\vec{K}_{i,s+1} \cdot \nabla_{p_i}\right) f_{s+1}\left(1, 2, ..., s+1\right),$$

$$(15.89)$$

where $s = 1, 2, ..., N$. These equations are certainly equivalent to the Liouville equation (or to the original Hamilton equations of the system), and cannot clarify the issue of irreversibility. As we have mentioned, there are many decoupling schemes, some of them quite ingenious, that are also unable to rigorously deal with irreversibility. In the next section, we sketch a decoupling that leads to Boltzmann's transport equation (and thus, to the H-theorem). Also, we give an example of another decoupling that leads to the Vlasov equation for a diluted plasma system (but presents time-reversal invariance).

The Boltzmann equation from the BBGKY hierarchy

Let us explicitly write the BBGKY equations for the reduced distribution densities of one and two particles,

$$\left(\frac{\partial}{\partial t} + \frac{1}{m}\vec{p}_1 \cdot \nabla_{r_1}\right) f_1(z_1, t) = -\int dz_2 \left(\vec{K}_{12} \cdot \nabla_{p_1}\right) f_2(z_1, z_2, t), \quad (15.90)$$

$$\left[\frac{\partial}{\partial t} + \frac{1}{m}\vec{p}_1 \cdot \nabla_{r_1} + \frac{1}{m}\vec{p}_2 \cdot \nabla_{r_2} + \frac{1}{2}\vec{K}_{12} \cdot (\nabla_{p_1} - \nabla_{p_2})\right] f_2(z_1, z_2, t)$$

$$= -\int dz_3 \left(\vec{K}_{13} \cdot \nabla_{p_1} + \vec{K}_{23} \cdot \nabla_{p_2}\right) f_3(z_1, z_2, z_3, t). \quad (15.91)$$

The first step consists in discarding collision terms in the second equation (on the basis of arguments that there are two distinct time scales, given by \vec{p}/m and \vec{K}, with collision times much shorter than time intervals between collisions). Thus, changing to coordinates such that $\vec{R} = (\vec{r}_1 + \vec{r}_2)/2$, $\vec{r} = \vec{r}_2 - \vec{r}_1$, $\vec{P} = \vec{p}_1 + \vec{p}_2$, and $\vec{p} = (\vec{p}_2 - \vec{p}_1)/2$, we write the following approximate form of the second equation

$$\left[\frac{\partial}{\partial t} + \frac{1}{M}\vec{P} \cdot \nabla_R + \frac{1}{\mu}\vec{p} \cdot \nabla_r + \vec{K}(\vec{r}) \cdot \nabla_p\right] f_2\left(\vec{P}, \vec{R}, \vec{p}, \vec{r}, t\right) = 0, \quad (15.92)$$

where $M = 2m$ is the total mass and $\mu = m/2$ is the reduced mass. Changing to the coordinate system of center of mass, so that $\vec{P} = 0$, we have

$$f_2\left(\vec{P}, \vec{R}, \vec{p}, \vec{r}, t\right) \rightarrow f_2(\vec{p}, \vec{r}, t), \quad (15.93)$$

in other words,

$$f_2\left(\vec{p} + \vec{K}(\vec{r})\,\delta t, \vec{r} + \frac{1}{\mu}\vec{p}\delta t, t + \delta t\right) = f_2(\vec{p}, \vec{r}, t), \quad (15.94)$$

which indicates that f_2 should reach equilibrium before f_1.

The first equation of the hierarchy may be written as

$$\left(\frac{\partial f_1}{\partial t}\right)_{col} = -\int dz_2 \vec{K}_{12} \cdot (\nabla_{p_1} - \nabla_{p_2}) f_2(z_1, z_2, t), \quad (15.95)$$

where, as a matter of convenience, we have added the gradient term with respect to \vec{p}_2, which vanishes anyway. Thus, we have

$$\left(\frac{\partial f_1}{\partial t}\right)_{col} = \frac{1}{m}\int_{r_0} dz_2 (\vec{p}_1 \cdot \nabla_{r_1} + \vec{p}_2 \cdot \nabla_{r_2}) f_2(z_1, z_2, t), \quad (15.96)$$

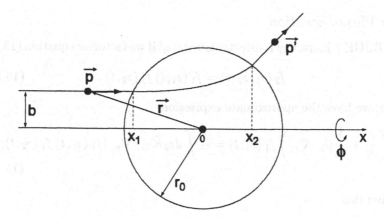

FIGURE 15.3. Relative coordinates to describe the collision between two particles (which are correlated within a sphere of radius r_o).

where we also assume that the interaction forces vanish for $|\vec{r_1} - \vec{r_2}| = r > r_o$. Using relative coordinates (see figure 15.3), and discarding gradient terms with respect to \vec{R}, we have

$$\left(\frac{\partial f_1}{\partial t}\right)_{col} = \frac{1}{m} \int d^3p_2 \int_{r<r_o} d^3r_2 \, (\vec{p_1} - \vec{p_2}) \cdot \nabla_r f_2$$

$$= \frac{1}{m} \int d^3p_2 \, |\vec{p_1} - \vec{p_2}| \int d\phi b db \int_{x_1}^{x_2} dx \frac{\partial}{\partial x} f_2. \quad (15.97)$$

Introducing the decoupling,

$$f_2(x_1) = f_1(\vec{p_1}) f_1(\vec{p_2}) \qquad (15.98)$$

and

$$f_2(x_2) = f_1(\vec{p_1'}) f_1(\vec{p_2'}), \qquad (15.99)$$

we finally write the Boltzmann equation,

$$\left(\frac{\partial f_1}{\partial t}\right)_{col} = \int d^3p_2 \int d\Omega \, |\vec{v_1} - \vec{v_2}| \, \sigma\,(\Omega) \, (f_1' f_2' - f_1 f_2). \quad (15.100)$$

As we have mentioned, the treatment in this section is based on the second edition of the text by Kerson Huang (*Statistical Physics*, John Wiley, New York, 1987). The interested reader may find a much more detailed discussion (of this and other topics of kinetic methods) in the book by H. J. Kreutzer, *Non-equilibrium thermodynamics and its statistical foundations*, Clarendon Press, Oxford, 1981.

The Vlasov equation

The BBGKY hierarchy is directly truncated if we factorize equation (15.90),

$$f_2(z_1, z_2, t) = f_1(z_1, t) f_1(z_2, t). \qquad (15.101)$$

Thus, we have the approximate expression

$$\left(\frac{\partial}{\partial t} + \frac{1}{m}\vec{p}_1 \cdot \nabla_{r_1}\right) f_1(z_1, t) = -\int dz_2 \vec{K}_{12} \cdot \nabla_{p_1}\left[f_1(z_1, t) f_1(z_2, t)\right]. \qquad (15.102)$$

Noting that

$$\int dz_2 \vec{K}_{12} \cdot \nabla_{p_1}\left[f_1(z_1, t) f_1(z_2, t)\right]$$

$$= \int dz_2 \left[-\nabla_{r_1} v\left(|\vec{r}_1 - \vec{r}_2|\right) \cdot \nabla_{p_1} f_1(z_1, t)\right] f_2(z_2, t)$$

$$= -\nabla_{r_1}\left\{\int dz_2 v\left(|\vec{r}_1 - \vec{r}_2|\right) f_1(z_2, t)\right\} \cdot \nabla_{p_1} f_1(z_1, t), \qquad (15.103)$$

we obtain the **Vlasov equation**,

$$\left(\frac{\partial}{\partial t} + \frac{1}{m}\vec{p}_1 \cdot \nabla_{r_1} - \left[\nabla_{r_1}\bar{u}(\vec{r}_1, t)\right] \cdot \nabla_{p_1}\right) f_1(z_1, t) = 0, \qquad (15.104)$$

where \bar{u} is an effective, or average, potential, given by

$$\bar{u}(\vec{r}_1, t) = \int dz_2 v\left(|\vec{r}_1 - \vec{r}_2|\right) f_1(z_2, t), \qquad (15.105)$$

which should be self-consistently calculated (since it determines, and at the same time depends on, the single-particle density f_1).

In contrast with the Boltzmann equation, the Vlasov equation is invariant under time reversal [since $f_1(\vec{r}, \vec{p}, t)$ and $f_1(\vec{r}, -\vec{p}, -t)$ obey the same equation]. Although it may account for some aspects of the behavior of a diluted plasma out of equilibrium, the Vlasov equation is unable to describe the irreversible time evolution of this plasma system toward its final macroscopic equilibrium state.

Exercises

1. Demonstrate relations (15.13) and (15.14).

2. From an elementary kinetic argument, and taking into account orders of magnitude only, show that the mean free path (between collisions) of molecules in a diluted gas is given by

$$\lambda = \frac{1}{\pi a^2 n},$$

where a is the molecular diameter and $n = N/V$ is the density of molecules. Obtain the order of magnitude of λ for hydrogen at the critical point, under normal conditions of pressure and temperature. What is the order of magnitude of λ for interstellar hydrogen (with a density of about one molecule per cubic centimeter)?

3. The typical velocity of molecules of an ideal gas is given by

$$\bar{v} = [\langle v^2 \rangle]^{1/2} = \left(\frac{3 k_B T}{m} \right)^{1/2}.$$

What is the order of magnitude of the time τ between collisions for hydrogen under normal conditions?

4. Use the formalism presented in the text to analyze the collisions between two molecules in order to show that the time τ between collisions, for a diluted gas of N particles, is given by the expression

$$\frac{N}{\tau} = \iint R d^3 \vec{v}_1 d^3 \vec{r}_1,$$

where

$$R = \int \frac{1}{4} a^2 d\Omega \int_{v_2} |\vec{v}_1 - \vec{v}_2| f_o(\vec{v}_1) f_o(\vec{v}_2) d^3 \vec{v}_2,$$

in which a is the molecular diameter, $d\Omega$ is an elementary solid angle, and $f_o(\vec{v})$ is the equilibrium Maxwell–Boltzmann distribution. Calculate the multiple integral to show that

$$\frac{1}{\tau} = 4a^2 n \left(\frac{\pi k_B T}{m} \right)^{1/2}.$$

Using the typical velocity,

$$\bar{v} = \left(\frac{3 k_B T}{m} \right)^{1/2},$$

show that

$$\lambda^{-1} = 4 \left(\frac{\pi}{3} \right)^{1/2} a^2 n,$$

where λ has the same order of magnitude as obtained in the previous exercise.

5. Consider a diluted gas of particles with electric charge q and mass m. In the absence on an applied electric field, the equilibrium velocity distribution is given by the usual Maxwell–Boltzmann expression,

$$f_o(\vec{v}) = \frac{N}{V}\left(\frac{m}{2\pi k_B T}\right)^{3/2} \exp\left(-\frac{mv^2}{2k_B T}\right).$$

In the presence of a **weak** uniform electric field \vec{E}, the distribution function can be derived from Boltzmann's transport equation in the collision-time approximation,

$$\left(\frac{\partial}{\partial t} + \frac{q}{m}\vec{E}\cdot\vec{\nabla}_v\right)f(\vec{v},t) = \left(\frac{\partial f}{\partial t}\right)_{coll} = -\frac{f-f_o}{\tau},$$

where τ is the characteristic time between collisions.

(a) As a first-order approximation in terms of the electric field, show that we can write

$$f(\vec{v},t) = f_o(\vec{v}) - \frac{\tau q}{m}\vec{E}\cdot\vec{\nabla}_v f_o(\vec{v}).$$

(b) Use this result to show that

$$\langle \vec{v}\rangle = \frac{\tau q}{m}\vec{E}.$$

Therefore,

$$\left\langle\vec{J}\right\rangle \equiv \langle qn\vec{v}\rangle = \sigma\vec{E},$$

where the conductivity σ is given by

$$\sigma = \frac{\tau q^2 n}{m}.$$

Note that for metallic electrons this result is absolutely preliminary (since we have to use the correct Fermi–Dirac distribution to describe equilibrium).

6. Obtain an expression for the function H associated with a diluted classical gas of N particles, in equilibrium within a container of volume V, at temperature T. What is the relation between H and the entropy of the gas?

7. Consider the Ehrenfest urn model at an initial situation such that $N_1 = 90$ and $N_2 = 10$ (where N_1 and N_2 are the numbers of balls inside boxes 1 and 2, respectively). As a function of time, the curve $n(t) = N_1(t) - N_2(t)$ intersects a certain value (for example, $n = 50$)

when it goes up, with increasing values of n, as well as when it goes down, with decreasing values of n. At first sight, this indicates that $n(t)$ has the same chance of either increasing or decreasing [as it comes from a certain initial value, say $n(0) = 80$]. In order to show that this is not true, consider $n(t)$ at three successive times, $n(t-1)$, $n(t)$, and $n(t+1)$. What is the probability that $n(t-1) < n(t) < n(t+1)$? What is the probability that $n(t-1) > n(t) > n(t+1)$? What is the probability that $n(t) > n(t-1)$ and $n(t) > n(t+1)$, that is, that $n(t)$ is a local maximum?

when it goes up, with increasing values of n, as well as when it goes down, with decreasing values of n. At first sight, this indicates that $\pi(t)$ has the same chance of either increasing or decreasing [as it comes from a certain initial value, say $\pi(0) = 80$]. In order to show that this is not true, consider $\pi(t)$ at three successive times, $\pi(t-1)$, $\pi(t)$, and $\pi(t+1)$. What is the probability that $\pi(t-1) > \pi(t) > \pi(t+1)$? What is the probability that $\pi(t-1) < \pi(t) < \pi(t+1)$? What is the probability that $\pi(t) > \pi(t-1)$ and $\pi(t) > \pi(t+1)$, that is, that $\pi(t)$ is a local maximum?

16
Nonequilibrium phenomena: II. Stochastic Methods

Instead of trying to present a sketch of the theory of **Markovian stochastic processes**, which is certainly beyond the scope of this text, we discuss some examples of the application of stochastic methods to analyze systems of physical interest. Initially, we introduce the **Langevin** and the **Fokker–Planck** equations for the Brownian motion, as investigated by Einstein and Smoluchowski during the early years of the twentieth century. We then present some probabilistic arguments to establish a **master equation**, which gives the time evolution of the probability of occurrence of the microscopic configurations of a physical system. However, as we have already pointed out, to obtain the transition probabilities of a master equation is usually as difficult as to satisfactorily decouple the BBGKY hierarchical set of kinetic equations.

In Chapters 13 and 14 we discussed several thermodynamic properties of the Ising model, which is a kind on nontrivial prototype of statistical models. *A priori*, the Ising model has no dynamics. Indeed, the Ising spin variables are real numbers (± 1) that are not subjected to the quantum commutation rules (for spin operators, as in the case of the Heisenberg Hamiltonian, we can certainly write down time evolution equations). However, we can still construct a master equation for the Ising model if we postulate some plausible transition probabilities (for example, if we make sure that the system evolves to a Gibbsian equilibrium state for sufficiently long time intervals). We then choose transition probabilities according to a proposal of Glauber to construct a **kinetic Ising model** (in one dimension, we obtain a number of exact results).

Finally, we use the example of the Ising model to introduce the **Metropolis dynamics**, which has been widely used in **Monte Carlo** dynamical simulations. With the popularization of computer equipment and techniques, numerical experiments began to play an increasingly important role in contemporary physics. Stochastic methods are the basis for the justification of most numerical simulations.

16.1 Brownian motion. The Langevin equation

The highly irregular motion of grains of pollen immersed in a fluid were discovered and investigated by the British botanist Robert Brown in 1827. Subsequent experimental investigations, supported by the developments of microscopy techniques, have given indications that this is a much more general phenomenon, taking place in suspensions of different sorts of microscopic particles in not so viscous fluids. According to Jean Perrin, the very small particles "come and go, go up and down, come back again, without any rest." The first theories of this Brownian motion, independently published by Einstein (1905) and Smoluchowski (1906), are a successful application of the atomistic ideas of the kinetic theory of gases. In the beginning of the twentieth century, the investigations of the Brownian motion were an important element for the establishment of the particle structure of matter, in opposition to the dominating vision of energetics.

If we assume that the Brownian motion is produced by the collisions between the tiny Brownian particles and the (much smaller) molecules of a fluid, we can write the **Langevin equation**,

$$m\frac{d\vec{v}}{dt} = \vec{F} - \alpha\vec{v} + \vec{F}_a(t),\tag{16.1}$$

where \vec{v} is the velocity of a particle of mass m, \vec{F} is the external force, α is the coefficient of viscous friction, and $\vec{F}_a(t)$ is a random (or stochastic) force related to the continuously colliding small molecules. Let us suppose that $\vec{F}_a(t)$ is independent of velocity and that it is changing very quickly with time (with respect to the characteristic time of the molecules).

In order to simplify the notation, let us look at a one-dimensional case, in zero external force (the three-dimensional generalization is immediate). Hence, we drop the vector notation, and have

$$m\frac{dv}{dt} = -\alpha v + F_a(t),\tag{16.2}$$

which is usually written as

$$\frac{dv}{dt} = -\gamma v + A(t),\tag{16.3}$$

where $\gamma = \alpha/m > 0$ and $A(t) = F_a(t)/m$. This is a stochastic (random) differential equation. To solve the problem, we have to find the probability density $p(v, t; v_o)$ that there is a solution between v and $v + dv$, at time t, such that $v = v_o$ at the initial time $t = 0$. Very close to the initial time, we have

$$p(v, t \to 0; v_o) \to \delta(v - v_o). \tag{16.4}$$

For a sufficiently long time $(t \to \infty)$, we are supposed to recover the equilibrium (Maxwell–Boltzmann) velocity distribution (for a one-dimensional gas), no matter the initial velocity v_o,

$$p(v, t \to \infty; v_o) \to \left(\frac{m}{2\pi k_B T}\right)^{1/2} \exp\left(-\frac{mv^2}{2k_B T}\right). \tag{16.5}$$

The generic solution of equation (16.3) is given by

$$v(t) = v_o \exp(-\gamma t) + \exp(-\gamma t) \int_0^t \exp(\gamma t') A(t') \, dt'. \tag{16.6}$$

Therefore, considering an ensemble of particles, we have the expected value of the velocity,

$$\langle v(t) \rangle = v_o \exp(-\gamma t) + \exp(-\gamma t) \int_0^t \exp(\gamma t') \langle A(t') \rangle \, dt'. \tag{16.7}$$

It is reasonable to assume that the expected value of the random forces vanishes,

$$\langle A(t) \rangle = 0, \tag{16.8}$$

so that we have an exponential decay of the initial velocity,

$$\langle v(t) \rangle = v_o \exp(-\gamma t). \tag{16.9}$$

To obtain the expected value of the quadratic deviations of the velocity, we write

$$\Delta v = v(t) - \langle v(t) \rangle = \exp(-\gamma t) \int_0^t \exp(\gamma t') A(t') \, dt'. \tag{16.10}$$

Therefore,

$$\left\langle (\Delta v)^2 \right\rangle = \exp(-2\gamma t) \int_0^t \int_0^t \exp(\gamma t' + \gamma t'') \langle A(t') A(t'') \rangle \, dt' dt''. \tag{16.11}$$

Now we assume that the random forces are such that the time correlation $\langle A(t') A(t'') \rangle$ is an even function of $t'' - t'$ and that it is appreciable in

the immediate vicinity of $t'' = t'$ only. We then write these correlations in terms of Dirac's δ-function,

$$\langle A(t') A(t'') \rangle = \tau \delta(t'' - t'), \tag{16.12}$$

Inserting this expression into equation (16.11), we have the expected value of the quadratic deviation,

$$\left\langle (\Delta v)^2 \right\rangle = \frac{\tau}{2\gamma} \left[1 - \exp(-2\gamma t) \right]. \tag{16.13}$$

To obtain an expression for the time parameter τ, we assume a Maxwellian velocity distribution in equilibrium (and the usual results of the classical equipartition theorem). Thus, in the $t \to \infty$ limit, we have

$$\left\langle (\Delta v)^2 \right\rangle = \frac{\tau}{2\gamma} \left[1 - \exp(-2\gamma t) \right] \to \frac{\tau}{2\gamma} = \frac{k_B T}{m}, \tag{16.14}$$

from which we obtain

$$\tau = \frac{2\gamma k_B T}{m}. \tag{16.15}$$

Therefore, we finally write the expected value of the quadratic deviation,

$$\left\langle (\Delta v)^2 \right\rangle = \frac{k_B T}{m} \left[1 - \exp(-2\gamma t) \right]. \tag{16.16}$$

Assuming that the probability distribution $p(v, t; v_o)$ is well represented by a Gaussian form, we use the first two moments to write

$$p(v, t; v_o) = \left[2\pi \left\langle (\Delta v)^2 \right\rangle \right]^{-1/2} \exp \left[-\frac{(v - \langle v \rangle)^2}{2 \left\langle (\Delta v)^2 \right\rangle} \right], \tag{16.17}$$

that is,

$$p(v, t; v_o) = \left\{ \frac{m}{2\pi k_B T \left[1 - \exp(-2\gamma t) \right]} \right\}^{1/2} \exp \left\{ -\frac{m \left[v - v_o \exp(-\gamma t) \right]^2}{2 k_B T \left[1 - \exp(-2\gamma t) \right]} \right\}. \tag{16.18}$$

It is easy to check that $p(v, t; v_o)$ tends indeed to the Maxwell–Boltzmann distribution, given by equation (16.5), as $t \to \infty$.

Now we obtain the probability distribution $p(x, t; x_o)$ for the displacements, $x = x(t)$, and make an attempt to establish a connection with the problem of the random walk as discussed in Chapter 1. Since $v = dx/dt$, we have

$$x(t) = x_o + \int_0^t v(t') \, dt', \tag{16.19}$$

which leads to

$$\langle x\,(t)\rangle = x_o + \int\limits_0^t \langle v\,(t')\rangle \, dt'. \tag{16.20}$$

From equation (16.9) for the expected value of the velocity, we have

$$\langle x\,(t)\rangle = x_o + \int\limits_0^t v_o \exp\left(-\gamma t'\right) dt' = x_o + \frac{v_o}{\gamma}\left[1 - \exp\left(-\gamma t\right)\right]. \tag{16.21}$$

To calculate the quadratic deviation, we write

$$[x\,(t)]^2 = x_o^2 + 2x_o \int\limits_0^t v\,(t')\,dt' + \int\limits_0^t\int\limits_0^t v\,(t')\,v\left(t'\right) dt' dt''. \tag{16.22}$$

Thus,

$$\left\langle [x\,(t)]^2 \right\rangle = x_o^2 + \frac{2x_o v_o}{\gamma}\left[1 - \exp\left(-\gamma t\right)\right] + \int\limits_0^t\int\limits_0^t \left\langle v\,(t')\,v\left(t'\right)\right\rangle dt' dt''.$$

$$\tag{16.23}$$

The calculation of the double integral is straightforward. From equations (16.9) and (16.10), we can write

$$\langle v\,(t')\,v\,(t'')\rangle \quad = \quad v_o^2 \exp\left(-\gamma t' - \gamma t''\right)$$

$$+ \exp\left(-\gamma t' - \gamma t''\right) \int\limits_0^{t'}\int\limits_0^{t''} \exp\left(\gamma y + \gamma z\right) \langle A\,(y)\,A\,(z)\rangle \, dy dz.$$

$$\tag{16.24}$$

Taking into account that $\langle A\,(y)\,A\,(z)\rangle = \tau\delta\,(z - y)$, we have

$$\int\limits_0^{t'}\int\limits_0^{t''} \exp\left(\gamma y + \gamma z\right) \langle A\,(y)\,A\,(z)\rangle \, dy dz = \tau \int\limits_0^{t'} dy \exp\left(2\gamma y\right)$$

$$\times \int\limits_{-y}^{t''-y} \exp\left(\gamma w\right)\delta\,(w)\,dw = \tau \int\limits_0^{t'} dy \exp\left(2\gamma y\right)\theta\,(t'' - y), \tag{16.25}$$

where $\theta(s)$ is the Heaviside function, $\theta(s) = 1$ for $s > 0$ and $\theta(s) = 0$ for $s < 0$. Thus, we have

$$\langle v(t') v(t'') \rangle = \exp\left(-\gamma t' - \gamma t''\right) \times \begin{cases} \left[v_o^2 + \frac{\tau}{2\gamma}\left(e^{2\gamma t'} - 1\right)\right]; & (t'' > t'), \\[2mm] \left[v_o^2 + \frac{\tau}{2\gamma}\left(e^{2\gamma t''} - 1\right)\right]; & (t'' < t'), \end{cases}$$

(16.26)

where $\tau = 2\gamma k_B T/m$, in agreement with equation (16.15). Inserting these expressions into equation (16.23), we have the expected value

$$\left\langle [x(t)]^2 \right\rangle = x_o^2 + \frac{2x_o v_o}{\gamma}\left[1 - \exp\left(-\gamma t\right)\right] + \frac{v_o^2}{\gamma^2}\left[1 - \exp\left(-\gamma t\right)\right]^2$$

$$+ \frac{k_B T}{m\gamma^2}\left[2\gamma t - 3 + 4\exp\left(-\gamma t\right) - \exp\left(-2\gamma t\right)\right],$$

(16.27)

from which we finally write

$$\left\langle (\Delta x)^2 \right\rangle = \langle x^2 \rangle - \langle x \rangle^2 = \frac{k_B T}{m\gamma^2}\left[2\gamma t - 3 + 4\exp\left(-\gamma t\right) - \exp\left(-2\gamma t\right)\right].$$

(16.28)

Now, assuming a Gaussian form, we have

$$p(x, t; x_o, v_o) = \left[2\pi \left\langle (\Delta x)^2 \right\rangle\right]^{1/2} \exp\left[-\frac{(x - \langle x \rangle)^2}{2\left\langle (\Delta x)^2 \right\rangle}\right],$$

(16.29)

where $\langle x \rangle$ and $\left\langle (\Delta x)^2 \right\rangle$ are given by equations (16.21) and (16.28), respectively.

These equations lead to interesting short and long-time limits:

(i) For $\gamma t \ll 1$, that is, for sufficiently short times, $t \ll 1/\gamma = m/\alpha$, we have

$$\langle x \rangle = x_o + v_o t - \frac{1}{2}v_o\gamma t^2 + O\left(t^3\right)$$

(16.30)

and

$$\langle x^2 \rangle = x_o^2 + 2x_o v_o t + v_o\left(v_o - x_o\gamma\right)t^2 + O\left(t^3\right),$$

(16.31)

so that $\left\langle (\Delta x)^2 \right\rangle = O\left(t^3\right)$. It should be pointed out that these results can be obtained from a very simple analysis of the motion, if we just use the laws of mechanics without the inclusion of a random force.

(ii) For $\gamma t \gg 1$, that is, for sufficiently long times, $t \gg 1/\gamma = m/\alpha$, we have

$$\left\langle (\Delta x)^2 \right\rangle \to \frac{2k_B T}{m\gamma} t \equiv 2Dt, \qquad (16.32)$$

where $D = k_B T/m\gamma$ is the diffusion coefficient. This celebrated result has been derived by Einstein, and experimentally checked by Jean Perrin. The dependence of the standard deviation of the displacements on \sqrt{t}, similar to the results for the problem of the random walk as analyzed in Chapter 1, indicates the stochastic nature of the Brownian motion. For sufficiently long times, we can write

$$p(x, t; x_o, v_o) \to (4\pi Dt)^{-1/2} \exp\left[-\frac{(x - x_o - v_o/\gamma)^2}{4Dt} \right], \qquad (16.33)$$

which is the solution of a **diffusion equation** in one dimension,

$$\frac{\partial p}{\partial t} = D \frac{\partial^2 p}{\partial x^2}, \qquad (16.34)$$

and points out the irreversible character of the Brownian motion.

16.2 The Fokker–Planck equation

Since the Fokker–Planck equation refers to the evolution of the probability distributions, we should make some comments on Markovian processes. Suppose that y_i is a random event at time t_i (y may refer, for example, to the set of variables that characterize a microscopic state of a given system). A sequence of random events, (y_1, t_1), (y_2, t_2), (y_3, t_3), ..., such that $t_1 < t_2 < t_3 < ...$, is called **Markovian** if the probability of occurrence of any of them depends only on the probability of occurrence of the immediately previous event (that is, if the probability of occurrence of any element of the sequence does not depend on the "previous history" of the system). All of the examples that we have been treating so far (and most of the problems of physical interest) belong to this class of Markovian problems.

Let us use the notation $P(y_F, t_F; y_I, t_I)$ for the (conditional) probability of occurrence of event y_F, at time t_F, known the occurrence of event y_I, at time t_I. The Markovian sequences obey the Chapman–Kolmogorov relation,

$$P(y_F, t_F; y_I, t_I) = \sum_k P(y_F, t_F; y_k, t_k) P(y_k, t_k; y_I, t_I), \qquad (16.35)$$

where the subscript k labels the intermediate events, between the initial time t_I and the final time t_F. In a continuous version, for probability densities, we write

$$p\left(y_F, t_F; y_I, t_I\right) = \int p\left(y_F, t_F; y_k, t_k\right) p\left(y_k, t_k; y_I, t_I\right) dy_k. \qquad (16.36)$$

Since the initial instant of time is arbitrary, the conditional probabilities should only depend on the difference between the final and the initial times of any process of physical relevance. Therefore, we can also write

$$p\left(y_F, t_F - t_I; y_I\right) = \int p\left(y_F, t_F - t_k; y_k\right) p\left(y_k, t_k - t_I; y_I\right) dy_k. \qquad (16.37)$$

Considering equation (16.37), we introduce the change of variables $t_F - t_I = t + \Delta t$ and $t_k - t_I = t$. Thus, we have

$$p\left(y_F, t + \Delta t; y_I\right) = \int p\left(y_k, t; y_I\right) p\left(y_F, \Delta t; y_k\right) dy_k. \qquad (16.38)$$

Now, we adapt this expression for the Brownian motion. Using the notation of last section for the distribution of velocities, we can write the following Chapman–Kolmogorov relation,

$$p\left(v, t + \Delta t; v_o\right) = \int\limits_{-\infty}^{+\infty} p\left(v', t; v_o\right) p\left(v, \Delta t; v'\right) dv', \qquad (16.39)$$

where the conditional probability $p\left(v, \Delta t; v'\right)$ may be interpreted as a kind of "transition probability" between two states with different velocities.

Let us go through some additional manipulations to write the Fokker–Planck equation. Introducing a well-behaved function $\phi\left(v\right)$, we can use equation (16.39) to write the integral form

$$\int\limits_{-\infty}^{+\infty} p\left(v, t + \Delta t; v_o\right) \phi\left(v\right) dv = \int\limits_{-\infty}^{+\infty}\!\!\!\int p\left(v', t; v_o\right) p\left(v, \Delta t; v'\right) \phi\left(v\right) dv' dv. \qquad (16.40)$$

Taylor-expanding the left-hand side of this expression, we have

$$\int\limits_{-\infty}^{+\infty} p\left(v, t + \Delta t; v_o\right) \phi\left(v\right) dv$$

$$= \left\langle \phi\left(v\right)\right\rangle + \Delta t \int\limits_{-\infty}^{+\infty} \frac{\partial p\left(v, t; v_o\right)}{\partial t} \phi\left(v\right) dv + ..., \qquad (16.41)$$

where the expected value of $\phi(v)$ is given by

$$\langle \phi(v) \rangle = \int_{-\infty}^{+\infty} p(v,t;v_o)\,\phi(v)\,dv. \tag{16.42}$$

Also, the expansion of $\phi(v)$ as a Taylor series leads to

$$\int_{-\infty}^{+\infty} p(v,\Delta t;v')\,\phi(v)\,dv = \int_{-\infty}^{+\infty} dv\,p(v,\Delta t;v')\,\big[\phi(v')$$

$$+\phi'(v')(v-v') + \frac{1}{2}\phi''(v')(v-v')^2 + \dots\Big]$$

$$= \phi(v') + \phi'(v')\,A(v')\,\Delta t + \frac{1}{2}\phi''(v')\,B(v')\,\Delta t + \dots, \tag{16.43}$$

where

$$A(v')\,\Delta t = \int_{-\infty}^{+\infty} p(v,\Delta t;v')\,(v-v')\,dv \tag{16.44}$$

and

$$B(v')\,\Delta t = \int_{-\infty}^{+\infty} p(v,\Delta t;v')\,(v-v')^2\,dv. \tag{16.45}$$

Hence, the right-hand side of equation (16.40) may be written as

$$\iint_{-\infty}^{+\infty} p(v',t;v_o)\,p(v,\Delta t;v')\,\phi(v)\,dv'dv$$

$$= \int_{-\infty}^{+\infty} p(v',t;v_o)\left[\phi(v') + +\phi'(v')\,A(v')\,\Delta t + \frac{1}{2}\phi''(v')\,B(v')\,\Delta t + \dots\right]dv'. \tag{16.46}$$

Performing some integrations by parts, we have

$$\int_{-\infty}^{+\infty} p(v',t;v_o)\,A(v')\,\phi'(v')\,dv' = -\int_{-\infty}^{+\infty} \phi(v')\,\frac{\partial}{\partial v'}\,[p(v',t;v_o)\,A(v')]\,dv' \tag{16.47}$$

and

$$\int\limits_{-\infty}^{+\infty} p\left(v',t;v_o\right) B\left(v'\right) \phi''\left(v'\right) dv' = \int\limits_{-\infty}^{+\infty} \phi\left(v'\right) \frac{\partial^2}{\partial \left(v'\right)^2} \left[p\left(v',t;v_o\right) B\left(v'\right)\right] dv'.$$

(16.48)

Inserting these expressions into equation (16.40), we finally have the Fokker–Planck equation,

$$\frac{\partial p\left(v,t;v_o\right)}{\partial t} = -\frac{\partial}{\partial v} \left[A\left(v\right) p\left(v,t;v_o\right)\right] + \frac{1}{2} \frac{\partial^2}{\partial v^2} \left[B\left(v\right) p\left(v,t;v_o\right)\right]. \quad (16.49)$$

The coefficients $A\left(v\right)$ and $B\left(v\right)$ can be chosen according to different criteria. For example, if we adopt the Gaussian form of the Langevin treatment of the Brownian motion, we have

$$A\left(v'\right) = \frac{1}{\Delta t} \int\limits_{-\infty}^{+\infty} \left(v - v'\right) p\left(v, \Delta t; v'\right) dv = -\gamma v' \quad (16.50)$$

and

$$B\left(v'\right) = \frac{1}{\Delta t} \int\limits_{-\infty}^{+\infty} \left(v - v'\right)^2 p\left(v, \Delta t; v'\right) dv = \frac{2k_B T\gamma}{m} = 2\gamma^2 D. \quad (16.51)$$

Therefore, the Fokker–Planck equation with the Langevin choice is given by

$$\frac{\partial p}{\partial t} = \gamma \frac{\partial}{\partial v}\left(vp\right) + \frac{k_B T\gamma}{m} \frac{\partial^2 p}{\partial t^2}, \quad (16.52)$$

which is known as the Ornstein–Uhlenbeck equation, and corresponds to a diffusion process in velocity space. With some additional algebraic effort, we can show that the probability distribution given by equation (16.18) is indeed a solution of this Fokker–Planck equation.

16.3 The master equation

In order to go beyond the Fokker–Planck treatment, we now introduce the **master equation** for the time evolution of stochastic Markovian processes. Let $P\left(y,t\right)$ be the probability of finding a system in the microscopic state y, at a time t. As an illustration, we write

$$\frac{\partial}{\partial t} P\left(y,t\right) = T_{in} - T_{out}, \quad (16.53)$$

where the rate of the probability change into state y is given by

$$T_{in} = \sum_{y'} P(y', t) \, w(y' \rightarrow y), \qquad (16.54)$$

where $w(y' \rightarrow y)$ may be regarded as the probability, during an elementary time interval, that the system changes from state y' to state y. Analogously, the rate of probability change off the state y is written as

$$T_{fora} = P(y, t) \sum_{y'} w(y \rightarrow y'). \qquad (16.55)$$

Therefore, we have the **master equation**

$$\frac{\partial}{\partial t} P(y, t) = \sum_{y'} [P(y', t) \, w(y' \rightarrow y) - P(y, t) \, w(y \rightarrow y')]. \qquad (16.56)$$

In problems of physical interest, we either calculate the probability rates, $w(y_1 \rightarrow y_2)$, from first principles (which is usually impossible, as we have already mentioned) or else resort to some plausible forms, which are consistent with the underlying physical features of the problem.

In a stationary state, the probability $P(y, t)$ is not an explicit function of time. Thus, in equilibrium, we have

$$\frac{\partial P}{\partial t} = 0. \qquad (16.57)$$

From equation (16.56), a **sufficient condition** for equilibrium is given by the **detailed balance principle**,

$$P(y', t) \, w(y' \rightarrow y) = P(y, t) \, w(y \rightarrow y'), \qquad (16.58)$$

for all states y and y'. This detailed balance equation, as the name indicates, has a simple intuitive meaning. In the stationary situation, we should have the same number of transitions from y to y' and in the opposite direction, from y' to y.

One of the most frequent strategies in this area, consists in choosing the rates $w(y_1 \rightarrow y_2)$ so that they obey the detailed balance condition in equilibrium. Thus, we write

$$P(y', t \rightarrow \infty) \, w(y' \rightarrow y) = P(y, t \rightarrow \infty) \, w(y \rightarrow y'). \qquad (16.59)$$

With this choice, which does not lead to unique expressions, we are absolutely sure that we do reach a final equilibrium state after a sufficiently long time. It is interesting to remark that this has been implicitly done in Chapter 1, for establishing a master equation associated with the random walk in one dimension, and in Chapter 15, to analyze the Ehrenfest urn model.

Example: Chemical kinetics

The treatment of fluctuations in chemical reactions provides quite intuitive applications of the stochastic formalism of the master equation. Consider a very simple example, given by the reaction

$$X \xrightarrow{k} Y, \tag{16.60}$$

where molecule X is transformed into molecule Y at the rate k (as in a radioactive decay). Call $P_n(t)$ the probability of finding n molecules of type X at time t. At the initial time, we have N_o molecules of type X, that is, $P_n(0) = \delta_{n,N_o}$. If we adopt the rate k for the molecular processes, we can write the master equation

$$\frac{\partial}{\partial t} P_n(t) = -knP_n(t) + k(n+1)P_{n+1}(t). \tag{16.61}$$

To obtain a solution for the master equation (16.61), we introduce the generating function

$$F(s,t) = \sum_{j=0}^{\infty} P_j(t) s^j, \tag{16.62}$$

where $|s| < 1$. Then, we have the partial differential equation,

$$\frac{\partial F}{\partial t} = k(1-s)\frac{\partial F}{\partial s}, \tag{16.63}$$

whose solution, with the initial condition

$$F(s,0) = \sum_{j=0}^{\infty} P_j(0) s^j = s^{N_o}, \tag{16.64}$$

is given by

$$F(s,t) = [1 + (s-1)\exp(-kt)]^{N_o}. \tag{16.65}$$

Now it is easy to obtain the time evolution of the expected values associated with the various powers of the number n of molecules of type X. Indeed, we have

$$\langle n \rangle = \sum_{n=0}^{\infty} nP_n(t) = \left[s\frac{\partial F}{\partial s}\right]_{s=1} = N_o \exp(-kt), \tag{16.66}$$

which is identical to the well-known law of mass action, given by the differential equation

$$\frac{d\langle n \rangle}{dt} = -k\langle n \rangle. \tag{16.67}$$

Hence, the law of mass action of chemical kinetics can be interpreted as an equation for the average number of particles, in the absence of statistical fluctuations. The quadratic deviation with respect to the mean is given by

$$\left\langle (\Delta n)^2 \right\rangle = \left\langle n^2 \right\rangle - \left\langle n \right\rangle^2 = \left[s\frac{\partial F}{\partial s} + s^2\frac{\partial^2 f}{\partial s^2} - \left(s\frac{\partial F}{\partial s} \right)^2 \right]_{s=1}$$

$$= N_o \left[1 - \exp\left(-kt\right) \right] \exp\left(-kt\right). \tag{16.68}$$

Therefore,

$$\frac{\sqrt{\left\langle (\Delta n)^2 \right\rangle}}{\left\langle n \right\rangle} \propto \frac{1}{\sqrt{N_o}}, \tag{16.69}$$

which indicates that typical fluctuations about the mean are very small (for a sufficiently big number of molecules of type X at the beginning).

Probabilistic justification of the master equation

To give a probabilistic justification of the master equation, let us consider the Chapman–Kolmogorov relation for a Markovian process, given by equation (16.36),

$$p\left(y_F, t_F; y_I, t_I\right) = \int p\left(y_k, t_k; y_I, t_I\right) p\left(y_F, t_F; y_k, t_k\right) dy_k. \tag{16.70}$$

Integrating over the initial variables, we have

$$\int dy_I p\left(y_F, t_F; y_I, t_I\right) = \int dy_I \int p\left(y_k, t_k; y_I, t_I\right) p\left(y_F, t_F; y_k, t_k\right) dy_k, \tag{16.71}$$

from which we can write

$$p\left(y_F, t_F\right) = \int p\left(y_k, t_k\right) p\left(y_F, t_F; y_k, t_k\right) dy_k, \tag{16.72}$$

where the conditional probability $p\left(y_F, t_F; y_k, t_k\right)$ may be interpreted as a transition probability.

Now we write $t_F = t_k + \Delta t$. Hence,

$$p\left(y_F, t_k + \Delta t\right) = \int p\left(y_k, t_k\right) p\left(y_F, t_k + \Delta t; y_k, t_k\right) dy_k. \tag{16.73}$$

We can also write the expressions

$$p\left(y_F, t_k + \Delta t\right) = p\left(y_F, t_k\right) + \frac{\partial p\left(y_F, t_k\right)}{\partial t_k} \Delta t + \ldots \tag{16.74}$$

and

$$p\left(y_F, t_k + \Delta t; y_k, t_k\right) = p\left(y_F, t_k; y_k, t_k\right) + \frac{\partial p\left(y_F, t_k; y_k, t_k\right)}{\partial t_k}\Delta t + ...,$$

(16.75)

where

$$p\left(y_F, t_k; y_k, t_k\right) = \delta\left(y_F - y_k\right).$$

(16.76)

Inserting these expressions into equation (16.73), we have

$$\frac{\partial p\left(y_F, t_k\right)}{\partial t_k} = \int p\left(y_k, t_k\right)\frac{\partial p\left(y_F, t_k; y_k, t_k\right)}{\partial t_k}dy_k +$$

(16.77)

Thus, we write

$$\frac{\partial p\left(y_F, t_k; y_k, t_k\right)}{\partial t_k} = w_{t_k}\left(y_k \rightarrow y_F\right) - \delta\left(y_k - y_F\right)\int w_{t_k}\left(y_k \rightarrow y\right) dy.$$

(16.78)

Finally, inserting into equation (16.77), we have the usual form of the master equation,

$$\frac{\partial p\left(y_F, t_k\right)}{\partial t_k} = \int \left[p\left(y_k, t_k\right) w_{t_k}\left(y_k \rightarrow y_F\right) - p\left(y_F, t_k\right) w_{t_k}\left(y_k \rightarrow y\right)\right] dy_k.$$

(16.79)

16.4 The kinetic Ising model: Glauber's dynamics

The one-dimensional kinetic Ising model, as proposed by Glauber, is one of the few nontrivial examples where dynamic properties can be exactly calculated. As we have already pointed out, the Ising model has no intrinsic dynamics. In contrast with quantum models, we cannot establish a Heisenberg equation for the time evolution of angular momentum operators (since the "spins" are just numbers, ± 1). However, if we adopt plausible forms for the transition probabilities between states, we can write a master equation for the time evolution of the probability of occurrence of a certain microscopic state.

To introduce Glauber's proposal, consider a very simple system, of a single isolated spin, in the absence of external field. Let $P\left(\sigma, t\right)$ be the probability of finding the spin in state $\sigma = \pm 1$ at time t. We then write the master equation

$$\frac{\partial}{\partial t}P\left(\sigma, t\right) = -\frac{1}{2}\alpha P\left(\sigma, t\right) + \frac{1}{2}\alpha P\left(-\sigma, t\right),$$

(16.80)

where $\alpha/2$ is the probability of transition per time from state σ to state $-\sigma$. In equilibrium, which should be reached after a very long time, $t \to \infty$, we have $\partial P(\sigma, t)/\partial t = 0$, which corresponds to equal probabilities of occupation of both spin states. Taking into account the normalization,

$$P(1,t) + P(-1,t) = 1, \tag{16.81}$$

we define the expected value of the spin variable,

$$\langle \sigma \rangle = m(t) = \sum_{\sigma} \sigma P(\sigma, t) = P(1,t) - P(-1,t). \tag{16.82}$$

Therefore, we have the differential equation

$$\frac{d}{dt} m(t) = -\alpha m(t), \tag{16.83}$$

whose solution,

$$m(t) = m(0) \exp(-\alpha t), \tag{16.84}$$

leads to an exponential decay, with the characteristic time α^{-1}, toward the equilibrium value, $m(\infty) = 0$.

In the more interesting case of a system of N spins, $P(\sigma_1, \sigma_2, ..., \sigma_N, t)$ is the probability of occurrence of the state $\vec{\sigma} = (\sigma_1, \sigma_2, ..., \sigma_N)$ at time t. Let us call $w_j(\sigma_j)$ the probability per time that the jth spin changes from σ_j to $-\sigma_j$. It is then immediate to generalize the previous treatment to write the master equation

$$\frac{\partial}{\partial t} P(\sigma_1, ..., \sigma_N, t) = \sum_{j=1}^{N} w_j(-\sigma_j) P(\sigma_1, ..., -\sigma_j, ..., \sigma_N, t)$$

$$- \sum_{j=1}^{N} w_j(\sigma_j) P(\sigma_1, ..., \sigma_j, ..., \sigma_N, t). \tag{16.85}$$

Using a more compact notation, it is convenient to write

$$\frac{\partial}{\partial t} P(\vec{\sigma}, t) = \sum_{j=1}^{N} [w_j(F_j\vec{\sigma}) P(F_j\vec{\sigma}, t) - w_j(\vec{\sigma}) P(\vec{\sigma}, t)], \tag{16.86}$$

where the "operator" F_j applied on state $\vec{\sigma}$ changes the sign of spin σ_j.

Now we consider a nearest-neighbor one-dimensional Ising model, given by the Hamiltonian

$$\mathcal{H} = -J \sum_{j=1}^{N} \sigma_j \sigma_{j+1}. \tag{16.87}$$

Glauber's proposal consists in the choice

$$w_j(\vec{\sigma}) = \frac{1}{2}\alpha\left[1 - \frac{1}{2}\gamma\sigma_j(\sigma_{j-1} + \sigma_{j+1})\right], \tag{16.88}$$

where the parameter γ is selected to fulfill the condition of detailed balance. It is interesting to look at some consequences of this choice:

(i) If the neighbors, σ_{j-1} and σ_{j+1}, are positive,

$$w_j = \frac{1}{2}\alpha\left[1 - \frac{1}{2}\gamma\sigma_j\right]. \tag{16.89}$$

(ii) If the neighbors are negative,

$$w_j = \frac{1}{2}\alpha\left[1 + \frac{1}{2}\gamma\sigma_j\right]. \tag{16.90}$$

(iii) If the neighbors have opposite signs,

$$w_j = \frac{1}{2}\alpha. \tag{16.91}$$

Therefore, $0 < \gamma < 1$ should correspond to ferromagnetic interactions, while $-1 < \gamma < 0$ corresponds to the antiferromagnetic case. In order to satisfy the detailed balance condition, we require that

$$w_j(F_j\vec{\sigma})P(F_j\vec{\sigma}, t \to \infty) = w_j(\vec{\sigma})P(\vec{\sigma}, t \to \infty). \tag{16.92}$$

Using a simplified notation,

$$P(\vec{\sigma}, t \to \infty) = P_o(\sigma_j) \quad \text{and} \quad P(F_j\vec{\sigma}, t \to \infty) = P_o(-\sigma_j), \tag{16.93}$$

we can write equation (16.92) in the form

$$\frac{w_j(\sigma_j)}{w_j(-\sigma_j)} = \frac{P_o(-\sigma_j)}{P_o(\sigma_j)}. \tag{16.94}$$

It is known that equilibrium is associated with Gibbs states, given by the (canonical) probability distribution

$$P_o(\sigma_j) = \frac{1}{Z}\exp\left[... + \beta J\sigma_{j-1}\sigma_j + \beta J\sigma_j\sigma_{j+1} + ...\right], \tag{16.95}$$

where Z is the canonical partition function. Therefore, inserting the expressions for the transition probabilities, given by equation (16.88), and the

equilibrium probabilities, given by equation (16.95), into equation (16.94), we have

$$\frac{w_j(\sigma_j)}{w_j(-\sigma_j)} = \frac{1 - \frac{1}{2}\gamma\sigma_j(\sigma_{j-1}+\sigma_{j+1})}{1 + \frac{1}{2}\gamma\sigma_j(\sigma_{j-1}+\sigma_{j+1})} = \frac{\exp\left[-\beta J\sigma_{j-1}\sigma_j - \beta J\sigma_j\sigma_{j+1}\right]}{\exp\left[\beta J\sigma_{j-1}\sigma_j + \beta J\sigma_j\sigma_{j+1}\right]},$$
(16.96)

from which it is not difficult to see that

$$\gamma = \tanh(2\beta J).$$
(16.97)

Thus, for the Ising chain, with nearest-neighbor interactions and in the absence of an applied field, the Glauber transition probabilities are given by

$$w_j(\sigma_j) = \frac{1}{2}\alpha\left[1 - \frac{1}{2}\sigma_j(\sigma_{j-1}+\sigma_{j+1})\tanh(2\beta J)\right].$$
(16.98)

As the expression $(\sigma_{j-1}+\sigma_{j+1})/2$ can assume the values $+1$, -1, and 0 only, we rewrite equation (16.98) in the form

$$w_j(\sigma_j) = \frac{1}{2}\alpha\left\{1 - \sigma_j\tanh\left[\beta J(\sigma_{j-1}+\sigma_{j+1})\right]\right\}.$$
(16.99)

Later, we will consider an Ising model on a d-dimensional lattice, in zero field, in order to show that a suitable choice of the Glauber transition probabilities, in agreement with detailed balance, is given by

$$w_j(\sigma_j) = \frac{1}{2}\alpha\left\{1 - \sigma_j\tanh\left[\beta J\sum_\delta \sigma_{j+\delta}\right]\right\},$$
(16.100)

where the sum is over the δ neighbors of site j. Although these choices may lead to the simplest expressions for the transition probabilities, by no means they are the only possibilities. In fact, there may be different forms of the transition probabilities, corresponding to different dynamic rules, all of them in agreement with the detailed balance principle (and thus leading to the same Gibbs equilibrium states).

The expected value of a spin variable on site k of an Ising chain is written as

$$m_k(t) = \langle\sigma_k\rangle = \sum_{\vec{\sigma}}\sigma_k P(\vec{\sigma},t).$$
(16.101)

Thus, using the master equation (16.86), we can write

$$\frac{\partial}{\partial t}m_k(t) = \sum_{\vec{\sigma}}\sigma_k\frac{\partial}{\partial t}P(\vec{\sigma},t)$$

$$= \sum_{\vec{\sigma}} \sigma_k \sum_{j=1}^{N} \left[w_j \left(F_j \vec{\sigma} \right) P \left(F_j \vec{\sigma}, t \right) - w_j \left(\vec{\sigma} \right) P \left(\vec{\sigma}, t \right) \right]. \tag{16.102}$$

Going through some straightforward algebraic manipulations, we have

$$\sum_{\vec{\sigma}} \sigma_k \sum_{j=1}^{N} w_j \left(F_j \vec{\sigma} \right) P \left(F_j \vec{\sigma}, t \right) = \sum_{\vec{\sigma}} \sigma_k \sum_{j=1}^{N} w_j \left(-\sigma_j \right) P \left(-\sigma_j, t \right)$$

$$= \sum_{\vec{\sigma}} \sigma_k w_k \left(-\sigma_k \right) P \left(-\sigma_k, t \right) + \sum_{\vec{\sigma}} \sum_{j \neq k} \sigma_k w_j \left(-\sigma_j \right) P \left(-\sigma_j, t \right)$$

$$= \sum_{\vec{\sigma}} -\sigma_k w_k \left(\sigma_k \right) P \left(\sigma_k, t \right) + \sum_{\vec{\sigma}} \sum_{j \neq k} \sigma_k w_j \left(\sigma_j \right) P \left(\sigma_j, t \right)$$

$$= \sum_{\vec{\sigma}} \sum_{j=1}^{N} \sigma_j w_j \left(\sigma_j \right) P \left(\sigma_j, t \right) - 2 \sum_{\vec{\sigma}} \sigma_k w_k \left(\vec{\sigma} \right) P \left(\vec{\sigma}, t \right). \tag{16.103}$$

Thus, equation (16.102) may be written as

$$\frac{\partial}{\partial t} m_k \left(t \right) = -2 \sum_{\vec{\sigma}} \sigma_k w_k \left(\vec{\sigma} \right) P \left(\vec{\sigma}, t \right), \tag{16.104}$$

which works for the Ising model in all dimensions.

To obtain some further results, we restrict the considerations to one-dimensional lattices with nearest-neighbor interactions. Inserting the transition probabilities, given by equation (16.98), into equation (16.104), we have

$$\frac{\partial}{\partial t} m_k \left(t \right) = -2 \sum_{\vec{\sigma}} \sigma_k \frac{1}{2} \alpha \left[1 - \frac{1}{2} \sigma_j \left(\sigma_{j-1} + \sigma_{j+1} \right) \tanh \left(2\beta J \right) \right] P \left(\vec{\sigma}, t \right)$$

$$= \alpha \sum_{\vec{\sigma}} \left[-\sigma_k + \frac{1}{2} \tanh \left(2\beta J \right) \left(\sigma_{j-1} + \sigma_{j+1} \right) \right] P \left(\vec{\sigma}, t \right). \tag{16.105}$$

Thus, in the particular case of the Ising chain with nearest-neighbor interactions, we can write the simple differential equation

$$\frac{\partial}{\partial t} m_k \left(t \right) = -\alpha m_k \left(t \right) + \frac{\alpha}{2} \tanh \left(2\beta J \right) \left[m_{k-1} \left(t \right) + m_{k+1} \left(t \right) \right], \tag{16.106}$$

where there are no coupling terms involving the expected values of either two or more spin variables. The absence of a hierarchy of coupled equations

for the moments (as in the case of the BBGKY hierarchy) allows the establishment of an exact solution for this problem. Unfortunately, however, there is no parallel to the exact solution of Onsager, for the much more interesting two-dimensional lattice, with a finite critical temperature.

If we assume periodic boundary conditions,

$$m_k(t) = m_{k+N}(t), \tag{16.107}$$

where N is an integer, we can use Fourier analysis to solve equation (16.106). Let us write

$$m_k(t) = \frac{1}{N} \sum_q \hat{m}_q(t) \exp(iqk), \tag{16.108}$$

where q belongs to the first Brillouin zone,

$$q = 0, \pm\frac{2\pi}{N}, \pm 2\frac{2\pi}{N}, ..., \pi, \tag{16.109}$$

for even N. Since the magnetization is always real, we have

$$\hat{m}_q(t) = \hat{m}_{-q}^*(t). \tag{16.110}$$

Hence, from equation (16.106), we have

$$\frac{\partial}{\partial t}\hat{m}_q(t) = -\alpha\hat{m}_q(t) + (\alpha\gamma\cos q)\,\hat{m}_q(t), \tag{16.111}$$

where $\gamma = \tanh(2\beta J)$. Thus,

$$\hat{m}_q(t) = A_q \exp\{-\alpha[1 - \gamma\cos q]\,t\}, \tag{16.112}$$

where A_q is a constant, given by the initial conditions, such that

$$A_q = A_{-q}^*. \tag{16.113}$$

Inserting into equation (16.108), we finally have

$$m_k(t) = \frac{1}{N} \sum_q A_q \exp\{-\alpha[1 - \gamma\cos q]\,t + iqk\}, \tag{16.114}$$

which is a solution of equation (16.106).

If we begin from a very simple set of initial conditions, $m_k(t = 0) = \delta_{k,0}$, it is easy to check that $A_q = 1$, for all q. Hence,

$$m_k(t) = \frac{1}{N} \sum_q \exp\{-\alpha[1 - \gamma\cos q]\,t + iqk\}. \tag{16.115}$$

In particular,

$$m_o(t) = \frac{1}{N} \sum_q \exp\{-\alpha[1 - \gamma\cos q]t\}$$

$$= \frac{1}{2\pi} \int dq \exp\{-\alpha[1 - \gamma\cos q]t\} \rightarrow \frac{1}{(2\pi\alpha\gamma t)^{1/2}} \exp[-\alpha(1-\gamma)t],$$

(16.116)

which decays to the equilibrium value, $m_o(t \rightarrow \infty) \rightarrow 0$, with the time constant $\tau^{-1} = \alpha(1 - \gamma)$. The magnetization per site is given by

$$m(t) = \frac{1}{N} \sum_{k=1}^{N} m_k(t) = \frac{1}{N} \exp[-\alpha(1-\gamma)t],$$

(16.117)

which decays with the same time constant.

These calculations provide an illustration of the stochastic methods. It is beyond the scope of this book to calculate other quantities of relevance for the kinetic Ising chain. It should be mentioned that it is relatively easy to investigate the dynamic properties of various correlation functions and to analyze this model in the presence of an applied field. The interested reader is advised to check the seminal paper by R. J. Glauber, J. Math. Phys. **4**, 294 (1963).

Now we present an approximate (mean-field) calculation for a three-dimensional Ising model, which does display a phase transition at a finite temperature.

Glauber's dynamics in the mean-field approximation

According to Glauber's proposal, the transition probability for the Ising model should be written as

$$w_j(\sigma_j) = \frac{1}{2}\alpha[1 - \sigma_j f(\{\sigma_{j+\delta}\})],$$

(16.118)

where the function f depends on the set of spins in the neighborhood of a certain site j. In equilibrium, we have

$$P_o(\sigma_j) = \frac{1}{Z} \exp\left[\ldots + \beta J \sigma_j \sum_\delta \sigma_{j+\delta} + \ldots\right].$$

(16.119)

Thus, from the condition of detailed balance, expressed by equation (16.94), we can write

$$\frac{w_j(\sigma_j)}{w_j(-\sigma_j)} = \frac{1 - \sigma_j f(\{\sigma_{j+\delta}\})}{1 + \sigma_j f(\{\sigma_{j+\delta}\})} = \exp\left[-2\beta J \sigma_j \sum_\delta \sigma_{j+\delta}\right],$$

(16.120)

from which we have

$$f\left(\{\sigma_{j+\delta}\}\right) = \tanh\left[\beta J \sum_\delta \sigma_{j+\delta}\right], \qquad (16.121)$$

which leads to the expression

$$w_j\left(\sigma_j\right) = \frac{1}{2}\alpha\left\{1 - \sigma_j\tanh\left[\beta J \sum_\delta \sigma_{j+\delta}\right]\right\}, \qquad (16.122)$$

as we have obtained before [see equation (16.100)].

The mean-field approximation consists in the assumption that

$$\sum_\delta \sigma_{j+\delta} \approx q\left\langle\sigma_{j+\delta}\right\rangle = qm\left(t\right), \qquad (16.123)$$

where q is the coordination number of the lattice, and the expected value $\left\langle\sigma_{j+\delta}\right\rangle = m\left(t\right)$ should not depend on the particular site j. Therefore, the transition probability (16.122) is given by

$$w_j\left(\sigma_j\right) = \frac{1}{2}\alpha\left\{1 - \sigma_j\tanh\left[\beta Jqm\left(t\right)\right]\right\}. \qquad (16.124)$$

Inserting this expression into equation (16.104), which has been obtained under very general grounds, we have the mean-field expression

$$\frac{\partial m}{\partial t} = -\alpha m + \alpha\tanh\left[\beta Jqm\right]. \qquad (16.125)$$

In the neighborhood of the critical point, we can write the expansion

$$\frac{\partial m}{\partial t} = \alpha\left\{-m + \beta Jqm + ...\right\} = -\alpha\left[1 - \frac{T}{T_c}\right]m + \qquad (16.126)$$

We then have the asymptotic behavior

$$m \approx m_o\exp\left[-\alpha\left(1 - \frac{T}{T_c}\right)t\right], \qquad (16.127)$$

which leads to the characteristic time

$$\tau = \frac{1}{\alpha}\left(1 - \frac{T}{T_c}\right)^{-1}, \qquad (16.128)$$

that diverges with the exponent $\gamma = 1$ as $T \to T_c$. This critical slowing down of the relaxation times drastically affects both experimental measurements and numerical simulations in the neighborhood of second-order phase transitions (although the experimental critical exponent is different from $\gamma = 1$).

16.5 The Monte Carlo method

The master equation provides a justification for the increasingly relevant **Monte Carlo methods**. In equilibrium statistical mechanics, we are interested in calculations of averages of the form

$$\langle A \rangle = \frac{\sum\limits_{C} A\,(C)\exp\left[-\beta \mathcal{H}\,(C)\right]}{\sum\limits_{C}\exp\left[-\beta \mathcal{H}\,(C)\right]}, \tag{16.129}$$

where the sum is over all of the microscopic configurations of a system associated with the Hamiltonian \mathcal{H}. For example, in the case of the Ising model on a lattice with N sites, this sum is over 2^N configurations. Thus, for large N, it is not practical to use this kind of expression to perform numerical calculations. The solution consists in performing the average over a duly selected, and much smaller, number of the more-representative equilibrium configurations of the system,

$$\langle A \rangle = \frac{1}{M}\sum_{i=1}^{M} A_i. \tag{16.130}$$

Of course, it is important to know under which circumstances we can really obtain the expected value of the quantity A from this arithmetic average over M representative configurations. Also, we have to know how to choose these representative configurations. Questions of this sort are answered by the Monte Carlo method.

According to the Monte Carlo method, we choose a sequence of independent configurations (which are called a Markov chain). Some initial configurations are far from equilibrium, but as a function of time we generate many more equilibrium configurations that should be used to perform the average of equation (16.130). As we have already discussed, if we call $w\,(y \to y')$ the probability of transition per unit time from configuration y to configuration y', we have the master equation

$$\frac{\partial}{\partial t}P\,(y,t) = \sum_{y'}\left[w\,(y' \to y)\,P\,(y',t) - w\,(y \to y')\,P\,(y,t)\right]. \tag{16.131}$$

In equilibrium, that is, after going through a sufficiently big number of members of the sequence, the probabilities should tend to the Gibbs values,

$$P_o\,(y) = \frac{1}{Z}\exp\left[-\beta \mathcal{H}\,(y)\right], \tag{16.132}$$

where Z is the canonical partition function (for a system in contact with a heat bath). As we have mentioned, a sufficient condition for equilibrium is

given by the detailed balance condition,

$$P_o(y) \, w(y \to y') = P_o(y') \, w(y' \to y). \qquad (16.133)$$

Therefore, the transition probabilities are chosen such that

$$\frac{w(y \to y')}{w(y' \to y)} = \exp(-\beta \Delta \mathcal{H}), \qquad (16.134)$$

where $\Delta \mathcal{H} = \mathcal{H}(y') - \mathcal{H}(y)$ is the energy difference between configurations y' and y.

Equation (16.134) does not yield a unique transition probability. Two frequent choices in Monte Carlo simulations are the Glauber algorithm,

$$w(y \to y') = \frac{1}{2\tau} \left[1 - \tanh\left(\frac{1}{2}\beta \Delta \mathcal{H}\right) \right], \qquad (16.135)$$

and the Metropolis prescription,

$$w(y \to y') = \begin{cases} \frac{1}{\tau} \exp(-\beta \Delta \mathcal{H}) & , \quad \Delta \mathcal{H} > 0, \\ \frac{1}{\tau} & , \quad \Delta \mathcal{H} < 0, \end{cases} \qquad (16.136)$$

where the time τ is interpreted as a "Monte Carlo step" (and we may choose $\tau = 1$).

Monte Carlo calculation for the Ising model

To give a practical example of the use of the Monte Carlo method, consider the Ising ferromagnet on a square lattice, given by the Hamiltonian

$$\mathcal{H} = -J \sum_{(ij)} \sigma_i \sigma_j, \qquad (16.137)$$

where $\sigma_i = \pm 1$, for $i = 1, 2, ..., N$, and the sum is over pairs of nearest-neighbor sites. The only parameter of this problem is the product βJ that defines the temperature. We may begin the calculations with a small lattice (4×4 or 6×6), with either free or periodic boundary conditions [see, for example, D. P. Landau, Phys. Rev. **B13**, 2997 (1976), about this specific problem]. In order to generate the Markov chain, we go through the following steps:

(i) Choose a particular configuration of the spin system (all spins $+$, for example).

(ii) Select any site of the lattice (in many simulations, we just choose a sequence of sites) and calculate $r = \exp(-\beta \Delta E)$, where $\Delta E =$

$E_f - E_i$ is the energy difference associated with the change of sign of the selected spin. Note that, in order to save time, it is convenient to keep a table of the (few) values of ΔE (since each spin interacts with just four neighbors).

(iii) Compare r with a random number z between 0 and 1 (an available random number generator is usually enough).

(iv) Change the sign of the spin if $r > z$.

(v) Keep the configuration that has been produced by this process, and (sequentially) choose another site to go back to the second step.

It is not difficult to be convinced that these procedures do indeed reproduce the transition probabilities associated with the Metropolis algorithm, given by equation (16.136). Now, if we discard a certain number of initial configurations, which still reflect the original initial state, we can use the remaining configurations to calculate the arithmetic average of the quantities of interest. For example, we can calculate the internal energy and the magnetization (in modulus) for different lattice sizes. Since it is not possible to simulate an infinite system, we also have to develop a series of techniques to calculate numerical errors and to obtain information from a systematic analyses of finite-size systems. As an exercise, we suggest a Monte Carlo simulation for the Ising ferromagnet on the square lattice (try to reproduce the results in the paper by D. P. Landau!).

Exercises

1. The one-dimensional version of the Langevin equation to describe a Brownian motion in the presence of an external field is written as

$$\frac{dv}{dt} = -\gamma v + f + A(t),$$

where f is proportional to a constant force. The initial conditions are given by $v(0) = 0$ and $x(0) = 0$. Suppose that $\langle A(t) \rangle = 0$ and that

$$\langle A(t') A(t'') \rangle = \tau \delta(t' - t''),$$

where τ is a time constant.

Obtain expressions for $\langle v(t) \rangle$ and $\left\langle (\Delta v)^2 \right\rangle$. Find the time constant τ as a function of temperature and the coefficient of viscous friction (which leads to the fluctuation-dissipation relation).

2. Show that the Fokker–Planck equation associated with the previous problem can be written as

$$\left(\frac{\partial}{\partial t} + v\frac{\partial}{\partial x} - \frac{f}{m}\frac{\partial}{\partial v}\right)p\left(x,v,t\right) = \frac{\partial}{\partial v}\left(A\frac{\partial}{\partial v} + Bv\right)p\left(x,v,t\right).$$

Obtain expressions for A and B.

3. Consider the reversible unimolecular reaction,

$$A \Longleftrightarrow B,$$

with different rates, k_1 and k_2, along the two directions. Initially, there are n_o molecules of type A and no molecules of type B. Write a master equation for the probability of finding n molecules of type A at time t. Obtain the time evolution of the expected number of molecules of type A.

4. The ferromagnetic Ising chain in the presence of an external field is given by the spin Hamiltonian

$$H = -J\sum_{i=1}^{N}\sigma_i\sigma_{i+1} - H\sum_{i=1}^{N}\sigma_i,$$

where $J > 0$ and $\sigma_i = \pm 1$. Write a master equation for the probability of finding this system in the microscopic state $\vec{\sigma}$ at time t. What is the form of the transition probabilities according to Glauber's dynamics? Obtain the time evolution of the expected value $m_k = \langle\sigma_k\rangle$.

5. Consider an Ising ferromagnet on a square lattice, with nearest-neighbor interactions, in the absence of an external field. Write a Monte Carlo program, based on the Metropolis algorithm, to determine the internal energy and the magnetization as a function of temperature. Very reasonable results can already be obtained for 6×6 or 8×8 lattices. Discuss the following ingredients of this Monte Carlo calculation: (i) the choice of the initial conditions; (ii) the choice of the boundary conditions; (iii) the number of Monte Carlo steps (time) to reach equilibrium; (iv) the effects of the finite size of the lattice. What are the errors involved in this numerical calculation?

6. Consider again a Monte Carlo simulation for the Ising ferromagnet on the simple square lattice, but from the point of view of Glauber's heat-bath algorithm. Choose an initial spin configuration C_o and let the system evolve from a configuration C_t at time t to a configuration $C_{t+\Delta t}$ at time $t + \Delta t$, according to the following steps: (i) choose a site i at random; (ii) compute the probability

$$p_i\left(t\right) = \frac{1}{2}\left[1 + \tanh\left(\beta J\sum_j S_j\right)\right],$$

where the sum runs over the nearest neighbors of site i; (iii) choose a random number $z_i(t)$ uniformly distributed between 0 and 1; (iv) update spin S_i according to the rule

$$S_i(t + \Delta t) = sgn [p_i(t) - z_i(t)],$$

and leave the other spins unchanged, $S_j(t + \Delta t) = S_j(t)$ for $j \neq i$. Repeat this procedure again and again, as in the previous Metropolis simulation.

(a) Obtain estimates for the internal energy and the magnetization. Compare with the results from the Metropolis algorithm.

(b) Check that the probability $P_t(C)$ of finding the system in the microscopic configuration $C \equiv \{S_i\}$ at time t satisfies the master equation

$$P_{t+\Delta t}(C) = \frac{1}{N} \sum_{i=1}^{N} \frac{1}{2} \left[1 + S_i \tanh \left(\beta J \sum_j S_j \right) \right]$$
$$\times [P_t(C) + P_t(C_i)],$$

where C_i is the configuration obtained from C by flipping spin S_i.

Appendices

A.1 Stirling's asymptotic series

In order to obtain Stirling's asymptotic formula, given by

$$n! \approx n^n e^{-n} (2\pi n)^{1/2}, \qquad (\text{A.1})$$

for large values of n, we write the identity

$$n! = \int_0^\infty x^n e^{-x} dx, \qquad (\text{A.2})$$

which can be easily checked by performing a sequence of integrations by parts. By changing variables, equation (A.2) can be written as

$$n! = n^{n+1} \int_0^\infty \exp\left[n \left(\ln y - y\right)\right] dy. \qquad (\text{A.3})$$

The problem is then reduced to the asymptotic calculation of the integral

$$I\left(n\right) = \int_0^\infty \exp\left[n f\left(y\right)\right] dy, \qquad (\text{A.4})$$

where

$$f\left(y\right) = \ln y - y. \qquad (\text{A.5})$$

Since $f(y)$ is maximum at $y = y_o = 1$, we can used Laplace's method to obtain the asymptotic form of $I(n)$. For large values of n, the contributions to the integral come from the immediate neighborhood of the maximum y_o.

According to Laplace's method, we perform a Taylor expansion of $f(y)$ about the maximum $y_o = 1$,

$$f(y) = -1 - \frac{1}{2}(y-1)^2 + \dots. \tag{A.6}$$

Now we discard higher-order terms and deform the contour of integration to include the entire real axis. Thus, we have

$$I(n) \approx \int\limits_{-\infty}^{+\infty} \exp\left[-n - \frac{n}{2}(y-1)^2\right] dy$$

$$= \exp(-n) \int\limits_{-\infty}^{+\infty} \exp\left(-\frac{n}{2}x^2\right) dx = \left(\frac{2\pi}{n}\right)^{1/2} \exp(-n). \tag{A.7}$$

Inserting into equation (A.3), we obtain Stirling's asymptotic formula, given by equation (A.1). It is easy to check numerically, even for relatively small values of n, that Stirling's formula is indeed an excellent approximation.

In most of the application in statistical physics, we are interested in the first few terms of the asymptotic series,

$$\ln n! = n \ln n - n + O(\ln n), \tag{A.8}$$

since the terms of order $\ln n$ can be dropped in the thermodynamic limit. For continuous functions $f(y)$, with a maximum at $y_o > 0$, it is easy to prove that

$$\lim_{n\to\infty} \frac{1}{n} \ln I(n) = f(y_o), \tag{A.9}$$

which gives a justification for the heuristic deduction of the Stirling approximation.

We now use identity (A.2) to define the **gamma function**,

$$\Gamma(n+1) = \int\limits_{0}^{\infty} x^n e^{-x} dx, \tag{A.10}$$

which is identical to $n!$ for integer and positive values of n. However, we can also consider a gamma function, defined by the integral equation (A.10), for all values of n (in particular, for half-integer values, as far as the integral makes sense). It is easy to see that $\Gamma(n) = (n-1)\Gamma(n-1)$, $\Gamma(1) = 1$, and $\Gamma(1/2) = \sqrt{\pi}$.

A.2 Gaussian integrals

The most frequently used Gaussian integral of statistical physics is given by

$$I_0\left(a\right) = \int\limits_{-\infty}^{+\infty} \exp\left(-ax^2\right) dx = \left(\frac{\pi}{a}\right)^{1/2}, \qquad (A.11)$$

with $a > 0$. To prove this result, let us write

$$\left[I_0\left(a\right)\right]^2 = \int\limits_{-\infty}^{+\infty}\int\limits_{-\infty}^{+\infty} \exp\left(-ax^2 - ay^2\right) dxdy. \qquad (A.12)$$

Using planar polar coordinates, $x = r\cos\theta$, $y = r\sin\theta$, we have

$$\left[I_0\left(a\right)\right]^2 = \int\limits_{0}^{+\infty}\int\limits_{0}^{2\pi} r\exp\left(-ar^2\right) drd\theta = 2\pi \int\limits_{0}^{+\infty} r\exp\left(-ar^2\right) dr = \frac{\pi}{a}, \quad (A.13)$$

from which we obtain equation (A.11).

Other Gaussian integrals of frequent use may be written as

$$I_n\left(a\right) = \int\limits_{0}^{+\infty} x^n \exp\left(-ax^2\right) dx, \qquad (A.14)$$

where $a > 0$ and $n \geq 0$. For $n = 1$, we have a trivial result,

$$I_1\left(a\right) = \int\limits_{0}^{+\infty} x\exp\left(-ax^2\right) dx = \frac{1}{2a}. \qquad (A.15)$$

For odd values of n, we just perform a sequence of integrations by parts. For $n = 2$, we have

$$I_2\left(a\right) = \int\limits_{0}^{+\infty} x^2 \exp\left(-ax^2\right) dx = -\frac{1}{2}\frac{d}{da}I_0\left(a\right) = \frac{1}{4a}\left(\frac{\pi}{a}\right)^{1/2}. \qquad (A.16)$$

For even values of n, we take successive derivatives with respect to the parameter a. Introducing the change of variables $ax^2 = y$, we can also write

$$I_n\left(a\right) = \frac{1}{2}a^{-(n+1)/2} \int\limits_{0}^{+\infty} y^{(n-1)/2}e^{-y}dy, \qquad (A.17)$$

that is,

$$I_n(a) = \frac{1}{2}a^{-(n+1)/2}\Gamma\left(\frac{n+1}{2}\right),\tag{A.18}$$

with the gamma function given by equation (A.10)

A.3 Dirac's delta function

Dirac's delta "function" is defined by

$$\delta(x) = \begin{cases} 0; & x \neq 0, \\ \infty; & x = 0, \end{cases}\tag{A.19}$$

so that

$$\int_{-\infty}^{+\infty} f(x)\,\delta(x)\,dx = f(0),\tag{A.20}$$

for all continuous functions $f(x)$. To give a better definition, we have to invoke the "theory of distributions," which is certainly beyond the scope of this book. By using some heuristic arguments, and from the standard rules of differential calculus, we can derive a number of properties of this δ-function.

It is interesting to construct sequences of highly concentrated functions, $\phi_n(x)$, with $n = 1, 2, 3, ...$, which are called δ-**sequences**, so that

$$\lim_{n\to\infty} \int_{-\infty}^{+\infty} f(x)\,\phi_n(x)\,dx = f(0).\tag{A.21}$$

For example, the following δ-sequences are useful in statistical physics:

(i) A pulse function,

$$\phi_n(x) = \begin{cases} n/2; & -1/n < x < 1/n, \\ \infty; & |x| \geq 1/n; \end{cases}\tag{A.22}$$

(ii) A Lorentzian form,

$$\phi_n(x) = \frac{n}{\pi}\frac{1}{1 + n^2 x^2};\tag{A.23}$$

(iii) A Gaussian form,

$$\phi_n(x) = \frac{n}{\sqrt{\pi}}\exp\left(-n^2 x^2\right).\tag{A.24}$$

These sequences can now be used to establish several properties of the δ-function. In particular, it is easy to see how to perform derivatives and integrals of δ-functions. On the basis of the definitions (A.19) and (A.20), we can write the differential form

$$\delta\left(x\right) = \frac{d}{dx}S\left(x\right), \tag{A.25}$$

where $S\left(x\right)$ is the step function,

$$S\left(x\right) = \left\{ \begin{array}{ll} 0; & x < 0, \\ 1; & x > 0. \end{array} \right. \tag{A.26}$$

For example, we can show that

$$\int\limits_{-\infty}^{+\infty} f\left(x\right)\left[\frac{d}{dx}\delta\left(x\right)\right]dx = -f'\left(0\right), \tag{A.27}$$

which gives a meaning to the derivative of the δ-function.

There is an integral representation of the δ-function, of enormous interest in statistical physics, given by

$$\delta\left(x\right) = \frac{1}{2\pi}\int\limits_{-\infty}^{+\infty} \exp\left(ikx\right)dk. \tag{A.28}$$

To present a justification for this integral representation, we consider the sequence

$$\phi_a\left(x\right) = \frac{1}{2\pi}\int\limits_{-\infty}^{+\infty} \exp\left(-a^2k^2 + ikx\right)dk, \tag{A.29}$$

where the factor $\exp\left(-a^2k^2\right)$, with $a \neq 0$, has been included to make sure that the integral converges. From a formal point of view, we are tempted to say that

$$\delta\left(x\right) = \lim_{a \to 0}\phi_a\left(x\right). \tag{A.30}$$

In fact, if we complete the square in the exponent of the integrand of equation (A.29), we have

$$\phi_a\left(x\right) = \frac{1}{2\pi}\exp\left(-\frac{x^2}{4a^2}\right)\int\limits_{-\infty}^{+\infty} \exp\left[-a^2\left(k - \frac{ix}{2a^2}\right)^2\right]dk. \tag{A.31}$$

Also, we can write

$$\int_{-\infty}^{+\infty} \exp\left[-a^2\left(k - \frac{ix}{2a^2}\right)^2\right] dk = \int_{-\infty-\frac{ix}{2a}}^{+\infty-\frac{ix}{2a}} \exp\left(-a^2 z^2\right) dz$$

$$= \int_{-\infty}^{+\infty} \exp\left(-a^2 z^2\right) dz = \frac{\sqrt{\pi}}{a}, \tag{A.32}$$

since the exponential function is analytic, and hence we can deform the contour in the complex plane. Thus, we have

$$\phi_a(x) = \frac{1}{2a\sqrt{\pi}} \exp\left(-\frac{x^2}{4a^2}\right). \tag{A.33}$$

For $n = 1/(2a)$, we recover equation (A.24), which is one of the best known support sequences associated with the delta function.

A.4 Volume of a hypersphere

The volume of a hypersphere of radius R in a space of dimension n is given by

$$V_n(R) = \int \cdots \int_{0 \le x_1^2 + \cdots + x_n^2 \le R^2} dx_1 \ldots dx_n. \tag{A.34}$$

Since $V_n(R)$ should be proportional to R^n, we can write

$$V_n(R) = A_n R^n, \tag{A.35}$$

where the prefactor A_n depends on the dimensionality of the space only. Thus, we have

$$\delta V_n(R) = n A_n R^{n-1} \delta R = S_n(R) \delta R, \tag{A.36}$$

where $S_n(R)$ is the area of the hypersphere of radius R. Using the notation of Chapter 4, we write the volume of the hyperspherical shell (of radius R and thickness δR, in a space of n dimensions),

$$\Omega_n(R; \delta R) = S_n(R) \delta R = C_n R^{n-1} \delta R, \tag{A.37}$$

where $C_n = n A_n$.

In order to calculate the coefficient C_n, we note that

$$\left[\int_{-\infty}^{+\infty} \exp\left(-ax^2\right) dx \right]^n = \int_{-\infty}^{+\infty} \cdots \int \exp\left(-ax_1^2 - \ldots - ax_n^2\right) dx_1 \ldots dx_n$$

$$= \left(\frac{\pi}{a}\right)^{n/2}. \qquad (A.38)$$

Also, we can write

$$\int_{-\infty}^{+\infty} \cdots \int \exp\left(-ax_1^2 - \ldots - ax_n^2\right) dx_1 \ldots dx_n = \int_0^\infty \exp\left(-aR^2\right) nA_n R^{n-1} dR$$

$$= \frac{nA_n}{2a^{n/2}} \int_0^\infty x^{\frac{n}{2}-1} e^{-x} dx = \frac{nA_n}{2a^{n/2}} \Gamma\left(\frac{n}{2}\right). \qquad (A.39)$$

Hence, we have

$$\left(\frac{\pi}{a}\right)^{n/2} = \frac{nA_n}{2a^{n/2}} \Gamma\left(\frac{n}{2}\right), \qquad (A.40)$$

from which we can write

$$C_n = nA_n = 2\frac{\pi^{n/2}}{\Gamma\left(\frac{n}{2}\right)}, \qquad (A.41)$$

that is,

$$\Omega_n\left(R; \delta R\right) = 2\frac{\pi^{n/2}}{\Gamma\left(\frac{n}{2}\right)} R^{n-1} \delta R. \qquad (A.42)$$

Note that, for $n = 3N$, we have

$$C_{3N} = 2\frac{\pi^{3N/2}}{\Gamma\left(\frac{3N}{2}\right)} \sim N^{-\frac{3}{2}N} b^N, \qquad (A.43)$$

where b is a constant.

A.5 Jacobian transformations

The mathematical properties of Jacobian forms are particularly useful to help dealing with the partial derivatives of interest in thermodynamics.

Given the functions $u = u(x, y)$ and $v = v(x, y)$, the Jacobian is defined as a determinant,

$$\frac{\partial(u, v)}{\partial(x, y)} = \begin{vmatrix} \dfrac{\partial u}{\partial x} & \dfrac{\partial u}{\partial y} \\ \dfrac{\partial v}{\partial x} & \dfrac{\partial v}{\partial y} \end{vmatrix}. \tag{A.44}$$

Therefore, the partial derivative of u with respect to x may be written in terms of a Jacobian,

$$\left(\frac{\partial u}{\partial x}\right)_y = \frac{\partial(u, y)}{\partial(x, y)}. \tag{A.45}$$

The following properties of Jacobians are used in many manipulations with partial derivatives:

$$(i) \quad \frac{\partial(u, v)}{\partial(x, y)} = -\frac{\partial(v, u)}{\partial(x, y)}. \tag{A.46}$$

$$(ii) \quad \frac{\partial(u, v)}{\partial(x, y)} = \frac{\partial(u, v)}{\partial(r, s)} \frac{\partial(r, s)}{\partial(x, y)}. \tag{A.47}$$

$$(iii) \quad \frac{\partial(u, v)}{\partial(x, y)} = \left\{\frac{\partial(x, y)}{\partial(u, v)}\right\}^{-1}. \tag{A.48}$$

As an example of the application of Jacobian forms, let us calculate the derivative $(\partial T/\partial p)_s$, related to the adiabatic compression of a simple fluid, in terms of better known thermodynamic derivatives (κ_T, c_p, α). We then write

$$\left(\frac{\partial T}{\partial p}\right)_s = \frac{\partial(T, s)}{\partial(p, s)} = \frac{\partial(T, s)}{\partial(T, p)} \frac{\partial(T, p)}{\partial(p, s)} = -\left(\frac{\partial s}{\partial p}\right)_T \frac{\partial(T, p)}{\partial(s, p)}$$

$$= -\left(\frac{\partial s}{\partial p}\right)_T \left\{\frac{\partial(s, p)}{\partial(T, p)}\right\}^{-1} = -\left(\frac{\partial s}{\partial p}\right)_T \left\{\left(\frac{\partial s}{\partial T}\right)_p\right\}^{-1}, \tag{A.49}$$

where the last two derivatives are written in terms on the independent variables T and p (that is, in the Gibbs representation). Considering the Gibbs free energy per particle, $g = u - Ts + pv$, we have

$$dg = -sdT + vdp, \tag{A.50}$$

from which we obtain two Maxwell relations,

$$-s = \left(\frac{\partial g}{\partial T}\right)_p, \quad \text{and} \quad v = \left(\frac{\partial g}{\partial p}\right)_T. \tag{A.51}$$

Therefore,

$$\left(\frac{\partial s}{\partial p}\right)_T = -\left(\frac{\partial v}{\partial T}\right)_p = -v\alpha. \tag{A.52}$$

From the definition of the specific heat at constant pressure,

$$c_p = T\left(\frac{\partial s}{\partial T}\right)_p, \tag{A.53}$$

we finally have

$$\left(\frac{\partial T}{\partial p}\right)_s = \frac{Tv\alpha}{c_p}. \tag{A.54}$$

A.6 The saddle-point method

The saddle-point method is used to obtain an asymptotic expression, for large N, of integrals of the form

$$I(N) = \int_C \exp\left[Nf(z)\right] dz, \tag{A.55}$$

where $f(z)$ is an analytic function and C is a given contour in the complex plane. The asymptotic expression of the real version of this integral, along the real axis, is obtained from a simple application of the Laplace method.

If we write $f(z) = u(x, y) + iv(x, y)$, the more-relevant contributions to $I(N)$ come from regions along the contour where $u(x, y)$ is very large. The main idea of the method consists in deforming the contour in order to go through the stationary point of u along a conveniently chosen direction (so that the problem is reduced to the calculation of a simple Gaussian integral).

From the Cauchy–Riemann conditions,

$$\frac{\partial u}{\partial x} = \frac{\partial v}{\partial y} \quad \text{and} \quad \frac{\partial u}{\partial y} = -\frac{\partial v}{\partial x}, \tag{A.56}$$

we can see that the real and imaginary parts of a complex function $f(z)$ obey the Laplace equation,

$$\frac{\partial^2 u}{\partial x^2} + \frac{\partial^2 u}{\partial y^2} = 0, \quad \text{and} \quad \frac{\partial^2 v}{\partial x^2} + \frac{\partial^2 v}{\partial y^2} = 0. \tag{A.57}$$

Therefore, any extremum of the surface $u(x, y)$,

$$\frac{\partial u}{\partial x} = \frac{\partial u}{\partial y} = 0, \qquad (A.58)$$

is a **saddle point** (that is, a minimum of u along a certain direction, and a maximum along the other direction, so that the surface $u(x, y)$ looks like a saddle or a mountain path). Furthermore, from the Cauchy–Riemann conditions, an extremum of $u(x, y)$ is an extremum of $v(x, y)$ as well. Thus, we have

$$f'(z_o) = 0, \qquad (A.59)$$

at the saddle-point z_o.

We now write a Taylor expansion about the saddle-point,

$$f(z) = f(z_o) + \frac{1}{2} f''(z_o)(z - zo)^2 + \dots. \qquad (A.60)$$

If we introduce the notation

$$f''(z_o) = \rho \exp(i\theta) \qquad (A.61)$$

and

$$(z - zo) = s \exp(i\phi), \qquad (A.62)$$

we then have

$$f(z) = f(z_o) + \frac{1}{2}\rho s^2 \exp[i(\theta + 2\phi)] + \dots, \qquad (A.63)$$

from which we write

$$u(x, y) \approx u(x_o, y_o) + \frac{1}{2}\rho s^2 \cos(\theta + 2\phi) \qquad (A.64)$$

and

$$v(x, y) \approx v(x_o, y_o) + \frac{1}{2}\rho s^2 \sin(\theta + 2\phi). \qquad (A.65)$$

In order to go through the saddle-point along the direction associated with the maximum of the real part, we make the choice $\cos(\theta + 2\phi) = -1$, which corresponds to $\sin(\theta + 2\phi) = 0$, that is, $v \approx v(x_o, y_o)$. Thus, along this deformed path of integration, we have $v(x, y) \approx v(x_o, y_o)$, which prevents the introduction of destructive oscillations associated with the presence of the factor $\exp(iNv)$ in the integrand. Therefore, we have the asymptotic form

$$I(N) \sim \exp[Nf(z_o)] \int\limits_{-\infty}^{+\infty} \exp\left[-\frac{1}{2}\rho s^2 N\right] \exp(i\phi)\, ds$$

$$= \exp\left[Nf\left(z_{o}\right)\right]\left(\frac{2\pi}{\rho N}\right)^{1/2}\exp\left(i\phi\right), \tag{A.66}$$

where $\rho = |f''(z_{o})|$, and $\phi = -\theta/2 \pm \pi/2$ gives the inclination of the contour (the sign depends on the direction along the contour). One of the first rigorous applications of the saddle-point method to a problem in statistical physics has been carried out by T. H. Berlin and M. Kac to obtain the exact solution of the spherical model [see the description by Marc Kac, in Physics Today **17**, 40 (1964)].

Example: Asymptotic calculation (for large N) of the integral

$$Z\left(N\right) = \frac{1}{2\pi i}\oint_{C}\exp\left[Nf\left(z\right)\right]dz, \tag{A.67}$$

where C is the unit circle enclosing the origin, and

$$f\left(z\right) = v\left(\frac{2\pi m}{\beta h^{2}}\right)^{3/2}z - \ln z \tag{A.68}$$

(see Chapter 7, Section 7.2D).
From the first derivative,

$$f'\left(z\right) = v\left(\frac{2\pi m}{\beta h^{2}}\right)^{3/2}-\frac{1}{z}, \tag{A.69}$$

we obtain the saddle point,

$$z_{o} = \frac{1}{v}\left(\frac{\beta h^{2}}{2\pi m}\right)^{3/2}. \tag{A.70}$$

Therefore,

$$f\left(z_{o}\right) = 1 + \ln v - \frac{3}{2}\ln\left(\frac{\beta h^{2}}{2\pi m}\right). \tag{A.71}$$

From the second derivative,

$$f''\left(z\right) = \frac{1}{z^{2}}, \tag{A.72}$$

we have

$$f''\left(z_{o}\right) = v^{2}\left(\frac{2\pi m}{\beta h^{2}}\right)^{3}. \tag{A.73}$$

Hence,

$$\rho = |f''\left(z_{o}\right)| = v^{2}\left(\frac{2\pi m}{\beta h^{2}}\right)^{3}, \tag{A.74}$$

and $\theta = 0$. We thus write

$$Z(N) \sim \frac{1}{2\pi i} \exp\left[Nf(z_o)\right] \left[\frac{2\pi}{Nv^2} \left(\frac{\beta h^2}{2\pi m}\right)^3\right]^{1/2} \exp(i\phi), \qquad (A.75)$$

where $\phi = -\theta/2 \pm \pi/2 = \pm \pi/2$. Noting that the choice $\phi = \pi/2$ corresponds to the correct direction along the deformed contour, we have the asymptotic expression

$$Z(N) \sim \left(2\pi Nv^2\right)^{1/2} \left(\frac{\beta h^2}{2\pi m}\right)^{3/2} \exp\left[Nf(z_o)\right], \qquad (A.76)$$

from which we calculate the limit

$$\lim_{N \to \infty} \frac{1}{N} \ln Z(N) = f(z_o) = 1 + \ln v - \frac{3}{2} \ln \left(\frac{\beta h^2}{2\pi m}\right), \qquad (A.77)$$

in agreement with the results obtained in Chapter 7.

A.7 Numerical constants

The following numerical constants (the error is in the last digit) are frequently used in statistical physics (values from the *American Institute of Physics Handbook*):

(i) Charge of the electron

 $e = 1.60217733 \times 10^{-19} \, \text{C}$

(ii) Speed of light in vacuum

 $c = 2.99792458 \times 10^8 \, \text{m s}^{-1}$

(iii) Planck's constant

 $h = 6.6260755 \times 10^{-34} \, \text{J s}$

(iv) Electron rest mass

 $m_e = 9.1093897 \times 10^{-31} \, \text{kg}$

(v) Proton rest mass

 $m_p = 1.6726231 \times 10^{-27} \, \text{kg}$

(vi) Avogadro's number

 $A = 6.0221367 \times 10^{23} \, \text{mol}^{-1}$

(vii) Boltzmann's constant

 $k_B = 1.3806568 \times 10^{-23} \, \text{J K}^{-1}$

(viii) Universal gas constant
$$R = 8.314510 \, \text{J} \, \text{mol}^{-1} \, \text{K}^{-1}$$

In statistical physics it is interesting to recall the order of magnitude of the following quantities:

- Angstrom: $1 \, \text{Å} = 10^{-10} \, \text{m}$

- Electron-volt: $1 \, \text{eV} \approx 1.6 \, \text{J}$

- Atmospheric pressure: $1 \, \text{atm} \approx 10^5 \, \text{Pa}$ (pascal)

- Kelvin degrees: $1 \, \text{eV} \rightarrow 10^4 \, \text{K}$

- Standard molar volume: $\approx 22.4 \, \text{l}$ at $0 \, ^\circ\text{C}$ and $1 \, \text{atm}$.

(viii) Universal gas constant

$$R = 8.31451\ \mathrm{J\,mol^{-1}\,K^{-1}}$$

In statistical physics it is interesting to recall the order of magnitude of the following quantities:

- Angstrom $1\ \text{Å} = 10^{-10}$ m
- Electron-volt $1\ \text{eV} \simeq 1.6\ \mathrm{J}$
- Atmospheric pressure $1\ \text{atm} \approx 10^{5}$ Pa (pascal)
- Kelvin degrees $1\ \text{eV} \sim 10^{4}$ K
- Standard molar volume $\approx 22.4\ l$ at 0°C and 1 atm.

Bibliography

Several references on more specific topics have already been given along the text. In this bibliography we list a number of works of a more general character that were influential to writing this book.

A. Basic texts

1. F. Reif, *Fundamentals of Statistical and Thermal Physics* (McGraw-Hill Book Company, New York, 1965). An excellent introductory text that contributed to innovating the teaching of thermal physics. It is very accessible but quite detailed to our taste. There is a short and simpler version (the fifth volume of the *Berkeley Physics Course*, McGraw-Hill, 1965).

2. H. B. Callen, *Thermodynamics* (John Wiley and Sons, New York, 1960). Outstanding didactical exposition of classical Gibbsian thermodynamics. There is a second edition that includes a long introduction to equilibrium statistical mechanics as well (John Wiley, 1985).

3. W. Feller, *An Introduction to Probability Theory and its Applications* (John Wiley and Sons, New York, 1968). The first volume in an excellent and accessible text on the theory of probabilities.

4. R. Kubo, *Statistical Mechanics* (North-Holland Publishing Company, Amsterdam, 1965). It contains a brief exposition of the theory, at a more advanced level. The impressive collection of (solved) problems is well-known among the students in the area.

5. K. Huang, *Statistical Mechanics* (John Wiley and Sons, New York, 1963). It covers many topics of interest, at a slightly more advanced level. It is still one of the most useful and popular graduate texts of statistical mechanics. There is a second edition, with the inclusion of some modern topics (John Wiley, 1987).

6. L. D. Landau and E. M. Lifshitz, *Statistical Physics* (Pergamon Press, Oxford, second edition, 1969). An excellent treatise of statistical physics, with the development of the theory, at a more advanced level, and many applications. It is part of the famous course of theoretical physics by Landau and Lifshitz.

7. R. C. Tolman, *The Principles of Statistical Mechanics* (Oxford University Press, London, 1939; there is a Dover edition of 1979). Classical and accessible text on the fundamentals of statistical mechanics.

8. J. W. Gibbs, *Elementary Principles in Statistical Mechanics* (Yale University Press, New Haven, 1902; there are several reprinted editions). The classical text of Gibbs is still very much up-to date and readable.

9. C. Kittel, *Thermal Physics* (John Wiley and Sons, New York, 1969). Introductory text, of heuristic character, at the same level of the present book (there is a more recent edition, in collaboration with H. Kroemer, published by W. H. Freeman and Sons, New York, 1980). It is quite different from a more compact and older text by C. Kittel, *Elementary Statistical Physics* (John Wiley and Sons, New York, 1958).

10. R. K. Pathria, *Statistical Mechanics* (Butterworth-Heinemann, Oxford, second edition, 1996). Slightly more advanced text, with many applications, and a strong didactical character.

11. G. H. Wannier, *Statistical Physics* (John Wiley and Sons, New York, 1966; there is a Dover edition of 1987). It contains many applications to condensed matter physics.

12. R. P. Feynman, *Statistical Mechanics, a Set of Lectures* (Addison-Wesley Publishing Company, New York, 1972). Class notes of R. P. Feynman. It contains stimulating discussions and many applications to quantum liquids and condensed matter problems.

13. M. Toda, R. Kubo, and N. Saito, *Statistical Physics I, Equilibrium Statistical Mechanics* (Springer-Verlag, New York, 1983). It is a slightly more advanced didactical text. There is a second volume on non-equilibrium phenomena, R. Kubo, M. Toda, and N. Hashitsuma, *Statistical Physics II, Non Equilibrium Statistical Mechanics* (Springer-Verlag, New York, 1985).

14. D. A. McQuarrie, *Statistical Mechanics* (Harper and Row Publishers, New York, 1976). Excellent graduate level text, with many applications to liquids and solutions, and a good treatment of nonequilibrium phenomena. It follows the line of the classical text of T. L. Hill, *Statistical Mechanics* (McGraw-Hill Book Company, New York, 1956).

15. Yu. B. Rumer and M. Sh. Ryvkin, *Thermodynamics, Statistical Physics, and Kinetics* (MIR, Moscow, 1980). Modern didactical text on the principles and applications of statistical physics. It is much longer and slightly more advanced than the present book.

16. D. Chandler, *Introduction to Modern Statistical Mechanics* (Oxford University Press, New York, 1987). Didactical modern text at a very accessible level. It includes discussions on phase transitions, the Monte Carlo method, and non-equilibrium phenomena.

17. L. E. Reichl, *A Modern Course in Statistical Physics* (University of Texas Press, Austin, 1980). Modern and complete text, at a more advanced level, under the influence of the Brussels school. It contains a detailed exposition of the theory and many applications.

18. B. Diu, C. Guthmann, D. Lederer, and B. Roulet, *Physique Statistique* (Hermann editors, Paris, 1989). Long and detailed text in French, with a didactical exposition of the theory and many applications.

19. E. J. S. Lage, *Física Estatística* (Fundação Calouste Gulbenkian, Lisbon, 1995). Text in Portuguese. It covers equilibrium and non-equilibrium statistical physics at an intermediate level.

B. Phase transitions and critical phenomena (chapters 12, 13 and 14)

Although somewhat outdated, the book by H. E. Stanley, *Introduction to Phase Transitions and Critical Phenomena* (Oxford University Press, New York, 1971), is very well written and still remains quite interesting. Prior to the proposals of the renormalization-group scheme, we also refer to the work of M. E. Fisher in Reps. Progr. Phys. **30**, 615 (1967). The texts of K. Huang (5) and L. E. Reichl (17) contain specific chapters on critical phenomena. The many volumes of the collection edited by C. Domb and M. S. Green (and nowadays by C. Domb and J. Lebowitz), *Phase Transitions and Critical Phenomena* (Academic Press, New York, from 1971), give excellent information on almost all of the important topics in this area.

There are some recent didactical texts on critical phenomena:

20. J. M. Yeomans, *Statistical Mechanics of Phase Transitions* (Clarendon Press, Oxford, 1992). Simple and didactical text with a good treatment of many interesting topics.

21. J. J. Binney, N. J. Dowrick, A. J. Fisher, and M. E. J. Newman, *The Theory of Critical Phenomena* (Clarendon Press, Oxford, 1992). Slightly more advanced text with a detailed discussion of the renormalization-group techniques.

The more mathematically inclined readers should check the following texts:

22. C. J. Thompson, *Classical Equilibrium Statistical Mechanics* (Clarendon Press, Oxford, 1988). Very accessible text, with many examples.

23. D. Ruelle, *Statistical Mechanics: Rigorous Results* (W. A. Benjamin, Amsterdam, 1969). Advanced text, of a mathematical character, and more difficult to read (the reader may first check an article by R. B. Griffiths in the first volume of the Domb and Green series).

C. Non-equilibrium phenomena (chapters 15 and 16)

We have already listed several texts that include adequate treatments of some topics pertaining to non-equilibrium phenomena: F. Reif (1), K. Huang (5), D. M. McQuarrie (14), and L. E. Reichl (17). To complete this list, we should mention some specific books:

24. H. Haken, *Synergetics. An Introduction: Non-equilibrium Phase Transitions and Self-Organization in Physics, Chemistry and Biology* (Springer-Verlag, New York, 1978). Accessible and inspiring book, including many interdisciplinary problems.

25. N. G. van Kampen, *Stochastic Processes in Physics and Chemistry* (North-Holland, Amsterdam, 1983). More advanced text on stochastic methods (master equations and applications).

Applications of Monte Carlo techniques to problems in statistical physics are discussed in the following introductory texts:

26. K. Binder and D. W. Heermann, *Monte Carlo Simulation in Statistical Physics* (Springer-Verlag, New York, 1988).

27. D. W. Heermann, *Computer Simulation Methods in Theoretical Physics* (Springer-Verlag, New York, 1986).

Index